Pattern Recognition with Fuzzy Objective Function Algorithms

ADVANCED APPLICATIONS IN PATTERN RECOGNITION
General editor: Morton Nadler

A STRUCTURAL ANALYSIS OF COMPLEX AERIAL PHOTOGRAPHS — Makoto Nagao and Takashi Matsuyama

PATTERN RECOGNITION WITH FUZZY OBJECTIVE FUNCTION ALGORITHMS — James C. Bezdek

Pattern Recognition with Fuzzy Objective Function Algorithms

James C. Bezdek
Utah State University
Logan, Utah

PLENUM PRESS • NEW YORK AND LONDON

Library of Congress Cataloging in Publication Data

Bezdek, James C. 1939–
Pattern recognition with fuzzy objective function algorithms.

(Advanced applications in pattern recognition)
Bibliography: p.
Includes indexes.
1. Pattern perception. 2. Fuzzy-algorithms. 3. Cluster analysis. I. Title. II. Series.
Q327.B48 001.53'4 81-4354
ISBN 0-306-40671-3 AACR2

A Division of Plenum Publishing Corporation
233 Spring Street, New York, N.Y. 10013

Printed in the United States of America

Foreword

The fuzzy set was conceived as a result of an attempt to come to grips with the problem of pattern recognition in the context of imprecisely defined categories. In such cases, the belonging of an object to a class is a matter of degree, as is the question of whether or not a group of objects form a cluster.

A pioneering application of the theory of fuzzy sets to cluster analysis was made in 1969 by Ruspini. It was not until 1973, however, when the appearance of the work by Dunn and Bezdek on the Fuzzy ISODATA (or fuzzy c-means) algorithms became a landmark in the theory of cluster analysis, that the relevance of the theory of fuzzy sets to cluster analysis and pattern recognition became clearly established.

Since then, the theory of fuzzy clustering has developed rapidly and fruitfully, with the author of the present monograph contributing a major share of what we know today. In their seminal work, Bezdek and Dunn have introduced the basic idea of determining the fuzzy clusters by minimizing an appropriately defined functional, and have derived iterative algorithms for computing the membership functions for the clusters in question. The important issue of convergence of such algorithms has become much better understood as a result of recent work which is described in the monograph. Further, farreaching generalizations of the original fuzzy c-means algorithms have been recently discovered that generate linear varieties of arbitrary dimension up to one less than the dimension of data space. These algorithms (and convex combinations of them) are discussed. Their potential for a wide range of applications is suggested; it seems but a matter of time until these applications are made.

Although the results of experimental studies reported in this book indicate that fuzzy clustering techniques often have significant advantages over more conventional methods, universal acceptance of the theory of fuzzy sets as a natural basis for pattern recognition and cluster analysis is not likely to materialize in the very near future. The traditional view, that probability theory is all that is needed to deal with imprecision, is too deeply entrenched to be abandoned without resistance. But those of us who believe that fuzziness and randomness are two distinct facets of uncertainty feel that it is only a matter of time before this dualistic view of imprecision becomes accepted as a useful tool in mathematical models of real processes. This monograph is an important step towards that end.

Berkeley, California L. A. Zadeh

Preface

It is tempting to assert that the basic aim of all science is the recognition of patterns. Scientists study observed groups of variables, trying to isolate and identify functional relationships—qualitative and quantitative. These associations provide mathematical models which are in turn used to infer objective properties of the process being modeled. Traditional approaches to this course of scientific investigation have been radically altered by modern data processing capabilities: we examine data and postulate interrelationships whose level of complexity would have been unthinkable even three decades ago. This capability has stirred a concomitant interest in the notion of precision: precision in nature, in the data we gather from nature, in our machine representation of the data, in the models we construct from the data, in the inferences we draw from the models, and, ultimately, in our philosophical perception of the idea itself.

This monograph has as its cornerstones the notions of *imprecision* and *information* in mathematical modeling. Lotfi Zadeh introduced in 1965 an axiomatic structure—the fuzzy set—which offers a new way to capture these two ideas mathematically. Many researchers assume that "fuzziness" either proposes to replace statistical uncertainty, or can, by a suitable change of view, be interpreted as the same thing. Neither of these presumptions is accurate! Fuzzy models are compatible but distinct companions to classical ones: sometimes more useful; sometimes equally appropriate; and sometimes unsuitable to represent the process under study.

Since their inception, fuzzy sets have advanced in a wide variety of disciplines: e.g., control theory, topology, linguistics, optimization, category theory, and automata. Nowhere, however, has their impact been more profound or natural than in pattern recognition, which, in turn, is often a basis for decision theory. In particular, my aim is to discuss the evolution of fuzziness as a basis for feature selection, cluster analysis, and classification problems. It is not my intent to survey a wide variety of research concerning these subjects; nor will the reader find collected here a group of related (but uncorrelated) papers. Instead, this volume is specifically organized around two fundamental structures—*fuzzy relations and partitions*—their theory, and their applications to the pattern recognition problems mentioned above. These same two structures also provide a basis for a variety of models and algorithms in group decision theory—a variety so great, in fact, that I have decided to include herein only those results concerning decision theory which bear directly on the pattern recognition problems just mentioned. Some of this material is already "obsolete"; it is included to make the presentation historically and chronologically continuous (presumably, this is the soundest pedagogical path to follow). My choice in this matter is motivated by a desire to collect and unify those concepts that appear to play a major role in current research trends. This choice may, of course, prove historically to have been shortsighted, and at the very least, will occasion the omission of many bright and innovative ideas, for which I apologize *a priori*. Moreover, some of the material included here is the center of great controversy at present: for this I do *not* apologize. This is not only predictable, but exactly as it should be. New ideas need to be challenged, sifted, modified, and—if necessary—discarded, as their scientific value becomes evident. Indeed, the *method* of science is to successively approximate physical processes by various models, always hoping that the next step improves the accuracy and usefulness of previous models. Whether fuzzy models assume a place in this iterative procedure remains to be seen. (I hope they do and, of course, believe they will.) In any case, I have little doubt that many of the results presented herein will have been subsumed by more general ones, or perhaps by alternatives of more evident utility, even before this book is published.

My purposes in writing this monograph are simple:

(i) to synthesize the evolution of fuzzy relations and partitions as axiomatic bases for models in feature selection, clustering, and classification;
(ii) to give a careful, unified development of the mathematics involved, to the extent that this is possible and desirable;
(iii) to demonstrate that this theory can be applied to real physical problems with measurable success.

About Theorems. Mathematical detail sometimes bewilders the uninitiated, who often cannot afford the time needed to master the underpinnings of concise, thorny proofs. On the other hand, the erection of any theory ultimately requires a foundation based on precise formulations of previous results. Accordingly, I have tried, at what seem to be crucial junctions, to divide proofs into two parts: first, an informal "analysis" of the meaning, application, and key ideas in the proof of the result; and then, a formal proof. My hope is that this diversion from the usual course will enable readers impatient for the applications to press on with an intuitive understanding of the theory, while at the same time not disappointing those with both the time and inclination to follow detailed arguments. The level of rigor will vary from none, to sloppy, to annoyingly painstaking, even within a single section—unmistakable signs of "growing pains" for any evolving subject area! We find art and science intermingled here: the art will diminish if the subject flourishes; nonetheless, its charm as a harbinger of the straight and narrow to follow remains intact. One of the nicest surprises ahead is that some of the most useful results have a very straightforward theoretical basis.

The Numbering System. The following abbreviations are used throughout:

A = Algorithm	E = Example	P = Proposition
C = Corollary	Fig. = Figure	S = Section
D = Definition	H = Homework	T = Theorem

Items are numbered consecutively within a section, which is the fundamental unit of organization. For example, (T5.2) is Theorem 2 of S5 (Section 5); Fig. 8.1 is Figure 1 of Section 8. Following the table of contents there is a list of special symbols and the page on which they are defined. The end of a proof or example is indicated by •. Each section closes with remarks and bibliographic comments. Bibliographic references appear in parentheses without decimals—(33) is reference 33; (3.3) is equation 3 of Section 3.

Exercises. Homework problems are given at the ends of sections. Altogether there are several hundred problems which draw upon and extend the text material in various ways. The range of difficulty is quite wide, with a view towards accommodating the needs of a variety of readers. Many problems (with answers) are simple calculations, applications or examples of definitions or formulas; these I have assigned as homework in beginning graduate courses in pattern recognition. Also included are many published theorems which are peripherally related to the central issues of the text; the more difficult ones are accompanied by references, and thus afford an excellent means for easing into literature of interest. (Exercises of this type often lead toward conjectures and discoveries of new facts.) Finally, I have included some problems for which answers are not known. This type of question makes an excellent class project, and it too can lead to exciting results, both within the classroom and without. Several exercises ask for a "handheld" run of a particular algorithm. This type of exercise strengthens one's understanding of the method in question. However, the algorithms discussed herein are intended to be coded as computer programs, and I have described them in a manner which seems most useful in this respect, rather than providing detailed listings, which are very seldom compatible with specific facilities.

Acknowledgments. Among the countless people to whom I am indebted for various reasons in connection with this volume, several must be mentioned explicitly: Lotfi Zadeh, Joe Dunn, Mort Nadler, Laurel Van Sweden, and L. S. Marchand have, in different ways, made this task both possible and pleasant.

Utah State University
Logan, Utah

James C. Bezdek

Contents

Notation

1

Models for Pattern Recognition

Section 1 (S1) describes specifically the problems to be discussed in succeeding chapters. In S2 a short analysis of the modeling process suggests that *information* and *uncertainty* will be key concepts in the development of new mathematical structures for pattern recognition. Fuzzy sets are introduced in the third section as a natural and tractable way to model physical situations which seem to require something more than is offered by deterministic or stochastic representations.

S1. Pattern Recognition

The term "pattern recognition" embraces such a vast and diversified literature that a definition of it always invites debate. Verhagen[111] presents a survey of *definitions* of pattern recognition which cites the difficulties involved in such an attempt. Nonetheless, it is pedagogically useful for us to begin by attempting to describe what pattern recognition entails. I think one can successfully defend a literal approach: quite simply, pattern recognition is *a search for structure in data*. Let us extract each of the major elements in this definition:

The Data

Techniques of pattern recognition are applicable to data drawn from virtually any physical process. The data may be qualitative, quantitative, or both; they may be numerical, pictorial, textural, linguistic, or any combination thereof. Examples are medical records, aerial photos, market trends, library catalogs, galactic positions, fingerprints, psychological profiles, cash flows, chemical constituents, demographic features, stock options, military decisions. Techniques for data processing we call the search. X will be our usual notation for a data set.

The Search

One of the most important and oft-used data analysis methods is the "eyeball" technique. Subjective assessment of data patterns has long been a method accepted by many traditional data analyzers; medical diagnosis is an excellent example of subjective pattern recognition. More reliable forms of data analysis have evolved with modern technology. For example, the primary goal of statistics is the identification of a probabilistic model which can provide a reasonable explanation of the process generating the data. Statistical analysis proceeds slowly by hand, more rapidly with hand calculators, and can be quite fast with modern computers. At present, statistical pattern recognition is a major technique in "the search." The computer itself, however, has encouraged us to search for data structures that are not intuitively probabilistic, nor even approximately stochastic. With this increased computational power there is a need to hypothesize axiomatic structures, models, and methods which enable our machines to transmit their findings to us in usable forms. The type of search performed depends not only on the data and our models, but upon the structure we expect to find.

The Structure

Presumably, data carry information about the process generating them. By structure, we mean the manner in which this information can be organized so that relationships between the variables in the process can be identified. Relationships may be *causal* (noninvertible associations), or simply *connective* (invertible associations). Representations of the recognized structure depend on the data, the method of search, and the model used. In terms of information, the data contain it, the search recognizes it, and the structure represents it. Different models convey different *amounts and types* of information concerning the process they represent.

The preceding paragraphs deal with generalities: what specifically is discussed below? Three main topics: clustering, classification, and feature selection. To describe them more accurately, let $X = \{x_1, x_2, \ldots, x_n\}$ be a data set of n items x_k, each of which is one observation from some physical process. The specific form of the observations is immaterial for the definitions below.

Cluster Analysis

Clustering in X has classically meant the identification of an integer c, $2 \leq c < n$, and a partitioning of X by c mutually exclusive, collectively exhaustive subsets of X (the "clusters"). Presumably, the members of each cluster bear more similarity to one another than to members of other clusters. At least, this will be true for *mathematical similarities* between the x_k's in some well-defined operational sense; one then hopes that the same substructure exists in the data-generating process itself. Cluster structure in X may reveal associations between *individuals* of a population. Associated with the clustering problem is, when not prespecified, identification of c, the most appropriate number of clusters in X. This important question we call the *cluster validity* problem: its resolution will be implicit in the directions taken in later chapters. Clustering, then, seeks associations between subsets of a sample.

Classification

Let S denote the data space from which X has been drawn, i.e., $X \subset S$. A classifier for S is a device or means whereby S itself is partitioned into c "decision regions." Explicit representation of these regions depends on the nature of S, the way in which the regions are formed, and the model we choose—in other words, on the data, the search, and the structure. The role played by a sample data set X from data space S in classifier design also depends on each of these factors. X is often used to "train" the classifier, that is, to delineate the decision regions in S. In classification then, we search for structure in an entire *data space*: it is possible—but not necessary—to conduct this search by first clustering in a sample data set X. The structure we find may enable us to classify subsequent observations rapidly and automatically. The purpose of classification may be as simple as construction of a taxonomy of the physical processes involved; or as complex as deciding a course of action based on identification of a distinguished individual. Classification attempts to discover associations between subclasses of a population.

Feature Selection

A common question that arises in connection with clustering and classification concerns the constitution of data space S: are the features (or characteristics) of data item $x_k \in X$ sufficiently representative of the physical process to enable us to construct realistic clusters or classifiers? In other words, do we have the "right" data space? Should some features of x_k be discarded? modified? enriched? transformed?

In feature selection, we search for *internal* structure in data items: the goal is to improve their usefulness as constituents of data sets for clustering and/or classification. In short, we hope to identify the "best" data space by looking for structure within its elements. Feature selection seeks associations between the characteristics of individuals in the sample and/or population.

To summarize, the main issues discussed below will be:

Feature selection:	The search for structure in data items, or observations $x_k \in X$
Cluster analysis:	The search for structure in data sets, or samples $X \subset S$
Classification:	The search for structure in data spaces, or populations S.

It is worth noting that these problems exhibit a beautiful (but confounding!) sort of interactive symmetry. For example; if we could choose "optimal" features, clustering and classification would be trivial; on the other hand, we often attempt to *discover* the optimal features by clustering the feature variables! Conversely, if we could design an "optimal" classifier, then the features selected would be immaterial. This kind of interplay is easy to forget—but important to remember!—when one becomes engrossed in the details of a specific problem, for a successful pattern recognizer is one who has matched his or her particular data to an algorithmic technique.

Medical diagnosis affords an excellent example of these ideas. A doctor wants to classify patients according to their actual diseases, in order to prescribe correct treatments. First, data are collected: symptoms are noted, tests performed. Each patient generates a record x_k. Then records are clustered—this group is observed to have hypertension due to primary aldosteronism, that group due to renovascular constriction. Finally, a classifier is designed—a new patient is identified with the doctor's decision rule and treated accordingly. All the elements of pattern recognition are here: data, search, structure. The doctor may wonder: Did I observe the right symptoms? Did the data obscure important facts? Are more tests needed? Is a bigger data space needed? Would a smaller data space suffice? Would a different clustering have been more accurate, useful, meaningful? Can the classifier be improved? And, perhaps, *can I use a computer to help me interpret the data and guide my analysis*?

Remarks

It will be our practice to close each section with some bibliographic comments and references germane to the topic at hand. A number of excellent literature surveys concerning various aspects of pattern recognition have been published recently. One of the earliest and most compelling is that of Nagy.[80] Other surveys of note include those of Ho and Agrawala,[58] Gose,[48] Levine,[73] and Kanal.[61] Bremerman,[24] and Grenander[50] give insightful discussion on the role of mathematics in pattern recognition. Encyclopedic bibliographies of books on pattern recognition and papers concerning classifier designs are available in Toussaint.[105] Published proceedings and collections of papers on various aspects of pattern recognition are numerous: (112) and (113) are of this type.

Among the textbooks and monographs concerning specific bodies of pattern recognition theory listed in (105) are several deserving special note. Clustering, for example, is ably represented by the books of Anderberg,[1] Tryon and Bailey,[109] and Hartigan.[57] Statistical pattern recognition techniques are discussed by Fu,[41] Meisel,[79] Sebestyen,[96] and Fukunaga.[42] Texts on general pattern recognition include those of Bongard,[25] Patrick,[81] Tou and Wilcox,[104] Tou and Gonzalez,[103] and Duda and Hart.[34] Of these, Duda and Hart is especially valuable because of its readable style and an exceptional collection of chronologically annotated references to the early literature in many fields. Books and papers of a more topical nature abound in virtually all data processing fields. Although they are far too numerous to mention individually, we shall occasionally draw attention to those which contribute in a specific way to the development or exemplification of our subject.

S2. Some Notes on Mathematical Models

Since the subject of this book is a new type of mathematical model, we begin with some questions and answers about models.

Where Do We Get Models? Models originate during the investigation of a physical process. For example: the arrival at consensus of a decision-making group; the effect of welfare on civil disobedience; the interaction of stress and strain in an airplane wing; the psychological effects of fluoridation on a community; the impact of stream pollution on a trout population. Physical processes such as these generate data. The data bear information relating the variables in the process. A mathematical model of the process provides a vehicle for organization, recognition, and representation of the information. Although many models arise in this way, it often happens that an existing model fulfills these functions quite ably for more than one process. Indeed, a "borrowed" model can—and often does—perform more successfully in applications other than the one for which it was originally devised!

What Does a Model Provide? First, the model must assimilate the data and make them compatible with the manipulative strategies to be used.

Many models are postulated in parametric form, and the data are used to identify the parameters. The identification procedure was called the search in S1. If possible, the model should be tested with new data. Finally, the model enables us to infer objective properties of the process being studied. The properties inferred enable us to understand, predict, and control the process: the role of the model is to convert information concealed in the raw data into usable forms.

How are Models Altered? The parameters of a given model, the type of model within a generic class, or the generic class itself can be changed. Each of these changes has an increasingly profound effect on the utility of the model. Moreover, as the mathematical structure of a given model evolves, its capacity to translate information changes. For example, mathematical abstractions in Lorentzian geometry provide very real insights about the physics of relativistic phenomena. In other words, models can be altered externally, by adopting a different philosophical view (and hence a different axiomatic structure), and altered internally, by discovering new facts about an existing model. Both of these avenues will be pursued in subsequent sections.

What about Different Models of the Same Process? Many examples suggest that a given process can be modeled *usefully* in several ways. Newtonian, relativistic, quantum, statistical, and continuum mechanics are different models of dynamic motion, each of which contributes a facet to our understanding of the same process. Different models may *compete*: in this case we need a measure of their relative efficiency and accuracy. Or, they may *supplement* one another: in this instance, we try to assess the overall quality of the augmented model. Each of these eventualities will arise in subsequent sections.

How Is an "Optimal" Model Chosen? This question occasions the need for caution. It is an important issue since the desire to optimize often motivates scientific progress. The word "optimal" implies the existence of a criterion or standard of comparison, and one must be certain that the criterion being used is a meaningful one in the context of its application. As an example: one often encounters the claim that a Bayesian classifier is optimal.[34] This is true, provided one accepts the Neyman–Pearson definition of probability of error (a very reasonable thing to do). A problem arises, however, when one attempts to apply this criterion as a measure of performance, because the theoretically optimal error rate cannot be calculated. Even if the asymptotic behavior of a given classifier were theoretically optimal in this sense (i.e., could achieve Bayesian performance if implemented infinitely often), it would be a mistake to conclude therefrom that its expected performance on finite data sets would be better than that of a different model whose asymptotic properties might be suboptimal relative to this index of error. The notion of measure of performance will be an integral part of our development, as it affords a basis for comparing different models of the same process.

Remarks

The choice of an existing model or construction of a new one depends on many factors. The type, availability, and means for processing data; technical support; economic constraints; time; the experience and ability of the investigator; and historical precedent all play roles in model building. Several of these aspects are analyzed in Kemeny and Snell,[63] a particularly readable classic introduction to some of the issues raised in this section. Other treatments of these ideas include the books of Maki and Thompson,[77] and Grossman and Turner.[51] Some of the models developed below exhibit *mathematically* similar structures to graph-theoretic types discussed in the excellent text by Roberts.[85]

Exercises

H2.1. Describe a physical process which has been modeled by competitive models of the same generic type.

H2.2. Describe a physical process which has been modeled by complementary models of the same generic type.

S3. Uncertainty

One of the amusing paradoxes of modern science is our attempt to capture *precisely* the amount and effects of uncertainty in mathematical models! As there are several sources of uncertainty, it is important for us to distinguish between them, because the type of uncertainty involved may alter our construction or choice of a model. In particular, we identify uncertainty due to

(i) Inaccurate measurements
(ii) Random occurrences
(iii) Vague descriptions

Undoubtedly there are other philosophically distinguishable categories, but these three provide for us an adequate way to describe the manifestation of uncertainty in deterministic, probablistic, and fuzzy models, respectively.

A process is *deterministic* if its outcome can—with absolute certainty—be predicted, upon replication of the circumstances defining it. For example, the amount of bacteria, say $y(t)$, present at time t is often assumed to obey the law of exponential growth:

$$y(t) = c\,e^{kt} \tag{3.1}$$

where c and k are, respectively, mathematical and physical parameters of model (3.1). Given sufficient data, parameters c and k are uniquely determined. Thus, an initial time and amount $[t_0, y(t_0) = y_0]$ and any subsequent time and amount $[t_*, y(t_*) = y_*]$ lead to the choices

$$\hat{k} = \frac{\ln(y_*/y_0)}{t_0 - t_*} \tag{3.1a}$$

$$\hat{c} = y_0\,e^{-kt_0} \tag{3.1b}$$

The function $y(t) = \hat{c}\,e^{\hat{k}t}$ presumably determines *unequivocally* the evolution of growth for all times $t \geqslant t_0$. It should be noted that the validity of this inference hinges on being able to duplicate the conditions used to evaluate $\hat{c}$ and $\hat{k}$ *exactly*. If this is not the case (and in an absolute sense, it never is!), we must satisfy ourselves that $\hat{c}\,e^{\hat{k}t}$ at least provides a reasonable approximation of the process—one which furnishes us with useful information concerning it. Here, then, uncertainty due to measurement errors causes uncertainty about $\hat{c}$ and $\hat{k}$.

Inexact measurements can cause uncertainty in models of physical processes which are ostensibly deterministic. Another type of uncertainty, quite different in concept, arises when the outcome of a physical process is believed to be *random*. In this case, there is an element of "chance" concerning the evolution of the process which is unaffected by environmental imprecision. For example, flipping a fair coin presumably has two equally likely, mutually exclusive outcomes (heads or tails) per toss, so that the evolution of a sequence of identical and independent trials of this experiment cannot be predicted with certainty. Models of processes which exhibit this kind of uncertainty are called *stochastic or probabilistic* models. Experience has taught us, for example, that a reasonable model of n Bernoulli trials (of which the coin flip experiment is a specific instance) is the binomial distribution

$$b(n, p; k) = \binom{n}{k} p^k (1-p)^{n-k} \tag{3.2}$$

wherein n is the number of trials; k is the number of "successes" in n trials; p is the probability of success on each trial; and $b(n, p; k)$ is the probability of exactly k successes in n trials. The parameters of (3.2) are n and p: n is known to the modeler; and p must be estimated. For a fair coin, we have a *prior* idea that $p = (1/2)$: in other types of Bernoulli trials, however (exactly two outcomes per independent trial, p fixed for all trials), the value of p need not be so obvious, and we must devise a means for estimating it, perhaps by some method of statistical inference. *This is an important point*: the accuracy and usefulness of probabilistic models is entirely dependent upon the VALUES—assigned by US—to the probabilities! In some cases these assignments are not controversial; but in many instances, the validity of probabilistic assignments is subject to question. We are not trying to detract here from the importance of stochastic models, which are usually the most accurate and realistic way to model random processes; we are emphasizing that they are, more often than not, only as good as the numbers *we supply* for them.

Finally, we observe that the assertions we can make about a sequence of Bernoulli trials afforded by (3.2) are quite different from inferences we can draw about the objective properties of bacterial growth via (3.1). Specifically, (3.2) enables us to predict the *likelihood* of observing exactly k successes in n Bernoulli trials. This estimate becomes increasingly accurate as n grows large; nonetheless, assertions about the actual occurrence and specific evolution of this event are not available from (3.2). The point here is that deterministic and stochastic models transmit different *kinds and amounts* of information about the processes they represent (even if they are models of the same process!).

Functional relationships such as (3.2) are sometimes called *stochastic laws*. While models of this type can be used to analyze the effects of uncertainty in deterministic models, it is important to distinguish the philosophical *source* of uncertainty in these two examples: in the first case, the process is assumed deterministic, but our ability to monitor it exactly is uncertain; in the second case, the outcome of the process itself is uncertain. *The importance of this distinction is that it dictates the assumptions supporting the mathematical structure of the model chosen to represent the process.* Can uncertainty arise in *other* ways? Of course! Consider the question, "is person x nearly two meters tall?" There is little doubt as to the response expected—yes, or no. But the adjective *nearly* causes uncertainty as to how one determines an appropriate response, i.e., we must devise first a model of "nearly two meters," and then elicit from it enough information to infer a decision. There is here an element of uncertainty which is not caused by measurement error, nor by random occurrence. It is unquestionably true that this dilemma can be circumvented by a suitable agreement concerning the meaning of "nearly two meters." That is, we might choose to *approximate* the situation by a deterministic or a stochastic model, either of which will provide a decision rule to use when confronted with $h(x)$, the observed height of x. However, neither of these choices alters the physical situation, which manifests a source of *nonstatistical uncertainty, or fuzziness*, quite unlike that encountered in the previous cases. Furthermore, the information which *can be* extracted from either of the preceding models may be intuitively incompatible with the information we would *like to use* in the decision process.

To be concrete about this point, let X represent a sample of n people, and let A_1 be the subset of X for which $h(x)$ is exactly two meters:

$$A_1 = \{x \in X | h(x) = 2\} \tag{3.3}$$

If we agree to say that x is nearly two meters if and only if x belongs to $A_1 (x \in A_1)$, then A_1 will be a very sparse set, its obvious shortcoming being that we cannot measure $h(x)$ exactly. To overcome this deficiency, consider the set

$$A_2 = \{x \in X | h(x) = 2 \pm 0.005\} \tag{3.4}$$

If membership in A_2 is equivalent to nearly two meters, the decision rule determined thereby will identify many people that *are* nearly two meters tall. However, the threshold ± 0.005 would exclude, for example, person y, whose observed height $h(y)$ is 2.0051 meters. In other words, we are forced by model A_2 to conclude that y is not nearly two meters tall—an inference which is seemingly at odds with physical actuality.

Now let us regard X as a sample of size n from S, the population of California; let $x \in X$ be the elementary event "observation of x"; let $h: X \to (0, \infty)$ be the random variable "height of x." Consider the event $1.995 \leq h(x) \leq 2.005$. We see that this event is just A_2 above, now cast in a probabilistic setting, and membership in A_2 is still equivalent to nearly two meters. Accordingly, with a suitable series of experiments followed by statistical inference about a probability model, we could assign to each $x \in X$ a probability, say $P(x)$, of being in A_2, and hence, we could estimate finally $\Pr(1.995 \leq h(x) \leq 2.005) = \Pr(x \in A_2)$. Suppose, for example, we are told that $\Pr(x \in A_2) = 0.95$. What does the stochastic model imply about $h(x)$? That there is, on the average, an excellent *chance*—viz., 95%—that $h(x) \in [1.995, 2.005]$. It is, nevertheless, just a chance! Would you risk the cost of a 3000-mile plane trip to scout the performance of x in a basketball game based on this information? Of course not. The point here is simple: $\Pr(x \in A_2) = 0.95$ does not preclude the *possibility* that $h(x) = 1.221$, the occurrence of which would not be appreciated by your athletic director. What went awry? The stochastic model *attaches* an element of chance to the statement "nearly two meters" when none is warranted! This fact should not necessarily prejudice our confidence in a probabilistic decision rule about the process: it *should*, however, suggest that yet a third type of model might be even more appropriate for the circumstances at hand. To this end Zadeh[(122)] suggested the following approach: since *set membership* is the key to our decisions, let us alter our notion of sets when the process suggests it, and proceed accordingly. Mathematical realization of this idea is simple, natural, and plausible. We let

$$A_3 = \{x | x \text{ is nearly two meters tall}\} \tag{3.5}$$

Since A_3 is not a conventional (hard) set, there is no set-theoretic realization for it. We can, however, imagine a function-theoretic representation, by a function, say $u_3: X \to [0, 1]$, whose values $u_3(x)$ give the *grade of membership* of x in the *fuzzy set* u_3. This is a natural generalization of the function-theoretic realization of sets A_1 and A_2 by their characteristic (or indicator) functions, say u_1 and u_2, respectively, where

$$u_1(x) = \begin{Bmatrix} 1; & x \in A_1 \\ 0; & \text{otherwise} \end{Bmatrix} \tag{3.6}$$

$$u_2(x) = \begin{Bmatrix} 1; & x \in A_2 \\ 0; & \text{otherwise} \end{Bmatrix} \tag{3.7}$$

Put another way, u_3 embeds the two-valued logic of $\{0, 1\}$ in the continuously valued logic $[0, 1]$. Although A_3 in (3.5) is conceptually a fuzzy "set," u_3—the function—is its only mathematical realization. The term "fuzzy *set*" is used for u_3 because hard sets (A_1 or A_2) and their characteristic functions (u_1 or u_2) *are* mathematically isomorphic (operationally equivalent). Just as in probability, the question of primary importance concerning u_3 is this: where do the numbers $u_3(x)$ come from? First, what do the numbers $u_3(x)$ purport to mean? In this example we want u_3 to measure the extent to which $h(x)$ is close to the number 2. Given this criterion, there are many u_3's which might satisfy our needs. For example, we might define a discrete fuzzy model such as

$$u_3(x) = \begin{Bmatrix} 1, & 1.995 \leq h(x) \leq 2.005 \\ 0.95, & 1.990 \leq h(x) < 1.995 \text{ or } 2.005 < h(x) \leq 2.010 \\ \vdots & \vdots \\ 0.05 & \cdots \\ \vdots & \vdots \end{Bmatrix} \tag{3.8}$$

Given this u_3, suppose we are told that $u_3(x) = 0.95$, what information is available? We can infer that x *is* (not, "in all probability, is!") nearly two meters tall, because u_3 in (3.8) prescribes our definition of nearly two meters. Moreover, we have *bounds* on $h(x)$. And finally, we have in this model the capacity for rendering qualitative judgements about *relative* heights. Thus as soon as we know that $u_3(y) = 0.65$, it is certain that

Table 3.1. Deterministic, Stochastic, and Fuzzy Memberships: An Example

		$u_1(\cdot)$	$u_2(\cdot)$	$u_3(\cdot)$
x	$h(x) = 2.000$	1	1	1
y	$h(y) = 1.997$	0	1	1
z	$h(z) = 1.994$	0	0	0.95

x, $u_3(x) = 0.95$, is more nearly two meters tall than y. To fix the differences between these three models we construct Table 3.1, comparing a few values of u_1, u_2, and u_3 on different points in S. Note that both the *type* and *amount* of information varies in the columns (or models) above. It seems fair to assert that column 3—the fuzzy model—maximizes the information contained about the process being represented, and consequently, optimizes our ability to utilize the model as a decision-making aid. In other words, the fuzzy model provides the most useful way to *define*—via the values $u_3(x)$—what "nearly two meters" means.

What philosophical distinction can be made between $u_3(x) = 0.95$, and $\Pr(x \in A_2) = 0.95$? We emphasize that *upon observation of* x [*and hence* $h(x)$], the prior probability $\Pr(x \in A_2) = 0.95$ becomes a posterior probability; either $\Pr(x \in A_2|x) = 1$ if $h(x) = 2 \pm 0.005$; or $\Pr(x \in A_2|x) = 0$. Notice, however, that $u_3(x)$—a measure of the extent to which $h(x)$ is nearly two meters, *remains 0.95 after observation.* This illustrates rather strikingly the philosophical distinction between probability and fuzziness.

A natural question that arises is: How could prior assignments for $u_3(x)$ be made? Indeed, one may argue that $u_3(x)$ is known only after observation of $h(x)$. The same question can be asked about $\Pr(x \in A_2)$! Our point here is that *once the models are constructed,* their usefulness and capabilities vary. The question: Where do we get the numbers? really means: How is the model parametrized? This will be an area of primary concern in subsequent sections. In general, membership values are assigned—just as probabilities are—in one of three ways: by the model-builder (subjective), by using data for estimation (relative frequency), or by the model (*a priori*).

As general properties of models often play a major role in their selection and use, membership assignments are examined more carefully after establishing some basic theoretical results concerning fuzzy sets.

Remarks

Philosophical objections may be raised by the logical implications of building a mathematical structure on the premise of fuzziness, since it seems (at least superficially) necessary to require that an object be or not be an element of a given set. From an aesthetic viewpoint, this may be the most satisfactory state of affairs, but to the extent that mathematical structures are used to model physical actualities, it is often an unrealistic requirement. Our discussion of uncertainty above is by no means exhaustive or definitive: it was given merely to illustrate this point—that fuzzy sets have an intuitively *plausible* philosophical basis. Once this is accepted, analytical and practical considerations concerning fuzzy sets are in most respects quite orthodox: the subject will develop along the same lines as would any new branch of applied mathematics. Perhaps the first paper explicitly linking the theories of fuzzy and statistical uncertainty is that of Zadeh.[(123)] Readers interested in pursuing theories of uncertainty more deeply cannot begin more efficiently than with the preface of the recent bibliography by Gaines and Kohout.[(45)] Sections 2.2–2.4 of this paper explicitly survey dozens of key references to imprecision and related topics.

Exercises

H3.1. Describe a physical process which can be modeled with a deterministic, stochastic, or fuzzy scheme.

H3.2. Give examples of physical processes which seem most aptly modeled by deterministic, stochastic, and fuzzy schemes, respectively.

H3.3. Let X be the set of real numbers "close to" $x_0 \in \mathbb{R}$. Postulate three fuzzy models of X, by specifying three membership functions for $x \in X$, say u_1, u_2, and u_3. Make u_1 discrete; u_2 continuous but not differentiable at x_0; and u_3 continuously differentiable at x_0. Describe how you would parametrize each function using data from the process.

H3.4. Repeat H3.3, letting u_1, u_2, and u_3 be probability density functions instead of membership functions.

H3.5. Postulate a deterministic model for the situation in H3.3.

2

Partitions and Relations

In this chapter the foundations for models of subsequent chapters are discussed. S4 contains some definitions and first properties of fuzzy sets. The important device of mathematical embedding first appears here. Interestingly enough, however, a property of the embedded structure (that a subset and its complement reproduce the original set under union) which is *not* preserved under embedment turns out to be one of its most useful aspects! In S5 partitions of finite data sets are given a matrix-theoretic characterization, culminating in a theorem which highlights the difference between conventional and fuzzy partitions. S6 is devoted to exploration of the algebraic and geometric nature of the partition embedment: the main results are that fuzzy partition space is compact, convex, and has dimension $n(c-1)$. S7 considers hard and fuzzy *relations* in finite data sets and records some connections between relations and partitions that tie the observational and relation-theoretic approaches together.

S4. The Algebra of Fuzzy Sets

For readers unfamiliar with mathematical shorthand, we define some symbols to be used hereafter. If X is any set and x is an element of X, we write $x \in X$. The logical connectives "implies," "is implied by," and "if and only if" are denoted, respectively, by $\Rightarrow$, $\Leftarrow$, and $\Leftrightarrow$. The algebra of the power set $P(X)$ of X—that is, of the set of (hard) subsets of X—is formulated in terms of some familiar operations and relations: let A, $B \in P(X)$;

Containment: $A \subset B \Leftrightarrow x \in A \Rightarrow x \in B$ (4.1a)

Equality: $A \doteq B \Leftrightarrow A \subset B$ and $B \subset A$ (4.1b)

Complement: $\tilde{A} \doteq \{x \in X \mid x \notin A\} \doteq X - A$ (4.1c)

Intersection: $A \cap B \doteq \{x \in X \mid x \in A \text{ and } x \in B\}$ (4.1d)

Union: $A \cup B \doteq \{x \in X \mid x \in A \text{ or } x \in B \text{ or both}\}$ (4.1e)

We indicate that A is a subset of X by writing $A \subset X$; the same A is an element (one member) of $P(X)$, so $A \subset X \Rightarrow A \in P(X)$. In (4.1c), $x \notin A$ means x is *not* in A. The complement, intersection, and union are new elements of $P(X)$, that is, new subsets of X, the symbol $\doteq$ meaning "is defined to be." It is extremely convenient for later sections to have at our disposal the symbols $\forall$ ("for all"); $\exists$ ("there exists"); and $\ni$ ("such that"). The set containing no elements is the empty set, denoted hereafter as $\varnothing$. The zero vector in vector spaces is denoted as $\boldsymbol{\theta}$.

The quintuple of primitive operations exhibited in (4.1) is indicated by ($\subset$, $=$, $\tilde{}$, $\cap$, $\cup$). Different algebraic structures can be erected by requiring stipulated behavior for various combinations of these operations applied to elements of $P(X)$, or more generally, to elements of *any* family $\mathcal{F}(X)$ of subsets of X. Although various generalizations of the material to follow can be given, the structure defined below is the one of primary interest for the applications to follow, because real data sets are with few exceptions finite. Subsequently, we assume X to be finite, its cardinality (the number of elements in it) being n, indicated by $|X| = n$. In this case, $|P(X)| = 2^n$, and $P(X)$ is the largest family of subsets of X. The pair $(X, P(X))$ is called a (finite) σ-algebra, because with (4.1) the elements of family $P(X)$ satisfy three properties:

$$\varnothing \subset P(X) \tag{4.2a}$$

$$A \in P(X) \Rightarrow \tilde{A} \in P(X) \tag{4.2b}$$

$$A, B \in P(X) \Rightarrow A \cup B \in P(X) \tag{4.2c}$$

A number of simple and useful consequences derive from (4.2). In particular, (4.2a) and (4.2b) $\Rightarrow X \in P(X)$. Further, (4.2b) and (4.2c) assure that $A \cap B \in P(X)$ whenever A, B are, via the De Morgan laws,

$$\widetilde{A \cup B} = \tilde{A} \cap \tilde{B} \tag{4.3a}$$

$$\widetilde{A \cap B} = \tilde{A} \cup \tilde{B} \tag{4.3b}$$

for all $A, B \in P(X)$. An easy inductive argument establishes generalizations of (4.3) to any finite number of factors, namely,

$$\widetilde{\bigcup_{i=1}^{n} A_i} = \bigcap_{i=1}^{n} \tilde{A}_i \tag{4.4a}$$

$$\widetilde{\bigcap_{i=1}^{n} A_i} = \bigcup_{i=1}^{n} \tilde{A}_i \tag{4.4b}$$

Since fuzzy sets are realized mathematically via *functions*, it is instructive to give a dual characterization of (X, $P(X)$, $\subset$, $=$, $\tilde{}$, $\cap$, $\cup$) in terms of characteristic functions. Let $A \in P(X)$. The function $u_A : X \to \{0, 1\}$ defined by

$$u_A(x) = \begin{Bmatrix} 1, & x \in A \\ 0, & \text{otherwise} \end{Bmatrix} \tag{4.5}$$

is the characteristic (or indicator) function of hard subset $A \subset X$. To each $A \in P(X)$ there corresponds a unique u_A: denote by $P(F)$ the set of all characteristic functions on X. Equation (4.5) provides a one-to-one correspondence between the elements of $P(X)$ and $P(F)$, an eventuality indicated more briefly by writing $P(X) \leftrightarrow P(F)$. The set-theoretic operations and relations in (4.1) are equivalent to the following function-theoretic operations defined for all u_A, u_B in $P(F)$ and all $x \in X$:

$$\text{Containment:} \quad u_A \leqslant u_B \Leftrightarrow u_A(x) \leqslant u_B(x) \tag{4.6a}$$

$$\text{Equality:} \quad u_A = u_B \Leftrightarrow u_A(x) = u_B(x) \tag{4.6b}$$

$$\text{Complement:} \quad \tilde{u}_A(x) = u_{\tilde{A}}(x) = 1 - u_A(x) \tag{4.6c}$$

$$\begin{aligned} \text{Intersection:} \quad u_{A \cap B}(x) &= (u_A \wedge u_B)(x) \\ &= \min\{u_A(x), u_B(x)\} \end{aligned} \tag{4.6d}$$

$$\begin{aligned} \text{Union:} \quad u_{A \cup B}(x) &= (u_A \vee u_B)(x) \\ &= \max\{u_A(x), u_B(x)\} \end{aligned} \tag{4.6e}$$

In (4.6d) and (4.6e) $\wedge$, $\vee$ are the minimum and maximum operators, respectively. With operations (4.6), $(X, P(F))$ is endowed with the σ-algebra structure of axioms (4.2). The empty set $\varnothing$ is the constant function $\mathbb{0} \in P(F)$, $\mathbb{0}(x) = 0 \ \forall \ x \in X$; the whole set X is the constant function $\mathbb{1} \in P(F)$, $\mathbb{1}(x) = 1 \ \forall \ x \in X$. Since $P(X) \leftrightarrow P(F)$ as sets, and $(\subset, =, \tilde{}, \cap, \cup)$, $(\leqslant, =, \tilde{}, \wedge, \vee)$ equip them with identical algebraic structures, we say they are "isomorphic," indicated more briefly by writing

$$(X, P(X), \subset, =, \tilde{}, \cap, \cup) \approx (X, P(F), \leqslant, =, \tilde{}, \wedge, \vee)$$

In other words, the behavior of subsets of X is entirely equivalent to the behavior of characteristic functions on X. This suggests that there is no harm (although it is a bit unusual) in calling $u_A \in P(F)$ a hard "set"; or conversely, regarding $A \in P(X)$ as a characteristic "function," because the set-theoretic and function-theoretic structures are indistinguishable. One reason for emphasizing this is that the set-theoretic characterization is "missing," so to speak, in the fuzzy sets generalization of $(X, P(F))$.

Following Zadeh's original idea,[122] we now formally define a *fuzzy subset of* X as a *function* $u : X \to [0, 1]$. A finite σ-algebra of fuzzy subsets of X can be constructed simply by imitating equations (4.6). If we denote by

$P_f(F)$ the set of all fuzzy subsets of X (equivalently, the set of all functions u defined on X and valued in $[0, 1]$), then $(X, P_f(F), \leq, =, \tilde{}, \wedge, \vee)$ is a finite σ-algebra in the sense of requirements (4.2). Moreover, all the characteristic functions on X are in $P_f(F)$, so the structure $(X, P(F))$ of hard subsets of X is embedded in the structure $(X, P_f(F))$ of fuzzy subsets of X. The De Morgan laws (4.3) and (4.4) carry easily to $(X, P_f(F))$, and the fuzzy σ-algebra behaves in most respects quite normally.

There is an interesting and important property of $P(F)$ which is not preserved in the embedment of $P(F)$ in $P_f(F)$. As a result of definitions (4.1c)–(4.1e), any hard set $A \subset X$ and its complement have the following properties:

$$A \cup \tilde{A} = X \tag{4.7a}$$

$$A \cap \tilde{A} = \varnothing \tag{4.7b}$$

Note that conditions (4.7) hold for $A = X \Rightarrow \tilde{X} = \varnothing$ and conversely. We preclude these trivial partitions by a third requirement:

$$\begin{Bmatrix} \varnothing \Subset A \Subset X \\ \varnothing \Subset \tilde{A} \Subset X \end{Bmatrix} \tag{4.7c}$$

where $\Subset$ denotes *proper* containment (precludes equality). Properties (4.7) assert that A and $\tilde{A}$ are collectively exhaustive (reproduce all of X under union); mutually exclusive (are disjoint under intersection); and nonempty. These are the familiar requirements for a *partition* of X; specifically, we call the pair $(A, \tilde{A})$ a *hard 2-partition of* X.

In terms of their function-theoretic duals, $\mathbb{0} \leftrightarrow \varnothing$ and $\mathbb{1} \leftrightarrow X$, properties (4.7) are equivalent to

$$u_A \vee \tilde{u}_A = \mathbb{1} \tag{4.8a}$$

$$u_A \wedge \tilde{u}_A = \mathbb{0} \tag{4.8b}$$

$$\mathbb{0} < u_A < \mathbb{1} \tag{4.8c}$$

Equations (4.8) are equalities about functions in $P(F)$, just as (4.7) concern equalities for sets in $P(X)$. (4.8a), for example, means that $\forall x \in X$,

$$(u_A \vee \tilde{u}_A)(x) = \max\{u_A(x), \tilde{u}_A(x)\} = \max\{u_A(x), 1 - u_A(x)\} = \mathbb{1}(x) = 1$$

because either $x \in A$ [so $u_A(x) = 1$], or $x \in \tilde{A}$ [so $u_A(x) = 0$ and $1 - u_A(x) = \tilde{u}_A(x) = 1$], so $\max\{1, 0\} = \max\{0, 1\} = 1$ for every $x \in X$. Property (4.8c) asserts that A is neither empty ($u_A > \mathbb{0}$), nor all of X ($u_A < \mathbb{1}$). Of course, (4.8c) occurs if and only if $\mathbb{0} < \tilde{u}_A < \mathbb{1}$, so the same is true for $\tilde{A}$.

Properties (4.7) or (4.8) do *not*, with definitions (4.6c)–(4.6e), hold for fuzzy subsets of X. To see this, consider the fuzzy subset

$$u_{0.5}(x) = 0.5 \qquad \forall x \in X$$

$$\Rightarrow \tilde{u}_{0.5}(x) = 1 - u_{0.5}(x) = 0.5 \qquad \forall x \in X$$

$$\Rightarrow u_{0.5} = \tilde{u}_{0.5}$$

That is, $u_{0.5}$ equals its complement in $P_f(F)$! From this it follows that both the union and intersection of $u_{0.5}$ with its complement are again $u_{0.5}$:

$$u_{0.5} \vee \tilde{u}_{0.5} = u_{0.5} \qquad (\text{not } \mathbb{1})$$

$$u_{0.5} \wedge \tilde{u}_{0.5} = u_{0.5} \qquad (\text{not } \mathbb{0})$$

Note, however, that the pointwise sum of $u_{0.5}$ and $\tilde{u}_{0.5}$ is 1: of course, this example is an extreme case, because $u_{0.5}$ is the unique element of $P_f(F)$ which is reproduced by complementation. The set $u_{0.5}$ illustrates rather dramatically that properties (4.7) or (4.8) do not generally obtain in $(X, P_f(F))$. In fact, *they will hold if and only if* u_A from $P_f(F)$ is actually in $P(F)$, i.e., u_A is a *hard* subset of X. For otherwise, there will be at least one $x^* \in X$ so that

$$0 < u_A(x^*) < 1 \Rightarrow 0 < \tilde{u}_A(x^*) < 1$$

$$\Rightarrow \begin{cases} 0 < (u_A \vee \tilde{u}_A)(x^*) < 1 \Rightarrow \mathbb{0} < (u_A \vee \widetilde{u_A}) < \mathbb{1} \\ 0 < (u_A \wedge \tilde{u}_A)(x^*) < 1 \Rightarrow \mathbb{0} < (u_A \wedge \widetilde{u_A}) < \mathbb{1} \end{cases}$$

Thus, (4.7) or (4.8) do *not* hold for $(X, P_f(F))$: the union of a fuzzy set and its complement need not recover all of X, nor need their intersection be empty.

This seemingly innocent difference—which arises as a result of definitions (4.6c)–(4.6e)—is sometimes misinterpreted as a defect of the fuzzy embedding. However, it is precisely this difference that can be exploited to great advantage in many instances. We emphasize the importance of this for subsequent sections by formally summarizing the discussion above as follows.

(P4.1) *Proposition 4.1.* No subset u_A of X which has at least one membership greater than zero and less than one can, with its complement $\tilde{u}_A$, form a hard 2-partition of X. Moreover, both the union and intersection of u_A and $\tilde{u}_A$ are nonempty, and satisfy

$$\mathbb{0} < (u_A \wedge \tilde{u}_A) \leq (u_A \vee \tilde{u}_A) < \mathbb{1} \tag{4.9}$$

Because the pattern recognition problems of S1 often require the partitioning of data sets, (P4.1) raises the following question: what requirements should we ask a fuzzy set and its complement to satisfy in order that it might be called a *fuzzy 2-partition of X*? The embedding principle suggests that any reasonable definition should reduce to equations (4.8) in case u_A is actually hard. The example above shows that neither (4.8a) nor (4.8b) can, in general, be satisfied by an arbitrary fuzzy set and its complement. There is, however, a form of this pair of conditions which always holds for hard 2-partitions of X that *is* equivalent to (4.8a) and (4.8b):

$$(u_A + \tilde{u}_A)(x) = u_A(x) + \tilde{u}_A(x) = u_A(x) + [1 - u_A(x)] = \mathbb{1}(x) = 1 \quad (4.10)$$

$\forall x \in X$ *and* all $u_A \in P_f(F)$. If u_A is a hard subset of X, $u_A(x) = 1 \Leftrightarrow \tilde{u}_A(x) = 0$ and conversely, $\forall x \in X$; from this (4.8a) and (4.8b) follow. If with this we require u_A and $\tilde{u}_A$ to be nonempty fuzzy subsets of X, we can formulate a definition which reduces to conditions (4.8) for hard subsets of X.

(D4.1) *Fuzzy 2-Partition.* Let X be any set, and $P_f(F)$ be the set of all fuzzy subsets of X. The pair $(u_A, \tilde{u}_A)$ is a *fuzzy 2-partition* of X if

$$u_A + \tilde{u}_A = \mathbb{1} \quad (4.11a)$$

$$\mathbb{0} < u_A < \mathbb{1} \quad (4.11b)$$

Since (4.11a) is, *by definition*, true for every $u_A \in P_f(F)$, and (4.11b) simply requires non-emptiness of u_A and $\tilde{u}_A$, (D4.1) may seem somewhat vacuous. It is made here for two reasons: (i) to illustrate in the simplest case that equations (4.11) reduce to (4.8) if $u_A \in P(F)$ is *hard*; and (ii) it suggests an appropriate (nontrivial) generalization when more than two partitioning subsets are involved. This we do in Section 5 below.

Remarks

Definitions (4.6) follow Zadeh's original 1965 paper.[(122)] Although he was well aware of the apparent anomaly that the *union* of u_A and $\tilde{u}_A$ as defined at (4.6e) need not reproduce all of X, it was Ruspini[(89)] who first observed that because the *sum* $(u_A + \tilde{u}_A)$ always did, this could be an advantage rather than a defect. A formal construction of the algebraic isomorphism between the finite σ-algebras $(X, P(X))$ and $(X, P(F))$ is given in (10). Kauffman[(62)] contains an excellent treatment of many other operations and relations which various authors have defined on $P_f(F)$; those which are germane to our exposition will be introduced as required. Many algebraically oriented papers with the flavor of lattice theory and multivalued logic have concerned themselves with extensions of Zadeh's original structure: among the most extensive are those of Goguen[(47)] and Klaua[(66)]—see (45) for a more complete discussion and literature survey. Another vein of generalization is from the viewpoint of category theory, the leading proponents of which are perhaps Arbib and Manes.[3] In (44) Gaines reviews a series of papers concerning the evolution of fuzzy logic stemming from a desire to circumvent some of the classic paradoxes of hard set theory. The reader interested in a clear and concise introduction to the rationale and construction of fuzzy sets as a basis for most applied work can do no better than Zadeh's original papers: (122) is particularly lucid in this respect.

Exercises

H4.1. Prove the De Morgan laws, (4.4), for hard sets.

H4.2. Prove the duals of (4.4), when hard sets are represented by their characteristic functions.

H4.3. Show that $(X, P_f(F), \leq, =, \bar{\ }, \wedge, \vee)$ is a finite σ-algebra, by verifying the appropriate generalizations of (4.2).

H4.4. Verify that the De Morgan laws (4.4) hold for fuzzy sets in $(X, P_f(F))$.

H4.5. Prove that the union $(u_i \vee u_j)$ of fuzzy sets u_i and u_j is the smallest fuzzy set containing u_i and u_j. That is, if $u_i, u_j \subset u_k$, then $(u_i \vee u_j) \subset u_k$ for all fuzzy sets u_k.[122]

H4.6. Prove that the intersection $(u_i \wedge u_j)$ of fuzzy sets u_i and u_j is the largest fuzzy set contained in both u_i and u_j.[122]

H4.7. Let a, b be real numbers. Show that
(i) $1 - (a \vee b) = (1 - a) \wedge (1 - b)$;
(ii) $1 - (a \wedge b) = (1 - a) \vee (1 - b)$;
(iii) $a \vee b = \frac{1}{2}(a + b - |a - b|)$;
(iv) $a \wedge b = \frac{1}{2}(a + b + |a - b|)$.

H4.8. Let $g: \mathbb{R} \to \mathbb{R}$; $a = g(x)$; $b = g(y)$, where $x, y \in \mathbb{R}$. Restate the results of H4.7 in terms of g.

H4.9. $A = [0, 5]$, $f: A \to [0, 1]$, $f(x) = e^{-x}$. $g: A \to [0, 1]$, $g(x) = (x/5)$. Sketch the graphs of the fuzzy sets $f, g, f \wedge g, f \vee g, \tilde{f}, \tilde{g}$. Find the set of points in A that have equal memberships in f and g; in $f \vee g$ and $f \wedge g$; in f and $\tilde{f}$; in g and $\tilde{g}$.

H4.10. Membership functions (fuzzy sets) can be expressed parametrically, just as probability density functions often are. Let α, β, γ be parameters, x the real variable. Two common types of parametric fuzzy sets are

(i)
$$u(x; \alpha, \beta, \gamma) = \begin{cases} 0, & x \leq \alpha \\ 2\left(\dfrac{x-\alpha}{\gamma-\alpha}\right)^2, & \alpha < x < \beta \\ 1 - 2\left(\dfrac{x-\gamma}{\gamma-\alpha}\right)^2, & \beta \leq x \leq \gamma \\ 1, & \gamma < x \end{cases}$$

(ii)
$$w(x;\beta,\gamma)=\begin{cases} u(x;\gamma-\beta,\gamma-(\beta/2),\gamma), & x<\gamma \\ 1-u(x;\gamma,\gamma+(\beta/2),\gamma+\beta), & \gamma\leqslant x \end{cases}$$

(iii) Graph u. What geometric property of the graph is parametrized by α, β, γ? [Answer: $\beta = (\alpha + \gamma)/2$ = point of inflection.]

(iv) Graph w. What do β and γ parametrize for this fuzzy set? (Answer: β = bandwidth between points of inflection; γ = unique maximum of w.)

(v) What families of probabilistic functions resemble u and w? Identify the statistical parameters of these functions corresponding to β for u; and to β and γ for w. (Answer: $u \sim$ normal CDF; $\beta \sim$ median; $w \sim$ normal PDF; $\gamma \sim$ mean; $\beta \sim$ standard deviation.)

(vi) Describe a physical process of which u or w might be a realistic representation.

H4.11. Let $L \in \mathbb{R}$ be fixed. Graph the fuzzy membership function $u(x) = \{1 + [(x - L)/(\lambda L)]^2\}^{-1}$, $x \in \mathbb{R}$, $0 < \lambda < 1$. This fuzzy set has been used to represent the instruction "go about L meters," with λ a distance-fixing parameter.

H4.12. Let $x \in \mathbb{R}$ $[u(x) = (1 + x^{-2})^{-1}; x \geqslant 0; u(x) = 0; x < 0]$. Graph the fuzzy set u. This function has been used to represent "numbers much greater than zero."

H4.13. Let B be the σ-algebra of Borel sets in $\mathbb{R}^n$, let p render $(\mathbb{R}^n, B, p)$ a probability measure space [cf. (86)], and let $u : \mathbb{R}^n \to [0, 1]$ be Borel measurable. Then the number [see (123)]

$$\Pr(u) = \underset{\mathbb{R}^n}{\int\int\cdots\int} u(x)\,dp$$

is called the probability of the fuzzy set ("event") u, because it is the expected value of the membership function of fuzzy set u. Show that for fuzzy sets u, v and $\{u_i\}$ of this kind,

(i) $u \subset w \Rightarrow \Pr(u) \leqslant \Pr(w)$

(ii) $\Pr(u \vee w) = \Pr(u) + \Pr(w) - \Pr(u \wedge w)$

(iii) $\Pr\left(\bigvee_{i=1}^{k} u_i\right) = \sum_{i=1}^{k} \Pr(u_i) - \sum\sum_{i\neq j} \Pr(u_i \wedge u_j) + \cdots + (-1)^k \Pr\left(\bigwedge_{i=1}^{k}\right)$

(iv) $\Pr\left(\bigvee_{i=1}^{\infty} u_i\right) \leqslant \sum_{i=1}^{\infty} \Pr(u_i)$

H4.14. In the context of H4.13, u is a *Borel fuzzy set* if $L_\alpha = \{\mathbf{x} \in \mathbb{R}^n \mid u(\mathbf{x}) \leqslant \alpha\} \in B\,\forall\alpha \in [0, 1]$. Show that the collection of Borel fuzzy sets is a σ-algebra with respect to the operations (4.6). Note that (4.2c) must hold for *any* family of Borel fuzzy sets.

S5. Fuzzy Partition Spaces

If X is a finite set, say $X = \{x_1, x_2, \ldots, x_n\}$, the generalization of equations (4.7) or (4.8) for any c hard subsets of X, c an integer, $2 \leq c \leq n$, is quite straightforward. Thus, a family $\{A_i : 1 \leq i \leq c\} \subset P(X)$ is a hard c-partition of X in case

$$\bigcup_{i=1}^{c} A_i = X \tag{5.1a}$$

$$A_i \cap A_j = \varnothing, \qquad 1 \leq i \neq j \leq c \tag{5.1b}$$

$$\varnothing \subsetneq A_i \subsetneq X, \qquad 1 \leq i \leq c \tag{5.1c}$$

Equations (5.1) reduce to (4.7) if $c = 2$. If $c = n$, each A_i is necessarily a singleton, $A_i = \{x_i\}\ \forall i$: since this is a trivial case, the range of c is usually $2 \leq c < n$. The function-theoretic duals of (5.1) are

$$\bigvee_{i=1}^{c} u_i = \mathbb{1} \tag{5.2a}$$

$$u_i \wedge u_j = \mathbb{0}, \qquad 1 \leq i \neq j \leq c \tag{5.2b}$$

$$\mathbb{0} < u_i < \mathbb{1}, \qquad 1 \leq i \leq c \tag{5.2c}$$

In (5.2) u_i is the characteristic function of A_i, that is,

$$u_{ik} \doteq u_i(x_k) = \begin{cases} 1, & x_k \in A_i \\ 0, & x_k \notin A_i \end{cases} \tag{5.3}$$

To be consistent with our previous discussion, the appropriate notation for u_i would be u_{A_i}; however, the less cumbersome symbol u_i will be used hereafter unless confusion would otherwise ensue. Furthermore, it is extremely convenient to denote $u_i(x_k)$ as u_{ik}, a practice we adhere to throughout. Equations (5.2) reduce to (4.8) in case $c = 2$. Finally, note that equations (4.11) generalize to

$$\sum_{i=1}^{c} u_i = \mathbb{1} \tag{5.4a}$$

$$\mathbb{0} < u_i < \mathbb{1} \qquad \forall i \tag{5.4b}$$

for any c-tuple $(u_1, u_2, \ldots, u_c)$ in the c-fold Cartesian product of $P(F)$ which forms a hard c-partition of X. The functional equality and inequalities in (5.4) can be stated in terms of their values over X as

$$\sum_{i=1}^{c} u_{ik} = 1, \qquad \forall x_k \in X \tag{5.5a}$$

$$\left\{\begin{matrix} \exists k \text{ so that } 0 < u_{ik} = 1 \\ \exists j \text{ so that } 0 = u_{ij} < 1 \end{matrix}\right\}, \qquad 1 \leq i \leq c \tag{5.5b}$$

Condition (5.5b) states that there is at least one, but not all n, of the x_k's in the ith cluster, for $1 \leq i \leq c$.

Equations (5.5), together with the knowledge that every u_{ik} is either zero or one, lead to a very useful alternative characterization of hard c-partitions of X—one that we use almost exclusively in later sections. It is the *matrix* representation: to this end, let V_{cn} denote the vector space of $(c \times n)$ real matrices over $\mathbb{R}$ (the reals), equipped with the usual scalar multiplication and vector addition (for matrices). Because of equations (5.5), a matrix $U = [u_{ik}]$ represents a hard c-partition of X when and only when its elements satisfy three conditions:

$$u_{ik} \in \{0, 1\}, \qquad 1 \leq i \leq c, \qquad 1 \leq k \leq n \tag{5.6a}$$

$$\sum_{i=1}^{c} u_{ik} = 1, \qquad 1 \leq k \leq n \tag{5.6b}$$

$$0 < \sum_{k=1}^{n} u_{ik} < n, \qquad 1 \leq i \leq c \tag{5.6c}$$

The collection of all matrices in V_{cn} whose entries satisfy (5.6) are isomorphic to the set of all ordered c-tuples $\{A_i\} \in [P(X)]^c$ of hard subsets of X which satisfy (5.1); and to the set of all ordered c-tuples $\{u_i\} \in [P(F)]^c$ of characteristic functions on X which satisfy (5.2), (5.4), or (5.5). Because of this isomorphism, we can call any of the three—that is, $\{A_i\}$, $\{u_i\}$, or U, a hard c-partition of X.

(D5.1) *Hard c-Partition.* $X = \{x_1, x_2, \ldots, x_n\}$ is any finite set; V_{cn} is the set of real $c \times n$ matrices; c is an integer, $2 \leq c < n$. *Hard c-partition space for* X is the set

$$M_c = \left\{ U \in V_{cn} \;\middle|\; u_{ik} \in \{0, 1\} \forall i, k;\ \sum_{i=1}^{c} u_{ik} = 1 \;\forall k;\ 0 < \sum_{k=1}^{n} u_{ik} < n \;\forall i \right\} \tag{5.7}$$

For convenience in referencing, we remind the reader how the entries of $U \in M_c$ are interpreted, and what each of the three conditions means. Row i of U, say $U_{(i)} = (u_{i1}, u_{i2}, \ldots, u_{in})$, exhibits (*values of*) the characteristic function of the ith partitioning subset of X: $u_{ik} = u_i(x_k)$ is one or zero, according as x_k is or is not in the ith subset; $\sum_i u_{ik} = 1 \;\forall k$ means each x_k is in exactly one of the c subsets; and $0 < \sum_k u_{ik} < n \;\forall i$ means that no subset is empty, and no subset is all of X: in other words, $2 \leq c < n$.

Although we have yet to discuss any specific algorithms, it is appropriate to note here that M_c is the set of admissible solutions for the conventional (hard) cluster analysis problem with respect to X (cf. S1 above). To illustrate, suppose X has three elements, each a vector of characteristics of a fruit as follows.

(E5.1) *Example 5.1.*

$$X = \{x_1 = \text{peach}, x_2 = \text{nectarine}, x_3 = \text{plum}\} \tag{5.8}$$

What are the hard 2-partitions of X? There are but three, up to an arrangement of rows:

$$U_1 = \overset{\begin{matrix} x_1 & x_2 & x_3 \end{matrix}}{\begin{bmatrix} 1 & 1 & 0 \\ 0 & 0 & 1 \end{bmatrix}}; \qquad \text{(or) } U_2 = \overset{\begin{matrix} x_1 & x_2 & x_3 \end{matrix}}{\begin{bmatrix} 1 & 0 & 0 \\ 0 & 1 & 1 \end{bmatrix}};$$

$$\text{(or) } U_3 = \begin{bmatrix} 1 & 0 & 1 \\ 0 & 1 & 0 \end{bmatrix} \tag{5.9}$$

Constraint (5.6b) rules out matrices such as

$$\overset{\begin{matrix} x_1 & x_2 & x_3 \end{matrix}}{\begin{bmatrix} 1 & 1 & 0 \\ 1 & 0 & 1 \end{bmatrix}}$$

because x_1 would then be in $A_1 \cap A_2$. Constraint (5.6c) rules out matrices such as

$$\overset{\begin{matrix} x_1 & x_2 & x_3 \end{matrix}}{\begin{bmatrix} 1 & 1 & 1 \\ 0 & 0 & 0 \end{bmatrix}}$$

for then $A_2 = \varnothing$; $A_1 = X$, and $c = 1$, not 2. •

(E5.1) was not accidental. Indeed, it exhibits a very disappointing constraint the mathematical model chosen places on the physical process it purports to represent, viz., that x_2 (the nectarine—a peach–plum HYBRID!), must be in (5.9), unequivocally classified with either x_1 (the peach), or x_3 (the plum). And in either case, the matrix used in (5.9) can give no indication of the mathematical (and, therefore, physical) dependence or relationship of x_2 to each of its progenitors. Surely this is counter to one's intuitive expectation for the features of x_2. Thus, the hard model has an intrinsic sort of "physical intractability" which is often at odds with properties of the real processes it purports to represent. Matrix U_3 represents the worst possible sort of behavior: the peach and plum are grouped together, while the nectarine becomes a separate class. We mention here—and exemplify later—that the discreteness of M_c endows it with analytical and algorithmic intractabilities as well.

The difficulties manifested by M_c are nicely circumvented by allowing fuzzy subsets to partition X. The formal definition, due to Ruspini,[89] generalizes equations (4.11) in exactly the same way that (D5.1) did, with the obvious exception that each u_{ik} be in [0, 1] rather than {0, 1}:

(D5.2) *Fuzzy c-Partition.* X is any finite set; V_{cn} is the set of real $c \times n$ matrices; c is an integer; $2 \leqslant c < n$. *Fuzzy c-partition space for* X is the set

$$M_{fc} = \left\{ U \in V_{cn} \;\middle|\; u_{ik} \in [0, 1] \; \forall i, k; \; \sum_{i=1}^{c} u_{ik} = 1 \; \forall k; \, 0 < \sum_{k=1}^{n} u_{ik} < n \; \forall i \right\} \tag{5.10}$$

Row i of a matrix $U \in M_{fc}$ exhibits (values of) the ith membership function (or ith fuzzy subset) u_i in the fuzzy c-partition U of X. Because each column sum is 1, the *total* membership of each x_k in X is still 1, but since $0 \leqslant u_{ik} \leqslant 1 \; \forall i, k$, it is possible for each x_k to have an otherwise arbitrary distribution of membership among the c fuzzy subsets $\{u_i\}$ partitioning X. There may, of course, be one or more columns of U which assign all of the membership of some x_k to a single u_i; indeed, M_c is clearly a finite subset of M_{fc}.

As an example, consider the situation in (E5.1), using as data the set X of equation (5.8). In view of the constraints for $U \in M_{fc}$, there are infinitely many fuzzy 2-partitions of X [including the three hard matrices at (5.9)]. A typical fuzzy 2-partition which exhibits quite nicely the potential utility of this imbedding is

$$\begin{matrix} & x_1 & x_2 & x_3 \\ U = & \left[\begin{matrix} 0.91 \\ 0.09 \end{matrix}\right. & \begin{matrix} 0.58 \\ 0.42 \end{matrix} & \left.\begin{matrix} 0.13 \\ 0.87 \end{matrix}\right] \end{matrix} \tag{5.11}$$

The membership values in (5.11) indicate that x_1(peach) and x_3(plum) have high affinities for different subclasses, while x_2(nectarine) has features which demand relatively higher *partial* membership in both of these fuzzy clusters—the situation we anticipate from an intuitive idea of the real relationship between members of the data. In other words, M_{fc} has a significantly higher potential than M_c for modeling the physical realities of data set X. Readers with real data sets will immediately ask: can partitions such as (5.11) be generated *from the data*? Yes! We discuss several algorithms in subsequent sections which do this.

This section ends with the generalization of (P4.1) which advertises the main difference between hard and fuzzy c-partitions of X:

(T5.1) *Theorem 5.1.* Suppose $U \in M_{fc}$ is a fuzzy c-partition of X as defined at (5.10). Then

$$\begin{Bmatrix} u_i \wedge u_j = u_{i \cap j} = \mathbb{0} \\ \text{for } all\ i \neq j \end{Bmatrix} \Leftrightarrow U \text{ is } hard, \text{ i.e., } U \in M_c \tag{5.12}$$

Analysis. (T5.1) says that *all* pairwise intersections in the partition are empty when and only when *membership sharing is not allowed*—that is, U is not fuzzy after all. To prove the theorem, look at the columns of U. Suppose

$$U^{(k)} = \overset{x_k}{\begin{pmatrix} u_{1k} \\ \cdot \\ u_{ik} \\ \vdots \\ u_{jk} \\ \cdot \\ u_{ck} \end{pmatrix}} \begin{matrix} \\ \\ \text{— row } i \\ \\ \text{— row } j \\ \\ \\ \end{matrix}$$

is the kth column of U. If any entry, say u_{ik}, is between zero and one, there must be another one, here u_{jk}, which is also. In every instance of this kind, clusters i and j are coupled (via intersection) by $\min\{u_{ik}, u_{jk}\} > 0$, so U is not hard.

Proof. ($\Rightarrow$) assume $u_i \wedge u_j > \mathbb{0}$. Then $\exists k$ such that $(u_i \wedge u_j)(x_k) = \min\{u_{ik}, u_{jk}\} > \mathbb{0}(x_k) = 0$. Thus u_{ik} and u_{jk} are both in $(0, 1)$, and U cannot be in M_c.

($\Leftarrow$) If $U \in M_c$, rows i and j are characteristic functions of the ith and jth hard subsets partitioning U. Thus, they must satisfy (5.2b), the left side of (5.12). •

The point of this result? There must be some "overlap" in at least one pair of fuzzy subsets in every truly fuzzy c-partition of X. This theorem indicates precisely the difference between M_c and M_{fc} which gives M_{fc} a significant advantage, both mathematically and physically, in many clustering and cluster validity problems.

(E5.2) *Example 5.2.* The membership function $u_1 \wedge u_2$ of the intersection of fuzzy clusters u_1 and u_2 in fuzzy partition (5.11) is

$$(u_1 \wedge u_2)(x_1) = \min(0.91, 0.09) = 0.09$$

$$(u_1 \wedge u_2)(x_2) = \min(0.58, 0.42) = 0.42$$

$$(u_1 \wedge u_2)(x_3) = \min(0.13, 0.87) = 0.13$$

Thus all three x_k's have membership in the fuzzy intersection: x_2 has the most, a fact in agreement with the discussion following (E5.1). •

Remarks

(T5.1) has yet to be fully exploited. The material above first appeared in (10), although it was known to both Ruspini[89] and Dunn.[35] M_{fc} arises in other mathematical contexts quite removed from the imbedding of M_c: for example, in certain parametric estimation problems of unsupervised learning,[28] and in Q-factor analysis.[22] This identical mathematical structure may lead to unsuspected cross-connections between ostensibly disparate areas.

Exercises

H5.1. $X = \{(1, 1), (1, 3), (10, 1), (10, 3), (5, 2)\}$. Exhibit the distinct hard 3-partitions of X as 3×5 matrices.

H5.2. Let

$$f(\mathbf{u}) = \sum_{i=1}^{c} (u_i)^2, \qquad \mathbf{u}^T = (u_1, u_2, \ldots, u_c) \in \mathbb{R}^c$$

If the numbers $\{u_i\}$ are unconstrained, what are the minimum and maximum values f can attain? If $\sum_{i=1}^{c} u_i = 1$ is required, what are the minimum and maximum values of f? [Answer: min = $(1/c)$ when $u_i = (1/c)\ \forall i$; max = 1 when some $u_i = 1$.]

H5.3. Let $U \in M_{fc}$, and

$$F(U) = \left(\sum_{k=1}^{n} \sum_{i=1}^{c} (u_{ik})^2 / n \right)$$

Use the results of H5.2 to show that[10]

(i) $(1/c) \leq F(U) \leq 1\ \forall U$

(ii) $F(U) = 1 \Leftrightarrow U$ is hard

(iii) $F(U) = (1/c) \Leftrightarrow u_{ik} = (1/c)\ \forall i, k$

H5.4. Does the function F in H5.3 depend monotonically on c?

S6. Properties of Fuzzy Partition Spaces

In this section we explore some algebraic and geometric properties of M_{fc}.

To begin, note that hard c-partition space M_c is finite but quite large for all but trivial values of c and n. In fact,

$$|M_c| = (1/c!)\left[\sum_{j=1}^{c} \binom{c}{j} (-1)^{c-j} j^n\right] \tag{6.1}$$

is the number of distinct ways to partition X into c nonempty subsets.[34] If, for example, $c = 10$ and $n = 25$, there are roughly 10^{18} distinct hard 10-partitions of the 25 points. Although finiteness is sometimes an advantage, it is clear that the size of M_c will impede search by exhaustion for "optimal" partitionings. Moreover, discreteness can be an analytical intractability.

M_c is properly imbedded in M_{fc}, an infinite set whose cardinality is the same as the reals. To see the geometry of this imbedding, we need some vector space ideas. Vectors are boldface (underlined in figures): thus, e.g., $\mathbf{x}_k \in \mathbb{R}^p$ denotes a p-dimensional column vector whose transpose is $\mathbf{x}_k^T = (x_{k1}, x_{k2}, \ldots, x_{kp})$. A set S in any vector space V is *convex* when S contains the line connecting every pair of points in S; that is, if $\mathbf{x}, \mathbf{y} \in S$, the entire line segment $\alpha\mathbf{x} + (1 - \alpha)\mathbf{y}$, $\alpha \in [0, 1]$, between $\mathbf{x}$ and $\mathbf{y}$ is also in S.

(D6.1) *Convexity.* V is a vector space with $S \subset V$.

$$S \text{ is convex} \Leftrightarrow \mathbf{x}, \mathbf{y} \in S \Rightarrow \alpha\mathbf{x} + (1 - \alpha)\mathbf{y} \in S \qquad \forall \alpha \in [0, 1] \tag{6.2}$$

Convexity carries information about the interior connectivity and shape of the boundary of S: loosely speaking, it prevents "holes" in the interior, "bays" in the boundary. Convex sets can be closed (contain all of their boundary); or open (contain only their interior); or neither: they can be bounded (every line segment has finite length); or unbounded. Some of these ideas are depicted graphically in Fig. 6.1.

If a set is *not* convex, the smallest convex superset of it is called its convex hull:

(D6.2) *Convex Hull.* V is a vector space with $S \subset V$; $\{C_i\}$ is the family of all convex sets such that $S \subset C_i \subset V$.

$$\text{conv}(S) = \bigcap_i C_i \tag{6.3}$$

Equation (6.3) allows us to interpret conv(S) as the "minimal convex completion" of S: of course, if S is convex to begin with, then $S = \text{conv}(S)$.

If a vector in $\mathbf{v} \in V$ can be written as the special linear combination

$$\mathbf{v} = \sum_{k=1}^{m_v} \alpha_k \mathbf{s}_k, \qquad \sum_{k=1}^{m_v} \alpha_k = 1, \qquad \alpha_k \geq 0 \; \forall k \tag{6.4}$$

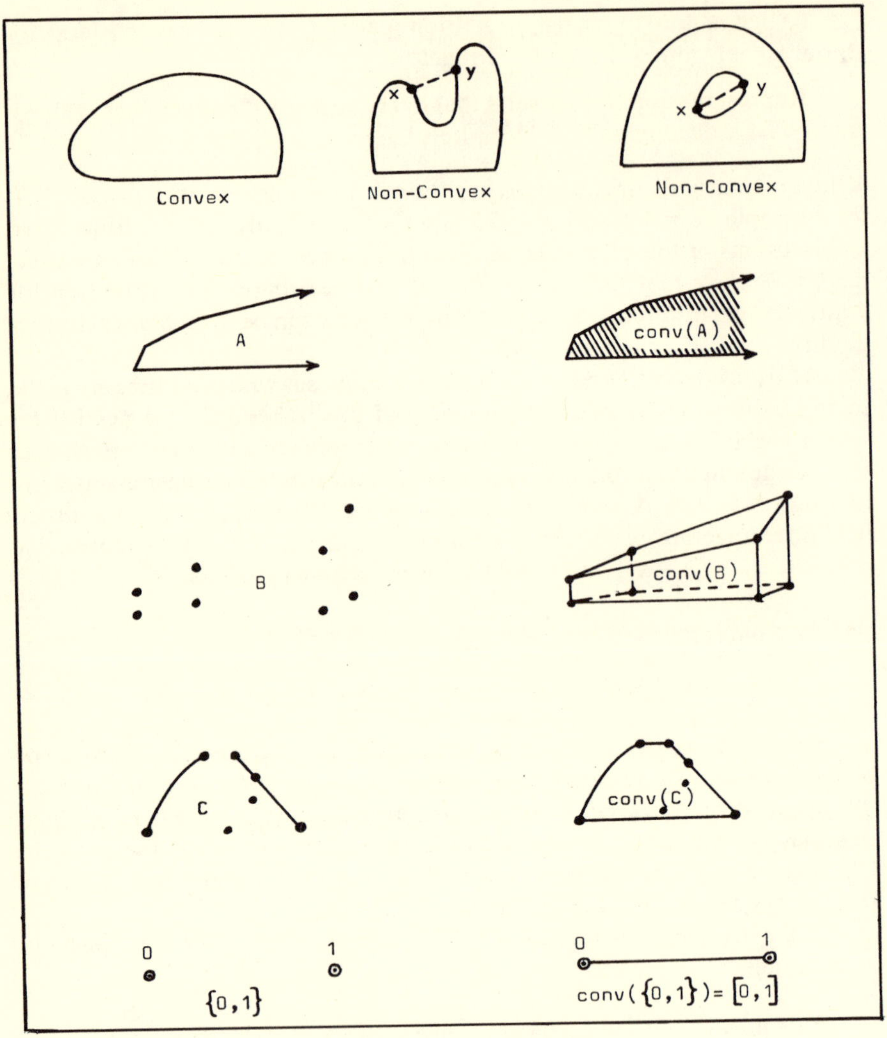

Figure 6.1. Convex sets and hulls.

of m_v vectors $\{\mathbf{s}_k\} \subset S$, $\mathbf{v}$ is called a convex combination of the $\{\mathbf{s}_k\}$. The convex hull of S is constructed as follows.

(T6.1) *Theorem 6.1.* V is a vector space with $S \subset V$:

$$\operatorname{conv}(S) = \left\{ \mathbf{v} \in V \,\middle|\, \mathbf{v} = \sum_{k=1}^{m_v} \alpha_k \mathbf{s}_k;\ \sum_{k=1}^{m_v} \alpha_k = 1;\ \alpha_k \geqslant 0;\ \mathbf{s}_k \in S \right\} \qquad (6.5)$$

Proof. Cf. Roberts and Varberg.[84] •

Some convex hulls are shown in Fig. 6.1.

Representation (6.5) asserts that every vector $\mathbf{v} \in \text{conv}(S)$ has (at least) one convex decomposition by finitely many $\mathbf{s}_k \in S$: the number of terms required (m_v) generally depends on $\mathbf{v}$. The most provocative example in Fig. 6.1 concerns {0, 1}, namely, that

$$[0, 1] = \text{conv}(\{0, 1\}) \tag{6.6}$$

From (6.6) one might suspect that $M_{fc} = \text{conv}(M_c)$, because (6.6) implies that for any integer p,

$$[0, 1] \times [0, 1] \times \cdots \times [0, 1] \doteq ([0, 1])^p$$

is the convex hull of $(\{0, 1\})^p$, where (x) denotes the Cartesian product [for any sets A and B, $A \times B = \{(a, b) | a \in A, b \in B\}$]. Thus,

$$([0, 1])^p = \text{conv}\big((\{0, 1\})^p\big) = \big(\text{conv}(\{0, 1\})\big)^p \tag{6.7}$$

$([0, 1])^3$, for example, is the unit cube in the first octant generated by its eight vertices, as in Fig. 6.2. Note that *each column* of hard 3-partitions is one of the three vertices $\{(1, 0, 0), (0, 1, 0), (0, 0, 1)\} \doteq B_3$; *each column* of a fuzzy 3-partition is one 3-vector from $\text{conv}(B_3)$, the "triangle" of Fig. 6.2a. In other words, column constraint (5.6b) means that $\text{conv}(B_3) \neq [0, 1]^3$, so M_{fc} is at best a proper subset of $([0, 1])^{cn}$. Row constraint (5.6c), in turn, prevents M_{fc} from being $(\text{conv}(B_c))^n$, B_c being the standard basis of $\mathbb{R}^c$. To see this, consider the following.

(E6.1) *Example 6.1.* The convex decomposition $\alpha_1 U_1 + \alpha_2 U_2$ of

$$U = \begin{bmatrix} 1 & 0.01 \\ 0 & 0.99 \end{bmatrix} = 0.99\begin{bmatrix} 1 & 0 \\ 0 & 1 \end{bmatrix} + 0.01\begin{bmatrix} 1 & 1 \\ 0 & 0 \end{bmatrix}$$

shows that something more than M_{f2} is needed to construct $\big(\text{conv}(B_2)\big)^2$, since $U_2 \in \big(\text{conv}(B_2)\big)^2$, but not M_2. $\sum_k u_{ik} < n$ means that no hard vertex can be repeated n times if it forms columns of $U \in M_c$. In this example, we cannot use $(1, 0)^T$ as successive columns of U_2 and have U_2 remain in M_2. •

Another way of saying this is: (5.6b) implies that the row rank of $U \in M_c$ is at most $(c - 1)$; (5.6c) implies that the column rank of $U \in M_c$ is at most $(n - 1)$. A convex fuzzy embedding which is, for *practical* purposes, the same as $M_c \subset M_{f_c}$ can be realized by uncoupling the columns of U. To do this, relax constraint (5.6c):

(D6.3) *Degenerate c-Partitions.* The supersets of M_c (D5.1) and M_{fc} (D5.2) obtained by allowing zero or one rows by relaxing (5.6c), i.e.,

$$0 \leqslant \sum_{k=1}^{n} u_{ik} \leqslant n \qquad \forall i \tag{6.8}$$

are denoted by M_{co} and M_{fco}, called, respectively, *degenerate* hard and fuzzy c-partition spaces for X.

(T6.2) *Theorem 6.2.* If B_c is the standard basis of $\mathbb{R}^c$, then

$$M_{fco} = \text{conv}(M_{co}) = (\text{conv}(B_c))^n \tag{6.9}$$

Analysis. This asserts that every $U \in M_{fco}$ has at least one convex decomposition by a finite number of (possibly) degenerate hard c-partitions of X. More to the point, since $M_{fc} \subset M_{fco}$, the same is true for *non*degenerate fuzzy c-partitions of X. The effect of (6.8) is to "uncouple" the *columns* of $U \in M_{co}$, allowing the *same* column to appear arbitrarily often. Proving (6.9) then reduces to demonstrating that a single fuzzy column can always be written as a convex sum of hard columns. Column constraint (5.6b) ensures this.

Proof. Let $B_c = \{\mathbf{e}_i \mid 1 \leqslant i \leqslant c\}$ denote the standard basis of $\mathbb{R}^c$, $\mathbf{e}_i^T = (0, 0, \ldots, 0, 1, 0, \ldots, 0)$, where the one occurs at the ith entry, $1 \leqslant i \leqslant c$. Single columns of $U \in M_{co}$ are just the $\mathbf{e}_i$'s. Now let $\mathbf{U}_{(k)}^T = (u_{1k}, u_{2k}, \ldots, u_{ck})$ be the kth column of $U \in M_{fco}$. Then

$$\mathbf{U}_{(k)} = \sum_{i=1}^{c} u_{ik}\mathbf{e}_i, \qquad \sum_{i=1}^{c} u_{ik} = 1, \qquad u_{ik} \geqslant 0 \tag{6.10}$$

so $\mathbf{U}_{(k)} \in \text{conv}(B_c)$. Since $\mathbf{U}_{(k)}$ is arbitrary, $\text{conv}(B_c)$ is exactly the set of c-vectors which can be the kth column of U. Under constraint (6.8), the columns of U are independent, so the n-fold Cartesian product of $\text{conv}(B_c)$ is M_{fco}. •

The geometric content of this theorem is illustrated in Fig. 6.2. As noted above, vertices $\mathbf{e}_i \in B_3$ are single columns of hard U's in M_{co}, while 3-vectors on the triangle $\text{conv}(B_3)$ are the possible columns of U's in M_{fco}. Geometrically, every $\mathbf{u}_k \in \text{conv}(B_3)$ can be written as a convex combination of the three vertices. Points in the interior will require three terms, points on the boundaries two.

Figure 6.2 suggests some further properties of M_{fco} that are useful later. Since M_{co} is finite, M_{fco} is called the convex polytope generated by M_{co}. M_{fco} is closed and bounded, and hence compact in V_{cn}. This, and the fact that M_{fco} is convex, ensure that M_{fco} is the convex hull of its extreme points [points

$$\text{Vertices} = \{\underline{v}_i \mid 1 \leqslant i \leqslant 8\}$$

$$(\,[0,1]\,)^3 = \text{conv}(\,\{\underline{v}_i\}\,)$$

$$\{\underline{v}_i\} = \text{ext}((\,[0,1]\,)^3)$$

$$\underline{u}_k \sim \left\{ \begin{array}{l} \underline{u}_k = u_{1k}\underline{e}_1 + u_{2k}\underline{e}_2 + u_{3k}\underline{e}_3 \\ 1 = u_{1k} + u_{2k} + u_{3k}\ ; \\ 0 \leqslant u_{ik} \leqslant 1 \end{array} \right\}$$

$$B_3 = \{\underline{e}_1,\underline{e}_2,\underline{e}_3\} = \text{column space for } M_{3o}$$

$$\text{conv}(B_3) = \text{column space for } M_{f3o}$$

$\underline{\bar{u}} = (1/3,1/3,1/3)$ is the centroid and is orthogonal to $\text{conv}(B_3)$.

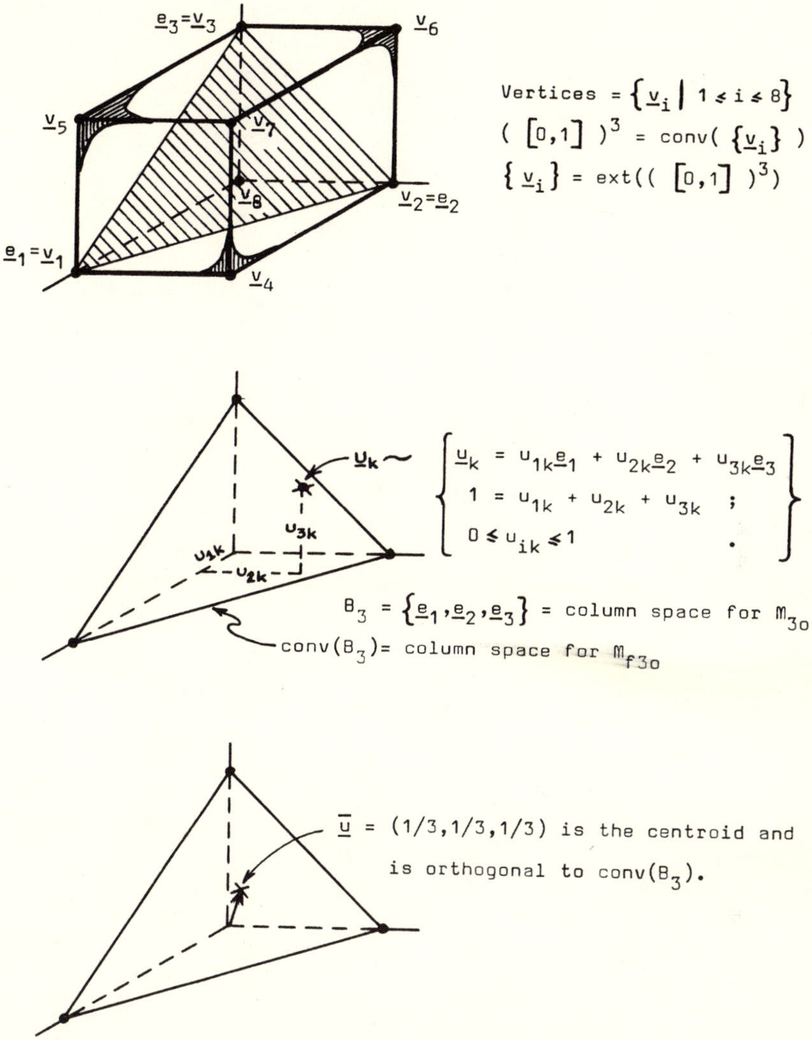

Figure 6.2. Column spaces for $c = 3$.

which cannot be written as $\alpha\mathbf{U} + (1 - \alpha)\mathbf{W}$ for any $\alpha \in (0, 1)$ and $\mathbf{U} \neq \mathbf{W}$]. Thus M_{co} are the extreme points of M_{fco} (in this case, the vertices of M_{fco}). We write $M_{co} = \text{ext}(M_{fco})$ to indicate this. Since $\mathbf{\bar{u}}^T = (1/c, 1/c, \ldots, 1/c)$ is the geometric centroid of $\text{conv}(B_c)$, the matrix $\mathbf{\bar{U}} = [1/c]$ with all cn entries $(1/c)$ is the geometric centroid of M_{fco}. Finally, it is geometrically clear that $\mathbf{\bar{u}}^T = (1/c, 1/c, \ldots, 1/c)$ is orthogonal (in the Euclidean sense) to $\text{conv}(B_c)$.

(T6.3) *Theorem 6.3.* M_{co}, M_{fco} as in (D6.3)

$$M_{fco} \text{ is closed and bounded (compact)} \tag{6.11a}$$

$$M_{co} = \text{ext}(M_{fco}) \tag{6.11b}$$

$$\bar{\mathbf{U}} = [1/c] \text{ is the centroid of } M_{fco} \tag{6.11c}$$

Proof. These results are simple consequences of (T6.2), together with some well-established general theorems on convexity [cf. (84), p. 84]. •

Yet another aspect of M_{fco} is its dimension. The *affine hull* of M_{co} is the set

$$\text{aff}(M_{co}) = \left\{ \mathbf{U} \in V_{cn} \;\middle|\; \mathbf{U} = \sum_{i=1}^{m_u} \alpha_i \mathbf{U}_i;\ \sum_{i=1}^{m_u} \alpha_i = 1;\ \alpha_i \in \mathbb{R} \right\} \tag{6.12}$$

$\text{Conv}(M_{co}) = M_{fco}$ is the subset of $\text{aff}(M_{co})$ with the added constraint $\alpha_i \geqslant 0\ \forall i$ on the weights $\{\alpha_i\}$ in (6.12). Since any set is affine if and only if it is the translate of a linear subspace, there is a subspace V_{co} of V_{cn} which is parallel

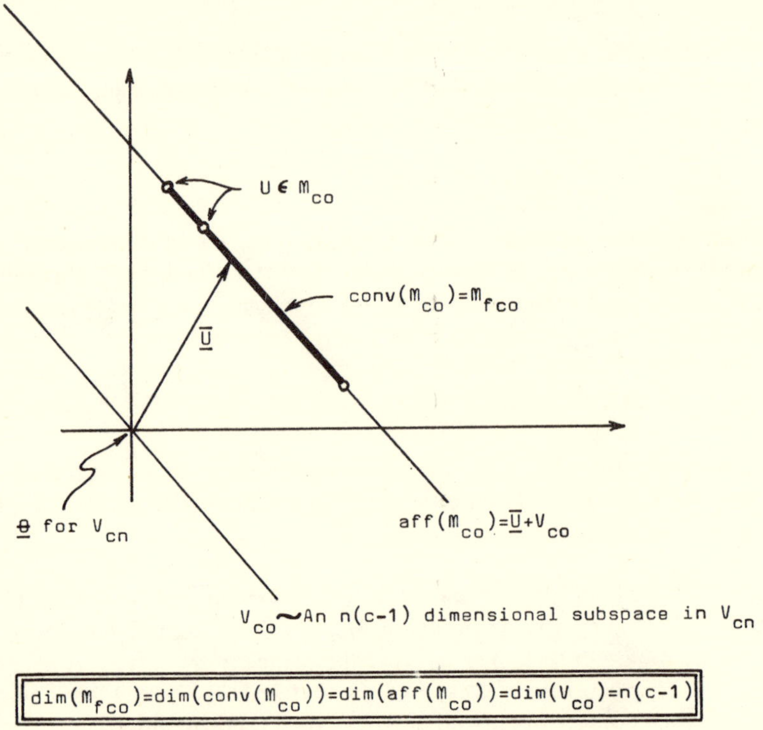

Figure 6.3. The dimension of fuzzy c-partition space.

to $\text{aff}(M_{co})$, and a $\mathbf{U}_0 \notin V_{co}$ so that $\mathbf{U}_o + V_{co} \doteq \{\mathbf{U}_o + \mathbf{W} \mid \mathbf{W} \in V_{co}\} = \text{aff}(M_{co})$. In fact, we may use $\bar{\mathbf{U}}$ for $\mathbf{U}_o$ (Fig. 6.3). Both $\text{conv}(M_{co})$ and $\text{aff}(M_{co})$ have as their $\dim(\text{conv}(M_{co})) = \dim(\text{aff}(M_{co})) = \dim(V_{co})$. Since $\dim(M_{fco}) < nc - 1$, none of these sets are really "flat" (in the hyperplane sense) as illustrated in (Fig. 6.3). M_{fco} may be envisioned as a multifaceted convex polytope in V_{cn}, symmetric about its centroid $\bar{\mathbf{U}}$, with flat *faces* connecting its vertices.

(T6.4) *Theorem 6.4.*

$$\dim(M_{fco}) = \dim(\text{conv}(M_{co})) = n(c-1) \tag{6.13}$$

Analysis. Look at Fig. 6.2: $\text{conv}(B_3)$ lies in a plane translated from $\mathbf{\theta}$ by $\bar{\mathbf{u}}^T = (1/3, 1/3, 1/3)$. This plane (also called a hyperplane) has dimension 2, so $\dim(\text{conv}(B_3)) \leqslant 2$. To guess the theorem, imagine first $\text{conv}(B_c)$. It lies in an affine set or hyperplane of dimension $(c-1)$, so $\dim(\text{conv}(B_c)) \leqslant c-1$. M_{fco} is made up of n copies of $\text{conv}(B_c)$; uncoupling columns via (6.8) makes the copies "independent," so each adds $(c-1)$ to the overall dimension. Thus for n columns, $n(c-1)$ is the answer, if $\dim(\text{conv}(B_c)) = c-1$.

Proof. Let $\langle \mathbf{x}, \mathbf{y} \rangle = \sum_{i=1}^{c} x_i y_i$ be the Euclidean inner product for vectors $\mathbf{x}, \mathbf{y} \in \mathbb{R}^c$; $\mathbf{x}$ and $\mathbf{y}$ are orthogonal if $\langle \mathbf{x}, \mathbf{y} \rangle = 0$. The hyperplane through any $\mathbf{x}_0 \in \mathbb{R}^c$ and orthogonal to $\mathbf{N} \in \mathbb{R}^c$ is the set

$$HP(\mathbf{x}_0; \mathbf{N}) = \{\mathbf{x} \in \mathbb{R}^c \mid \langle \mathbf{x} - \mathbf{x}_0, \mathbf{N} \rangle = 0\} \tag{6.14}$$

$\text{HP}(\mathbf{x}_0; \mathbf{N})$ is the translate by $\mathbf{N}$ of a linear subspace of dimension $c-1$, so $\dim(\text{HP}(\mathbf{x}_0, \mathbf{N})) = c-1$. Now consider the hyperplane in $\mathbb{R}^c$ through $\mathbf{e}_1^T = (1, 0, \ldots, 0)$ and orthogonal to $\bar{\mathbf{u}}^T = (1/c, 1/c, \ldots, 1/c)$. A vector $\mathbf{u}_k^T = (u_{1k}, u_{2k}, \ldots, u_{ck})$ belongs to $\text{HP}(\mathbf{e}_1; \bar{\mathbf{u}})$ if and only if

$$\langle \mathbf{u}_k - \mathbf{e}_1, \bar{\mathbf{u}} \rangle = \sum_{i=1}^{c} (u_{ik} - e_{i1})(\bar{u}_i) = [(u_{1k} - 1) + u_{2k} + \cdots + u_{ck}]/c = 0$$

or equivalently, $\sum_{i=1}^{c} u_{ik} = 1$. Therefore, $\text{conv}(B_c) \subset \text{HP}(\mathbf{e}_1, \bar{\mathbf{u}})$, and $\dim(\text{conv}(B_c)) \leqslant c-1$. Now $\{\mathbf{e}_1 - \mathbf{e}_j \mid 1 \neq j\}$ are $c-1$ independent vectors in $\text{conv}(B_c)$, so $\dim(\text{conv}(B_c)) = c-1$. Finally,

$$\dim(M_{fco}) = \dim(\text{conv}(B_c))^n = \dim(\underbrace{\text{conv}(B_c) \times \cdots \times \text{conv}(B_c)}_{n \text{ times}})$$

$$= \sum_{k=1}^{n} \dim(\text{conv}(B_c)) = n(c-1).$$

This proves (6.13). •

Note that even though each column of U lies in a hyperplane in $\mathbb{R}^c$, U itself is *not* in a hyperplane in V_{cn}, for its dimension would then be $nc - 1$, not $n(c-1)$. What is there in (6.13), aside from its mathematical content, that might interest practitioners? This: Caratheodory proved that whenever $\dim(\text{conv}(S)) = m$, every $\mathbf{v} \in \text{conv}(S)$ has a convex decomposition by *at most* $m + 1$ factors [cf. (84), p. 76]. This means that *every* fuzzy c-partition of X can be factored (in principle at least!) as

$$U = \sum_{k=1}^{n(c-1)+1} \alpha_k U_k, \qquad \sum_k \alpha_k = 1, \qquad \alpha_k \geqslant 0, \qquad U_k \in M_{co} \tag{6.15}$$

The sequence (α_k, U_k) in (6.15) is a nonnested set of *hard* c-partitions of X, and because the $\{\alpha_k\}$ are convex weights, it is natural to conjecture that α_k indicates the "percentage" of fuzzy U that U_k accounts for. Convex decomposition then, is one way to produce hard clusterings in X.

Remarks

Two avenues are opened by the results of this section. First, properties of fuzzy partition space are useful in further mathematical considerations: e.g., (6.11a) ensures that continuous functions attain maxima and minima on M_{fco}. On the other hand, the convex decomposition alluded to above has a physical interpretation that may be directly applicable to (real data) clustering problems. Each of these paths shall be traveled below.

The material of S6 evokes some interesting questions. When is $U \in M_{fc}$ uniquely decomposable? What algorithms produce decompositions? When are all the hard factors nondegenerate? If the decomposition is not unique, how many are there? What physical interpretation and utility does a specified decomposition have? Zadeh in (122) defined fuzzy convex sets—an intriguing idea that has subsequently received surprisingly little attention, in view of the general utility of convexity in more conventional settings. Chang[(26)] extends the idea of fuzzy convexity to Euclidean spaces. Roberts and Varberg[(84)] is an excellent general reference on convex analysis. T6.4 has a longer proof in reference (21), which contains some additional examples concerning convexity. Clustering by decomposition of $U \in M_{fc}$ is not confined to *convex* factorizations: several "affinity" decompositions due to Backer[(4)] will be discussed in Chapter 5.

Exercises

H6.1. Derive (6.1), the cardinality of M_c [see Feller, W., *An Introduction to Probability Theory and Its Applications*, Vol. 1, Wiley, New York (1958), p. 58].

H6.2. Compute the number of distinct hard c-partitions of the data set X in H5.1 for $c = 1, 2, 3, 4, 5$.

H6.3. Prove that each of the sets described is convex:
(i) $\{(x, y) \in \mathbb{R}^2 | x^2 + y^2 < 9\} = B(\boldsymbol{\theta}, 3)$
(ii) $\{(x, y) \in \mathbb{R}^2 | x^2 + y^2 \leq 9\} = \bar{B}(\boldsymbol{\theta}, 3)$
(iii) $\{(x, y) \in \mathbb{R}^2 | |x| + |y| \leq 9\}$
(iv) $\{\mathbf{x} \in \mathbb{R}^p | \|\mathbf{x}\| \leq r\} = \bar{B}(\boldsymbol{\theta}, r)$

H6.4. Let $A, B \in \mathbb{R}^p$, $A + B = \{\mathbf{a} + \mathbf{b} | \mathbf{a} \in A, \mathbf{b} \in B\}$, $\alpha \in \mathbb{R}$, $\alpha A = \{\alpha\mathbf{a} | \mathbf{a} \in A\}$. Show that, if A, B are convex, then
(i) αA is convex
(ii) $A + B$ is convex
(iii) $B = \lambda B + (1 - \lambda)B \; \forall \lambda \in [0, 1]$

H6.5. Prove (6.4), assuming (6.2) defines conv(S).

H6.6. Prove that the intersection of any collection of convex sets is again convex.

H6.7. Let

$$S_2 = \left\{ U \in V_{22} \,\middle|\, \sum_{i=1}^{2} u_{ik} = \sum_{k=1}^{2} u_{ik} = 1; u_{ik} \geq 0 \right\}$$

S_2 is the set of *doubly stochastic* 2×2 real matrices. Show that
(i) S_2 is convex in V_{22}.

(ii) $U_1 = \begin{bmatrix} 1 & 0 \\ 0 & 1 \end{bmatrix}$ and $U_2 = \begin{bmatrix} 0 & 1 \\ 1 & 0 \end{bmatrix}$

are extreme points of S_2.
(iii) Every $U \in S_2$ can be written as $\lambda U_1 + (1 - \lambda)U_2$ for some λ in [0, 1]. (Hint: show that U_1 and U_2 are the *only* extreme points of S_2.)
(iv) What is the dimension of S_2?
(v) Is S_2 a "straight line" in V_{22}?

H6.8. Graph the convex hull of the data set X in H5.1.

H6.9. Let $S^X = \{f : X \to S\}$, X any set, S any subset of a vector space V. S^X is a vector space with pointwise functional addition and scalar multiplication: $(\mathbf{f} + \mathbf{g})(\mathbf{x}) = \mathbf{f}(\mathbf{x}) + \mathbf{g}(\mathbf{x})$; $(\alpha\mathbf{f})(\mathbf{x}) = \alpha\mathbf{f}(\mathbf{x})$. Show that S is convex in $V \Leftrightarrow S^X$ is convex in V^X.[10]

H6.10. If A, $B \subset V$, V any vector space, X any set, show that $B = \text{conv}(A) \Leftrightarrow B^X = \text{conv}(A^X)$.[10]

H6.11. V and W are vector spaces; $A \subset B \subset V$; $S \subset T \subset W$. Let $L: V \to W$ be a linear map, and let $S = L[A]$; $T = L[B]$. Show that[10]
(i) B is convex $\Leftrightarrow T$ is convex
(ii) $B = \text{conv}(A) \Leftrightarrow T = \text{conv}(S)$

H6.12. Combine the results of problems H6.9–H6.11 to establish (6.9).

H6.13. $U = \begin{bmatrix} 0.90 & 0.80 & 0.30 & 0.40 & 0.05 \\ 0.10 & 0.20 & 0.70 & 0.60 & 0.95 \end{bmatrix}$

Find a five-term convex decomposition of U by factoring at each step the maximum possible coefficient from the remainder matrix. (Answer: $\alpha_1 = 0.60$, $\alpha_2 = 0.20$, $\alpha_3 = 0.10$; $\alpha_4 = 0.05$; $\alpha_5 = 0.05$: these coefficients are unique; the $\{U_k\} \subset M_{2_0}$ are not.)

H6.14. A fuzzy subset u of $X \subset \mathbb{R}^p$ is a *convex fuzzy set* if the *hard* subsets $L_\alpha = \{\mathbf{x} \in X \mid u(\mathbf{x}) \geq \alpha\}$ are convex in $\mathbb{R}^p$ $\forall \alpha \in (0, 1]$. Prove that u is convex in this sense if and only if

$$u(\mathbf{x}) \wedge u(\mathbf{y}) \leq u(\lambda \mathbf{x} + (1 - \lambda)\mathbf{y})$$

for all $\mathbf{x}, \mathbf{y} \in X$ and all $\lambda \in [0, 1]$.[122] Note that X is necessarily convex in $\mathbb{R}^p$.

H6.15. Suppose u, w to be convex fuzzy subsets of $X \subset \mathbb{R}^p$. Prove that the fuzzy set $u \wedge w$ is also a convex fuzzy set.[122]

H6.16. Let u be a convex fuzzy subset of $\mathbb{R}^p$, and suppose $\mathbf{v}, \mathbf{x}_1, \ldots, \mathbf{x}_r$ to be vectors in $\mathbb{R}^p$ in the domain of u. Show that

$$\bigwedge_{i=1}^{r} u(\mathbf{x}_i) \leq u(\mathbf{v})$$

whenever $\mathbf{v}$ is a convex combination of the $\{\mathbf{x}_i\}$.

H6.17. The *fuzzy convex hull* of a fuzzy subset $u : \mathbb{R}^p \to [0, 1]$ is defined in (26) as the smallest convex fuzzy subset of $\mathbb{R}^p$ containing u. That is, if $u_{\text{ch}} : \mathbb{R}^p \to [0, 1]$ is the (membership function of the) fuzzy convex hull of u, then $u(\mathbf{x}) \subset u_{\text{ch}}(\mathbf{x}) \subset w(\mathbf{x})$ $\forall w$ so that w is a fuzzy convex subset of $\mathbb{R}^p$. Let $\mathbf{v} = (\mathbf{v}_0, \mathbf{v}_1, \ldots, \mathbf{v}_p)$, with $\mathbf{v}_i \in \mathbb{R}^p$ $\forall i$, and for each $\mathbf{x} \in \mathbb{R}^p$, let

$$S(\mathbf{x}) = \left\{ \mathbf{v} \,\middle|\, \mathbf{x} = \sum_{i=0}^{p} \alpha_i \mathbf{v}_i ; 0 \leq \alpha_i \leq 1; \sum_{i=0}^{p} \alpha_i = 1 \right\}$$

$$u_{\text{ch}}(\mathbf{x}) = \sup_{\mathbf{v} \in S(\mathbf{x})} \left\{ \prod_{i=0}^{p} u(\mathbf{v}_i) \right\}$$

(i) Show that u_{ch} is a fuzzy convex subset of $\mathbb{R}^p$.
(ii) Show that u_{ch} is the fuzzy convex hull of u.

H6.18. Let u, w be fuzzy convex sets in $\mathbb{R}^p$, with $(u \vee w)$ their union. Prove that the fuzzy convex hull of $(u \vee w)$ is given by

$$(u \vee w)_{\text{ch}}(\mathbf{x}) = \sup_{\mathbf{v} \in T(\mathbf{x})} \{u(\mathbf{v}_1) \wedge w(\mathbf{v}_2)\}$$

where $\forall \mathbf{x}$,

$$T(\mathbf{x}) = \{(\mathbf{v}_1, \mathbf{v}_2) \in \mathbb{R}^{2p} \mid \mathbf{x} = \lambda \mathbf{v}_1 + (1 - \lambda)\mathbf{v}_2 ; 0 \leq \lambda \leq 1\}.$$

S7. Fuzzy Relations

So far, we have assumed that the data gathered have the form of a sample of size n of observations $\{x_k\} = X$. Data in this form can be clustered into c-partitions of X using various algorithms, and when this is done, the observations are grouped according to some clustering criterion. In a hard c-partition, the *relationship* between pairs of points in $X \times X$ is unequivocal: x_i and x_j are fully related $\Leftrightarrow$ they belong to the same cluster; and otherwise, they are total "strangers." For matrices in M_c then, there is a natural equivalence between hard c-partitions and hard equivalence relations which is delineated in (D7.1) below. The connection between fuzzy c-partitions and fuzzy relations is not so transparent; some initial facts concerning this are developed here. A second reason for introducing relation-theoretic structures is that many real processes generate raw data in precisely this form. That is, rather than a sample of n observations $\{x_k\}$, we often measure a set of n^2 values (or an $n \times n$ data matrix) which represent relationships (similarities, associations, etc.) between pairs (x_i, x_j). In this instance, the x_k's may not be recorded, and the raw data are an $n \times n$ relation matrix. With a view towards algorithms designed for this situation, we present some preliminary material on relational structures.

Given $X = \{x_1, x_2, \ldots, x_n\}$, a generalized (binary) relation in X is a map on $X \times X$ into $\mathbb{R}$, say, a function $r: X \times X \to \mathbb{R}$. A hard relation in X is a hard subset of $X \times X$ ($r: X \times X \to \{0, 1\}$); a fuzzy relation in X is a fuzzy subset of $X \times X$ ($r: X \times X \to [0, 1]$). In every case, the relation r has an equivalent description in terms of an $n \times n$ relation matrix $R = [r_{ij}] \doteq [r(x_i, x_j)]$, whose ijth entry prescribes the strength or value of the relationship between x_i and x_j. Each hard c-partition of X induces on $X \times X$ a unique hard equivalence relation:

(D7.1) *Hard Equivalence Relation.* $X = \{x_1, x_2, \ldots, x_n\}$ is any finite set. An $n \times n$ matrix $R = [r_{ij}] = [r(x_i, x_j)]$ is a hard equivalence relation on $X \times X$ if

$$r_{ii} = 1, \qquad 1 \leq i \leq n \tag{7.1a}$$

$$r_{ij} = r_{ji}, \qquad 1 \leq i \neq j \leq n \tag{7.1b}$$

$$\left.\begin{matrix} r_{ij} = 1 \\ r_{jk} = 1 \end{matrix}\right\} \Rightarrow r_{ik} = 1 \qquad \forall i, j, k \tag{7.1c}$$

These three properties of R are called *reflexivity*, *symmetry*, and *transitivity*, respectively. The set of all such relations we denote by

$$ER_n = \{R \in V_{nn} \mid r_{ij} \text{ satisfies (7.1) } \forall i, j\} \tag{7.2}$$

Each matrix in ER_n corresponds to a unique hard c-partition $U \in M_c$, by putting $r_{ij} = 1 \Leftrightarrow x_i$ and x_j belong to the same hard subset of U, and $r_{ij} = 0$ otherwise. Accordingly, hard c-partitioning algorithms for X induce hard equivalence relations on $X \times X$, and conversely.

(E7.1) *Example 7.1.*

$$U = \begin{array}{c} \begin{array}{ccccc} x_1 & x_2 & x_3 & x_4 & x_5 \end{array} \\ \begin{bmatrix} 1 & 1 & 0 & 1 & 0 \\ 0 & 0 & 1 & 0 & 1 \end{bmatrix} \end{array} \leftrightarrow R = \begin{array}{c} \begin{array}{ccccc} x_1 & x_2 & x_3 & x_4 & x_5 \end{array} \\ \begin{bmatrix} 1 & 1 & 0 & 1 & 0 \\ 1 & 1 & 0 & 1 & 0 \\ 0 & 0 & 1 & 0 & 1 \\ 1 & 1 & 0 & 1 & 0 \\ 0 & 0 & 1 & 0 & 1 \end{bmatrix} \end{array}$$

To see that $M_c \approx ER_n$, note that permuting columns 3 and 4 of U and relabeling the data $(x'_3 = x_4, x'_4 = x_3)$ yields

$$U' = \begin{array}{c} \begin{array}{ccccc} x'_1 & x'_2 & x'_3 & x'_4 & x'_5 \end{array} \\ \left[\begin{array}{ccc|cc} 1 & 1 & 1 & 0 & 0 \\ 0 & 0 & 0 & 1 & 1 \end{array} \right] \end{array} \leftrightarrow R' = \begin{array}{c} \begin{array}{ccccc} x'_1 & x'_2 & x'_3 & x'_4 & x'_5 \end{array} \\ \left[\begin{array}{ccc|cc} 1 & 1 & 1 & 0 & 0 \\ 1 & 1 & 1 & 0 & 0 \\ 1 & 1 & 1 & 0 & 0 \\ \hline 0 & 0 & 0 & 1 & 1 \\ 0 & 0 & 0 & 1 & 1 \end{array} \right] \end{array}$$

The indicated partitioning of U' and R' makes this isomorphism quite transparent.   •

The situation for hard equivalence relations is summarized in the following theorem.

(T7.1) *Theorem 7.1.* M_c is hard c-partition space for X (5.7), ER_n are the hard equivalence relations in X (7.2). Then

$$M_c \approx ER_n \qquad (7.3)$$

The objection to hard c-partitions illustrated in (E5.1) surfaces again in (E7.1). The entries of R (or R') imply that observations 1, 2, and 4 are fully related to one another, as are 3 and 5; but no member of either cluster is related (at all!) to members of other equivalence classes. In view of this, it is natural to consider the consequence in *relation space* of allowing fuzzy clusters in *partition space*. In other words, fuzzy clusters in X correspond intuitively to fuzzy relations in $X \times X$. The *mathematical* correspondence between M_c and $M_{fco} = \text{conv}(M_{co})$ on the one hand, and between M_c and ER_n on the other, suggests that the convex hull of ER_n might be isomorphic to M_{fco}. This is not the case; one difficulty lies with an appropriate generalization of transitivity (7.1c), which is unique for hard relations, but has taken several proposed forms in the fuzzy case. Another problem is that different mappings on M_{fc} produce sets of fuzzy relations with different—yet both useful and appealing properties—in V_{nn}. Moreover, the mappings proposed to date have been noninvertible. To close this section we continue (E5.1) in the following example.

(E7.2) *Example 7.2.* The hard 2-partition U of (E5.1) is associated with the matrix $R_1 \in ER_3$ shown below:

$$U_1 = \begin{bmatrix} 1 & 1 & 0 \\ 0 & 0 & 1 \end{bmatrix} \leftrightarrow R_1 = \begin{bmatrix} 1 & 1 & 0 \\ 1 & 1 & 0 \\ 0 & 0 & 1 \end{bmatrix}$$

One fuzzy relation R which can be constructed from the fuzzy 2-partition U at (5.11) using the formula $R = (U^T(\sum \wedge)U)$ defined in (20) is

$r_{ij} = \sum_{k=1}^{c} u_k(x_i) \wedge u_k(x_j)$

$$U = \begin{bmatrix} 0.91 & 0.58 & 0.13 \\ 0.09 & 0.42 & 0.87 \end{bmatrix} \rightarrow R = \begin{bmatrix} 1 & 0.67 & 0.22 \\ 0.67 & 1 & 0.55 \\ 0.22 & 0.55 & 1 \end{bmatrix}$$

From R we infer that x_1 and x_2 are more related ($r_{12} = 0.67$) to one another than are x_2 and x_3 ($r_{23} = 0.55$); and that x_1 and x_3 (the peach and the plum) are rather weakly related ($r_{13} = 0.22$). The entries of R do seem to characterize the information in U about the data $\{x_k\}$ in a different way. [Cf. comments following (5.11).]

Remarks

The point of this section is that data of two kinds may present themselves: observations, or relations between pairs of observations. The choice of treatment of data in terms of hard c-partitions or hard equivalence relations is a matter of convenience, since the two models are fully equivalent (philosophically and mathematically). Fuzzy relations and partitions are philosophically *similar* in some sense: their mathematical structures are not isomorphic. This circumstance provides a natural division between the theory of fuzzy partitions and relations, but one should bear in mind that there is an underlying connection (quite concrete in the hard case; implicit in the fuzzy case) between partitions and relations in X. Specific references will appear at more appropriate junctures: A nice general reference containing much of the early work on fuzzy relations is the book of Kaufmann.[62]

Exercises

H7.1. For each hard 2-partition of the data set X in H5.1, exhibit the unique equivalence relation in V_{55} to which it corresponds via (7.2).

H7.2. Find a convex decomposition of the fuzzy 2-partition U in E7.2 using hard 2-partitions (at least one such decomposition exists). Try to decompose the fuzzy relation R induced from U in E7.2 with hard equivalence relations [such a decomposition is *not* guaranteed; see (20)].

3

Objective Function Clustering

S8 illustrates some of the difficulties inherent with cluster analysis; its aim is to alert investigators to the fact that various algorithms can suggest radically different substructures in the same data set. The balance of Chapter 3 concerns objective functional methods based on fuzzy c-partitions of finite data. The nucleus for all these methods is optimization of nonlinear objectives involving the weights $\{u_{ik}\}$; functionals using these weights will be differentiable over M_{fc}—but not over M_c—a decided advantage for the fuzzy embedding of hard c-partition space. Classical first- and second-order conditions yield iterative algorithms for finding the optimal fuzzy c-partitions defined by various clustering criteria.

S9 describes one of Ruspini's[89] algorithms, and compares its performance to the hard c-means or classical within-groups sum-of-squared-errors method. S10 discusses an algorithm based on a mixed fuzzy statistical objective due to Woodbury and Clive[118] and reports an application of it to the diagnosis of tetralogy of Fallot (four variants of a congenital heart disease). The infinite family of fuzzy c-means algorithms is defined in S11 and is compared (numerically) to several earlier methods. S12 contains a proof of convergence for fuzzy c-means; and S13 illustrates its usefulness for feature selection with binary data by examining a numerical example concerning stomach disorders due to hiatal hernia and gallstones.

S8. Cluster Analysis: An Overview

In this chapter, we assume that the important question of feature extraction—which characteristics of the physical process are significant indicators of structural organization, and how to obtain them—has been answered. Our point of departure is this: given $X = \{x_1, x_2, \ldots, x_n\}$, find an integer c, $2 \leqslant c < n$, and a c-partition of X exhibiting categorically homogeneous subsets! The most important requirement for resolving this issue is a suitable measure of "clusters"—what *clustering criterion* shall be used?

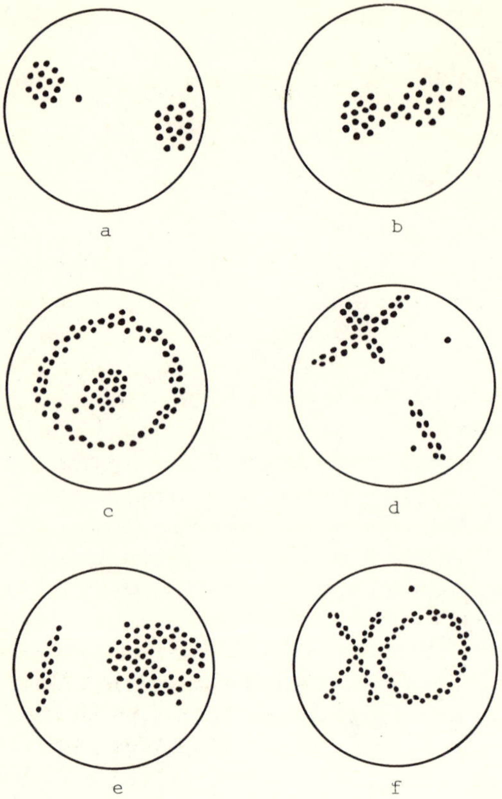

Figure 8.1. Some clustered data sets in the plane.

Specifically, what mathematical properties—e.g., distance, angle, curvature, symmetry, connectivity, intensity—possessed by members of the data should be used, and in what way, to identify the clusters in X? Since each observation can easily have several hundred dimensions, the variety of "structures" is without bound. It is clear that (i) *no* clustering criterion or measure of similarity will be universally applicable, and (ii) selection of a particular criterion is at least partially subjective, and always open to question.

Regardless of the actual dimension of data space, clustering models are usually interpreted geometrically by considering their action on two- or three-dimensional examples. Some of the difficulties inherent in trying to formulate a successful clustering criterion for a wide variety of data structures are illustrated in Fig. 8.1.

The ideal case—compact, well-separated, equally proportioned clusters (Fig. 8.1a)—is seldom encountered in real data. More realistically, data sets may embody a mixture of shapes (spherical, elliptical); sizes (intensities, unequal numbers of observations); and geometries (linear, angular, curved). Even in $\mathbb{R}^2$, further difficulties can arise if X is translated, dilated, compressed, rotated, or even reordered!

To illustrate how problematical the choice of a classification criterion can be, Fig. 8.2 depicts typical (hard) clusters obtained by processing two data sets with (i) a distance-based objective function algorithm (the within-group sum-of-squared-error (WGSS) criterion), and (ii) a distance-based graph-theoretic method (the single-linkage algorithm). Figure 8.2 should warn potential clusterers that adherence to a single method may lead to extremely untenable solutions. Behavior of this sort can occur *even when both algorithms use the same distances*, i.e., the same measure of similarity between points in X, but with different clustering criteria! We emphasize here that *similarity measures* are the building blocks for *clustering criteria*. In the simplest models, a measure of similarity can serve as a criterion of validity; but, more generally, the same measure can be used with various criteria to yield different models and results for the same data set.

In view of the perplexities discussed above, it is clear that *successful* cluster analysis ultimately rests not with the computer, but with the investigator, who is well advised to use some empirical hindsight concerning the physical process generating X to temper algorithmically suggested solutions. Specification of a similarity measure and/or clustering criterion is not enough. The method used must be matched to the data.

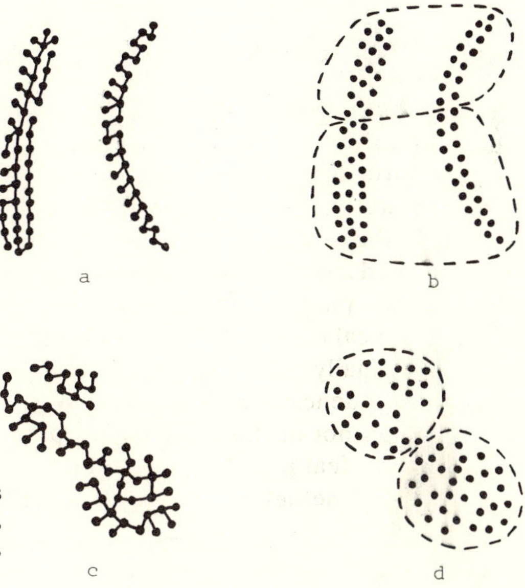

Figure 8.2. Cluster perversity: success and failure. (a) single linkage succeeds, (b) WGSS fails, (c) single linkage fails, (d) WGSS succeeds.

No matter how clustering methods are categorized, some exceptions will ensue. A reasonable system might be to separate them first by axiomatic basis into deterministic, stochastic, or fuzzy; and then by type of clustering criterion. Roughly speaking, there are three types of criteria for each axiomatic class: hierarchical, graph theoretic, and objective functional. Similarity measures provide a further subdivision of methodologies. Moreover, there are instances where two apparently quite dissimilar methods, developed independently from differing viewpoints, turn out to be the same technique. Nonetheless, it seems useful to provide a short description of each major type, because the mathematical structure of models of the different types can be quite diverse.

Hierarchical Methods

This group of methods had its origins in taxonomic studies. There are agglomerative (merging) and divisive (splitting) techniques. In both cases, new clusters are formed by reallocation of membership of one point at a time, based on some measure of similarity. Consequently, a hierarchy of nested clusters—one for each c—is generated. These techniques have as their forte conceptual and computational simplicity; they appear suitable when data substructure is dendritic in nature.

Graph-Theoretic Methods

In this group, X is regarded as a node set, and edge weights between pairs of nodes can be based on a measure of similarity between pairs of nodes. The criterion for clustering is typically some measure of connectivity or bonding between groups of nodes; breaking edges in a minimal spanning tree to form subgraphs is often the clustering strategy. Techniques of this kind are well suited to data with "chains," or pseudolinear structure. For example, the single-linkage technique should be quite adaptable to the data of Figs. 8.1d and 8.2a. Mixed data structures, e.g., Figs. 8.1c, 8.1e, and 8.1f, may cause pure graph-theoretic methods some difficulty. Data with hyperelliptical clusters, noise, and bridges, such as in Figs. 8.1a, 8.1b, and 8.2c, are usually badly distorted by graph-theoretic models because of their chaining tendencies; this failure is illustrated in Fig. 8.2c. Moreover, graph methods do not ordinarily generate paradigmatic representatives of each subclass—a feature of considerable value when designing prototype classifiers. Nonetheless, this group of methods is being studied intensively.

Objective Function Methods

These methods ordinarily allow the most precise (but not necessarily more valid) formulation of the clustering criterion. For each c, a criterion or objective function measures the "desirability" of clustering candidates. Typically, local extrema of the objective function are defined as optimal clusterings. For example, if one takes as the similarity measure Euclidean distance for data ($\mathbf{x}_k \in \mathbb{R}^d$), and as a measure of cluster quality the overall within-group sum of squared errors (WGSS) (between $\mathbf{x}_k$'s and "prototypical" $\mathbf{v}_i$'s), the objective function is the sum of squared errors. This clustering criterion is called a minimum variance objective: we shall meet many infinite families of fuzzy generalizations of it in this volume. Until recently, criterion function algorithms were regarded as most applicable to data such as are illustrated in Figs. 8.1a and 8.2c, where clusters are basically hyperspherical and of roughly equal proportions. Note that WGSS, for example, is an effective criterion in precisely the instance where graph-theoretic criteria are not (cf. Fig. 8.2). There are some fuzzy methods which try to combine the best features of each type; these are taken up in Chapter 5.

We reiterate that different similarity measures, clustering criteria, and axiomatic structures lead to astonishingly disparate structural interpretations of the same data set. Geometric insight concerning the applicability of a particular choice—so invaluable for two- and three-dimensional data—must be augmented by algebraic and analytic means for higher-dimensional situations.

Remarks

There is a long and distinguished literature on hierarchical clustering due to its usefulness in numerical taxonomy. In addition to the general references given in S1, we single out here the books of Sneath and Sokal[(99)] and Jardine and Sibson[(59)] as especially pertinent general references. Some key papers in this area include those of Lance and Williams,[(69,70)] Johnson,[(60)] and Wishart.[(116)] The single-linkage method was shown by Gower and Ross[(49)] to be hierarchical in nature. Other key papers on graph-theoretic techniques include those of Zahn,[(124)] Ling,[(74)] Chen and Fu,[(27)] and Balas and Padberg.[(7)] Ling contains excellent criticisms and examples of the failure of objective functions in specific instances, as does Duda and Hart.[(34)]

Fuzzy sets as a basis for clustering were first suggested by Bellman, Kalaba, and Zadeh.[(9)] Shortly thereafter, some initial attempts were reported by Wee, [(114)] Flake and Turner,[(39)] and Gitman and Levine.[(46)] The first systematic exposition of fuzzy clustering appears to have been the work of Ruspini, which is discussed and referenced in S9 below.

S9. Density Functionals

For the balance of Chapter 3 we assume data set X to be a subset of the real p-dimensional vector space $\mathbb{R}^p$: $X = \{\mathbf{x}_1, \mathbf{x}_2, \ldots, \mathbf{x}_n\} \subset \mathbb{R}^p$; each $\mathbf{x}_k = (x_{k1}, x_{k2}, \ldots, x_{kp}) \in \mathbb{R}^p$ is called a *feature or pattern vector*; and x_{kj} is the jth *feature or characteristic* of observation $\mathbf{x}_k$. The first fuzzy objective-function method we discuss is due to Ruspini.[89] Let d_{jk} denote the "distance" in $\mathbb{R}^p$ between $\mathbf{x}_j$ and $\mathbf{x}_k$; we assume that $\forall\ \mathbf{x}_j$ and $\mathbf{x}_k$ in $\mathbb{R}^p$ this function satisfies

$$\begin{Bmatrix} d_{jk} \doteq d(\mathbf{x}_j, \mathbf{x}_k) \geq 0 \\ d_{jk} = 0 \Leftrightarrow \mathbf{x}_j = \mathbf{x}_k \end{Bmatrix} \tag{9.1a}$$

$$d_{jk} = d_{kj} \tag{9.1b}$$

In other words, $d: X \times X \to \mathbb{R}^+$ is positive-definite and symmetric, but does not necessarily obey the triangle inequality (so d is not necessarily a metric on $\mathbb{R}^p$). Functions that satisfy (9.1) are often called *measures of dissimilarity*. A similarity measure s can be constructed from d by first normalizing d, say to $d^*: X \times X \to [0, 1]$, and then defining $s_{ij} = 1 - d_{ij}^*$: it is mathematically immaterial which point of view one adopts.

Ruspini[89] considered a number of clustering criteria incorporating the distances $\{d_{jk}\}$ and fuzzy c-partitions of X. We discuss below one of the most successful ones, leaving interested readers to pursue the references for related algorithms. To this end, let $J_R: M_{fco} \to \mathbb{R}^+$ be defined as

$$J_R(U) = \sum_{j=1}^{n} \sum_{k=1}^{n} \left\{ \left[\sum_{i=1}^{c} \sigma(u_{ij} - u_{ik})^2 \right] - d_{jk}^2 \right\}^2 \tag{9.2}$$

where σ is a real constant and $2 \leq c < n$ is fixed *a priori*. J_R is the clustering criterion; d the measure of (dis-) similarity. Ruspini interprets J_R as a measure of cluster quality based on local density, because J_R will be small when the terms in (9.2) are individually small: in turn, this will occur when close (pairs of) points have nearly equal fuzzy cluster memberships in the c u_i's in U.

Optimal fuzzy c-partitionings of X are taken as local minima of J_R. Ruspini's algorithms are based on several general results concerning the constrained optimization problem

$$\underset{U \in M_{fco}}{\text{minimize}} \{J_R(U)\} \tag{9.3}$$

The main result is contained in the following theorem.

(T9.1) *Theorem 9.1.* Let $J : M_{fco} \to \mathbb{R}^+$ be differentiable on M_{fco}. For fixed $U \in M_{fco}$, define

$$I_k = \{i | u_{ik} > 0\} \qquad \forall k \tag{9.4a}$$

$$\beta_{ik} = -\left(\frac{\partial J(U)}{\partial u_{ik}}\right) \qquad \forall i, k \tag{9.4b}$$

$$\beta_{Mk} = \bigvee_{i=1}^{c} \beta_{ik} \qquad \forall k \tag{9.4c}$$

$$\beta_{mk} = \bigwedge_{i=1}^{c} \beta_{ik} \qquad \forall k \tag{9.4d}$$

If $\beta_{Mk} = \beta_{mk}$ for $k = 1, 2, \ldots, n$, then U is a constrained stationary point of J, that is, for all possible directions $\{r_j(\mathbf{x}_k)\}$ such that

$$\sum_{i=1}^{c} r_i(\mathbf{x}_k) = 0 \qquad \forall k \tag{9.4e}$$

$$r_i(\mathbf{x}_k) \geqslant 0 \qquad \text{for } i \notin I_k \tag{9.4f}$$

$$\sum_{i=1}^{c} r_i^2(\mathbf{x}_k) = 1 \qquad \forall k \tag{9.4g}$$

it is true that

$$\sum_{k=1}^{n} \sum_{i=1}^{c} r_i(\mathbf{x}_k)\beta_{ik} \leqslant 0 \tag{9.5}$$

Analysis. This theorem simply invokes the classical first-order conditions on the gradient of J with respect to U which are necessary for stationary points.[55] The importance of the hypotheses as stated lies in their usage for defining algorithm (A9.1) below. The proof itself is straightforward, being computationally direct.

Proof (*Ruspini*[89]). Assume the hypotheses. Then (9.4c) and (9.4d) imply that

(A) $$\beta_{ik} = \beta_{mk} = \beta_{Mk} \qquad \forall i \in I_k$$

and

(B) $$\beta_{ik} \leqslant \beta_{Mk} \qquad \forall i \notin I_k$$

Let

$$\alpha_k = \sum_{i \notin I_k} r_i(\mathbf{x}_k) \geqslant 0$$

the inequality due to (9.4f). In view of (9.4e) and (9.4g), we have

$$-\alpha_k = \sum_{i \in I_k} r_i(\mathbf{x}_k)$$

This and (A) yield

$$\text{(C)} \qquad \sum_{i \in I_k} r_i(\mathbf{x}_k)\beta_{ik} = -\alpha_k \beta_{Mk}$$

Likewise, from (B)

$$\text{(D)} \qquad \sum_{i \notin I_k} r_i(\mathbf{x}_k)\beta_{ik} \leqslant \alpha_k \beta_{Mk}$$

Equations (C) and (D) now yield

$$\sum_{i=1}^{c} r_i(\mathbf{x}_k)\beta_{ik} \leqslant \beta_{Mk}(\alpha_k - \alpha_k) = 0$$

Summing this over k now yields (9.5). •

To implement the gradient method for iterative optimization of J_R, the following corollary is needed.

(C9.1) *Corollary 9.1.* If for some t, $1 \leqslant t \leqslant n$, $\beta_{mt} < \beta_{Mt}$, then taking the directions

$$r_m(\mathbf{x}_t) = -2^{1/2}/2 \tag{9.6a}$$

$$r_M(\mathbf{x}_t) = 2^{1/2}/2 \tag{9.6b}$$

$$r_i(\mathbf{x}_k) = 0 \qquad \forall i \neq m, M; \qquad \forall k \neq t \tag{9.6c}$$

yields

$$\sum_{k=1}^{n} \sum_{i=1}^{c} r_i(\mathbf{x}_k)\beta_{ik} > 0$$

With these results in mind, we now define the basic algorithm of Ruspini for approximation of local minima of J_R by the gradient method: In this and subsequent iterative algorithms, superscript (l) represents iterate number l in a Picard loop, starting with $l = 0$ for the initial guess.

(A9.1) *Algorithm 9.1* (*Ruspini*[89])

(A9.1a) Fix c, $2 \leqslant c < n$, and a dissimilarity measure d. Initialize $U^{(0)} \in M_{fc}$, Then at step $l : l = 0, 1, \ldots$:

(A9.1b) Calculate $\sigma^{(l)}$ (using least squares in one variable).

(A9.1c) For a fixed column k, compute $\{\beta_{ik}^{(l+1)}\}$ and find $\beta_{Mk}^{(l+1)}, \beta_{mk}^{(l+1)}$.

(A9.1d) If $\beta_{Mk}^{(l+1)} = \beta_{mk}^{(l+1)}$: if $k = n$, go to (A9.1f); if $k < n$, put $k = k + 1$ and return to (A9.1c).

(A9.1e) If $\beta_{Mk}^{(l+1)} > \beta_{mk}^{(l+1)} + \varepsilon_1$: (i) choose directions for column k via (9.6); update memberships $\{u_{ik}^{(l)}\}$ to $\{u_{ik}^{(l+1)}\}$ using the cube root strategy reported in (89), p. 345; if $k = n$, go to (A9.1f); if $k < n$, set $k = k + 1$ and return to (A9.1c).

(A9.1f) Compare $U^{(l)}$ to $U^{(l+1)}$ in a convenient matrix norm: if $\|U^{(l+1)} - U^{(l)}\| \leq \varepsilon_L$ stop; otherwise, return to (A9.1b), with $l = l + 1$.

A number of particulars concerning (A9.1) are left to the references. One must have a tie-breaking rule in case $\beta_{Mk}^{(l)}$ or $\beta_{mk}^{(l)}$ is not unique; the updating procedure in step (A9.1e) is quite involved; and sizes of local error (ε_1), loop error (ε_L), and a measure of closeness for matrices in V_{cn} ($\|U^{(l+1)} - U^{(l)}\|$) must be chosen. The advance on direction selected and column updates are controlled by a cubic equation because J_R is a polynomial of degree 4 in the length of the correction.

Although (T9.1) and its corollary ensure that the sequence $\{U^{(l)} | l = 0, 1, 2, \ldots\}$ converges to a stationary point of J_R [because (A9.1) is a valid gradient method of iterative optimization], a number of important provisos which apply to many algorithms of this type must be made. Specifically:

(i) Stationary points of any objective function are not necessarily local minima.

(ii) There is no assurance that even a *global* optimum of any objective function is a "good" clustering of X.

(iii) Different choices for algorithmic parameters (here, $U^{(0)}$, d, ε_1, ε_L, $\|\cdot\|$) may yield different "optimal" partitionings of X.

(iv) There may be a different number of clusters (c) which result in a "better" partitioning of X. In fact, there may be more than one value of c for which reasonable substructural interpretations exist!

All of these facts underly the cluster validity question—have we identified *meaningful* substructure in X? If varying $U^{(0)}$ or c produces different "optimal" clusters, how shall the "most optimal" U be chosen? Presumably, the smallest value of $J_R(\hat{U})$ identifies $\hat{U}$ as best. But (ii) points up the fact that goodness of $\hat{U}$ ultimately rests with the mathematical properties measured by J_R: unfortunately, it is usually rather easy to find a simple two-dimensional data set for which any global criterion function optimizes at a ridiculous "clustering" solution for X (cf. Fig. 8.2b)! This is perhaps the main criticism leveled at *all* criterion-function algorithms. On the other hand, there is no doubt that any objective function performs well on *certain* kinds of patterns—the real trick is to find out which algorithm is suitable for data generated from a specific process. If the objective function itself is not enough, a cluster validity functional with which to test optimal U's is needed. This aspect of clustering is discussed in later sections.

To see how algorithm (A9.1) works, we give the following example.

(E9.1) *Example 9.1* (*Ruspini*[89]). Data set X consists of the 15 points in $\mathbb{R}^2$ listed in column 2 of Table 9.1 and illustrated graphically in Fig. 9.1. This pattern has the general structure of Fig. 8.1b. Data points (2, 2), (3, 2), and (4, 2) form a bridge or neck between the wings of the butterfly. From a physical standpoint, these vectors are analogous in concept to the nectarine in (E5.1); the left wing would correspond to, say, plums; and the right wing to peaches. Another possible interpretation of this pattern is that points in the wings were drawn from two fairly distinct classes; points in the neck are noise. The results of applying (A9.1) to X with $c = 2$ are listed as membership functions in Table 9.1; the memberships of fuzzy cluster u_1 are reported in Fig. 9.1. Measure d was the Euclidean metric, and $U^{(0)} \in M_{fco}$ was obtained by a preprocessing method reported in (89). Memberships seem to be assigned in an intuitively reasonable fashion. Plots of u_1 and u_2 (shown as continuous although they are really discrete) as functions of the horizontal variate illustrate the way in which fuzzy subsets of X "soften" the bound-

Table 9.1. The Butterfly: (E9.1) and (E9.2)

Data		Membership functions			
		$J_R \simeq$ (A9.1)		$J_W \simeq$ (A9.2)	
k	$\mathbf{x}_k$	u_{1k}	u_{2k}	u_{1k}	u_{2k}
1	(0, 0)	0.99	0.01	1.0	0.0
2	(0, 2)	0.98	0.02	1.0	0.0
3	(0, 4)	0.99	0.01	1.0	0.0
4	(1, 1)	0.86	0.14	1.0	0.0
5	(1, 2)	0.85	0.15	1.0	0.0
6	(1, 3)	0.86	0.14	1.0	0.0
7	(2, 2)	0.67	0.33	1.0	0.0
8	(3, 2)	0.50	0.50	0.0	1.0
9	(4, 2)	0.33	0.67	0.0	1.0
10	(5, 1)	0.14	0.86	0.0	1.0
11	(5, 2)	0.15	0.85	0.0	1.0
12	(5, 3)	0.14	0.86	0.0	1.0
13	(6, 0)	0.01	0.99	0.0	1.0
14	(6, 2)	0.02	0.98	0.0	1.0
15	(6, 4)	0.01	0.99	0.0	1.0

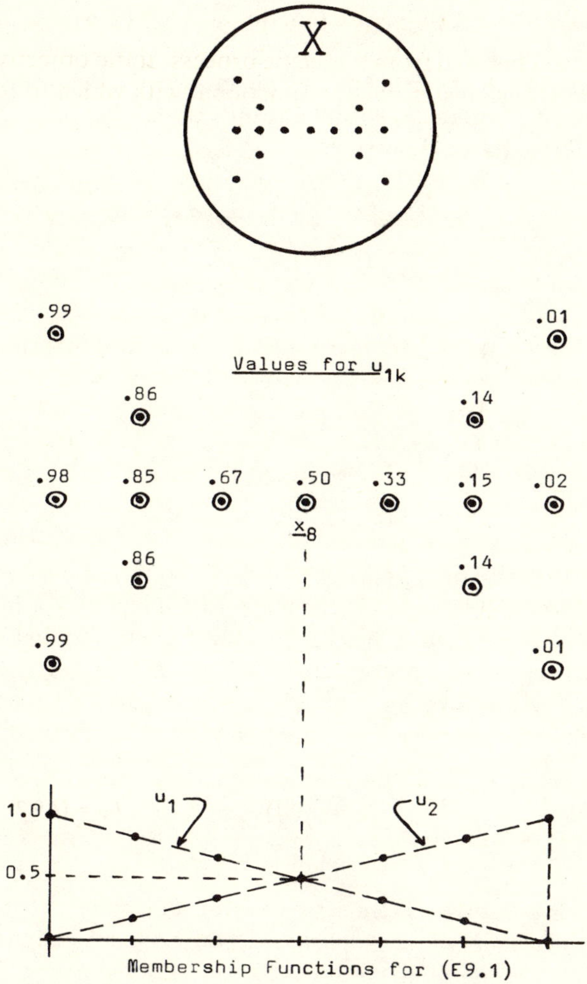

Figure 9.1. The butterfly, (E9.1): membership assignments using Ruspini's algorithm (A9.1).

aries required by hard characteristic functions [cf. (E9.2)]. u_1 and $u_2 = 1 - u_1 = \tilde{u}_1$ are nearly linear functions: note that they are symmetric with respect to $\mathbf{x}_8$ because the data themselves have this property. Point $\mathbf{x}_8 = (3, 2)$, the geometric centroid of X, garners membership of 0.5 in each fuzzy cluster: as one progresses away from this point (towards the "core" of each wing), memberships become more and more distinct. Thus, in this instance fuzzy clustering provides a very natural description of two-cluster substructure in X. •

The classical WGSS objective functional is perhaps the most extensively studied clustering criterion which generates hard clusters in X. Its fuzzy generalization is taken up later: we introduce the method here to provide a basis for comparison with the results of (E9.1). Towards this end, let $J_W : M_{co} \times \mathbb{R}^{cp} \to \mathbb{R}^+$ be defined as

$$J_W(U, \mathbf{v}) = \sum_{k=1}^{n} \sum_{i=1}^{c} u_{ik}(d_{ik})^2 \tag{9.7a}$$

where

$$d_{ik} = d(\mathbf{x}_k, \mathbf{v}_i) = \|\mathbf{x}_k - \mathbf{v}_i\| = \left[\sum_{j=1}^{p} (x_{kj} - v_{ij})^2\right]^{1/2} \tag{9.7b}$$

$$\mathbf{v} = (\mathbf{v}_1, \mathbf{v}_2, \ldots, \mathbf{v}_c) \in \mathbb{R}^{cp}, \mathbf{v}_i \in \mathbb{R}^p \forall i \tag{9.7c}$$

and

$$U = [u_{ik}] \in M_{co} \text{ is } \textit{hard} \tag{9.7d}$$

Vector $\mathbf{v}$ is a set of c prototypical "cluster centers," $\mathbf{v}_i$ the cluster center for hard cluster $u_i \in U$, $1 \leq i \leq c$. The measure of dissimilarity in J_W is the Euclidean norm metric. Note, however, that dissimilarity between *single* data points $\mathbf{x}_k$ and cluster prototypes $\mathbf{v}_i$ *not* necessarily in X is what J_W assesses. Since $u_{ik} = u_i(\mathbf{x}_k) = 1 \Leftrightarrow \mathbf{x}_k \in u_i$, and is zero otherwise, an equivalent way to write (9.7a) that exhibits clearly the geometric property J_W gauges is

$$J_W(U, \mathbf{v}) = \sum_{i=1}^{c} \left(\sum_{\mathbf{x}_k \in u_i} \|\mathbf{x}_k - \mathbf{v}_i\|^2 \right) \tag{9.7a'}$$

Since $u_{ik}(d_{ik})^2$ is the squared (Euclidean) error incurred by representing $\mathbf{x}_k$ by $\mathbf{v}_i$, it is also a measure of local density. J_W will be small when the points in each hard u_i adhere tightly (have small d_{ik}'s) to their cluster center $\mathbf{v}_i$. The ith term in (9.7a') is the sum of within-cluster squared errors over all the $\mathbf{x}_k$'s $\in u_i$; the sum over i in (9.7a') then renders J_W the overall or total within-group sums of squared errors associated with the pair $(U, \mathbf{v})$. This provides for J_W a geometrically appealing rationale, but one should remember that Euclidean geometry is essentially hyperspherical in $\mathbb{R}^P$, i.e., tight clusters will be "round" (because the unit ball is) in the Euclidean topology. If clusters in $\mathbb{R}^P$ are not shaped this way, J_W is a poor criterion (cf. Fig. 8.2).

J_W also measures a statistical property of the pair $(U, \mathbf{v})$: to describe it, we define the scatter matrices of Wilks as follows (see Duda and Hart,[34] p. 221).

(D9.1) *Definition 9.1 (Hard Scatter Matrix).* Assume $X = \{\mathbf{x}_1, \mathbf{x}_2, \ldots, \mathbf{x}_n\} \subset \mathbb{R}^p$ and $(U, \mathbf{v}) \in M_c \times \mathbb{R}^{cp}$. We define the following.

$$\text{Centroid of cluster } u_i: \quad \mathbf{v}_i = \sum_{k=1}^{n} u_{ik}\,\mathbf{x}_k \Big/ \sum_{k=1}^{n} u_{ik} \tag{9.8a}$$

$$\text{Scatter matrix for } u_i: \quad S_i = \sum_{k=1}^{n} u_{ik}(\mathbf{x}_k - \mathbf{v}_i)(\mathbf{x}_k - \mathbf{v}_i)^T \tag{9.8b}$$

$$\text{Within-cluster scatter matrix:} \quad S_W = \sum_{i=1}^{c} S_i \tag{9.8c}$$

Because the norm used in (9.7) is Euclidean, one can check that the trace of S_W is just J_W:

$$\mathrm{Tr}(S_W) = J_W(U, \mathbf{v})$$

Since $\mathrm{Tr}(S_W)$ is proportional to the sum of variances in the p coordinate directions, minimization of J_W amounts to minimizing this measure of variance, hence the name "minimum variance partitioning problem" for

$$\underset{M_c \times \mathbb{R}^{cp}}{\text{minimize}} \{J_W(U, \mathbf{v})\} \tag{9.9}$$

Accordingly, minima of J_W have both statistical and geometric appeal as clustering solutions.

Finding optimal pairs $(U, \mathbf{v})$ for J_W is not as easy as one might guess. The difficulty stems from M_c, which is finite but huge [cf. (6.1)]. One of the great advantages of fuzziness is differentiability in the u_{ik}'s over M_{fc}, whereas such is not the case for J_W over M_c. The point, of course, is that first-order necessary conditions for fuzzy criteria can often be found with gradients, as, e.g., in (T9.1) above. One of the most popular algorithms for approximating minima of J_W is iterative optimization; the algorithm described below is variously called the (hard) c-means or basic ISODATA method.

(A9.2) *Algorithm 9.2 [Hard c-Means, (HCM), Duda and Hart*[34]]

(A9.2a) Fix c, $2 \leq c < n$, and initialize $U^{(0)} \in M_c$. Then at step l, $l = 0, 1, 2, \ldots$:

(A9.2b) Calculate the c mean vectors $\{\mathbf{v}_i^{(l)}\}$ with (9.8a) and $U^{(l)}$.

(A9.2c) Update $U^{(l)}$: the new memberships must be calculated as follows $\forall i$ and k

$$u_{ik}^{(l+1)} = \begin{cases} 1, & d_{ik}^{(l)} = \min\limits_{1 \leq j \leq c} \{d_{jk}^{(l)}\} \\ 0, & \text{otherwise} \end{cases} \tag{9.10}$$

(A9.2d) Compare $U^{(l)}$ to $U^{(l+1)}$ in a convenient matrix norm: if $\|U^{(l+1)} - U^{(l)}\| \leq \varepsilon_L$ stop; otherwise, set $l = l + 1$ and return to (A9.2b).

From an intuitive standpoint, (A9.2) is quite reasonable: guess c hard clusters; find their centroids, reallocate cluster memberships to minimize squared errors between the data and current prototypes; stop when looping ceases to lower J_W. Since the hard space M_c is discrete, the notion of local minimum is not defined for J_W, so convergence of (A9.2) is itself an undefined notion.

The necessity of computing $\{\mathbf{v}_i^{(l)}\}$ with (9.8a) for fixed $U^{(l)}$ is easy to establish—set the gradient of J_W with respect to each $\mathbf{v}_i$ equal to zero. Dunn[35] established algebraically the necessity of nearest prototype rule (9.10) for membership updates; an alternate analytic proof is given in S11. A tie-breaking rule is needed in case $\min_j \{d_{jk}^{(l)}\}$ is nonunique: the usual one is to arbitrarily assign $\mathbf{x}_k$ to the first cluster at which $d_{jk}^{(l)}$ minimizes in the indexed sequence over j (this hardly ever occurs in practice!). We call this a "singularity" for (A9.2). Note that *one* d_{ik} *can be zero*: singularity in column k for hard c-means implies a *nonunique minimum*, zero or otherwise, for the $\{d_{jk}\}$. Initialization should specify c *distinct* clusters, for otherwise, nondistinctness will be propagated through subsequent iterations.

Duda and Hart call (A9.2) the basic ISODATA method because it forms the core of a greatly embellished clustering strategy called ISODATA by its inventors (Ball and Hall[8]). Note that one could initialize the cluster centers $\{\mathbf{v}_i^{(0)}\}$ in (A9.2), thereby starting the loop for hard c-means a "half-step" out of phase with the method listed: moreover, one can choose to compare $\|\mathbf{v}_i^{(l+1)} - \mathbf{v}_i^{(l)}\|$ to ε_L in a norm on $\mathbb{R}^p$, rather than use the termination criterion of (A9.2d). The matter of initialization is immaterial; the termination criterion *does* affect the amount of storage and number of iterations required for fixed ε_L. These details are best decided with an application in hand. Since (A9.2) is essentially a hill-climbing technique, it is sensitive to provisos (ii)–(iv) following (A9.1).

(E9.2) *Example 9.2.* The data of (E9.1) were processed with hard 2-means, using for $U^{(0)}$ the matrix

$$U^{(0)} = \begin{bmatrix} 1 & 1 & 1 & 1 & 1 & 0 & 0 & 0 & 0 & 0 & 0 & 0 & 0 & 0 & 0 \\ 0 & 0 & 0 & 0 & 0 & 1 & 1 & 1 & 1 & 1 & 1 & 1 & 1 & 1 & 1 \end{bmatrix}.$$

The termination criterion was the max norm on $V_{2,15}$, viz., $\|U^{(l+1)} - U^{(l)}\| = \max_{i,k} \{|u_{ik}^{(l+1)} - u_{ik}^{(l)}|\} \leq \varepsilon_L$. Choosing $\varepsilon_L < 1$ requires $U^{(l+1)} = U^{(l)}$ in order to stop. In this example, $\varepsilon_L = 0.01$ was used, and (A9.2) terminated in four iterations at

$$U^{(3)} = U^{(4)} = \begin{bmatrix} 1 & 1 & 1 & 1 & 1 & 1 & 1 & 0 & 0 & 0 & 0 & 0 & 0 & 0 & 0 \\ 0 & 0 & 0 & 0 & 0 & 0 & 0 & 1 & 1 & 1 & 1 & 1 & 1 & 1 & 1 \end{bmatrix}$$

which is listed in Table 9.1, in columns 5 and 6, for ease of comparison. These membership functions are also displayed graphically in Fig. 9.2 without iterate superscripts. Note that u_1 and u_2 cannot be symmetric with respect to $\mathbf{x}_8$, because $\mathbf{x}_8$ must belong *entirely* to one cluster or the other,

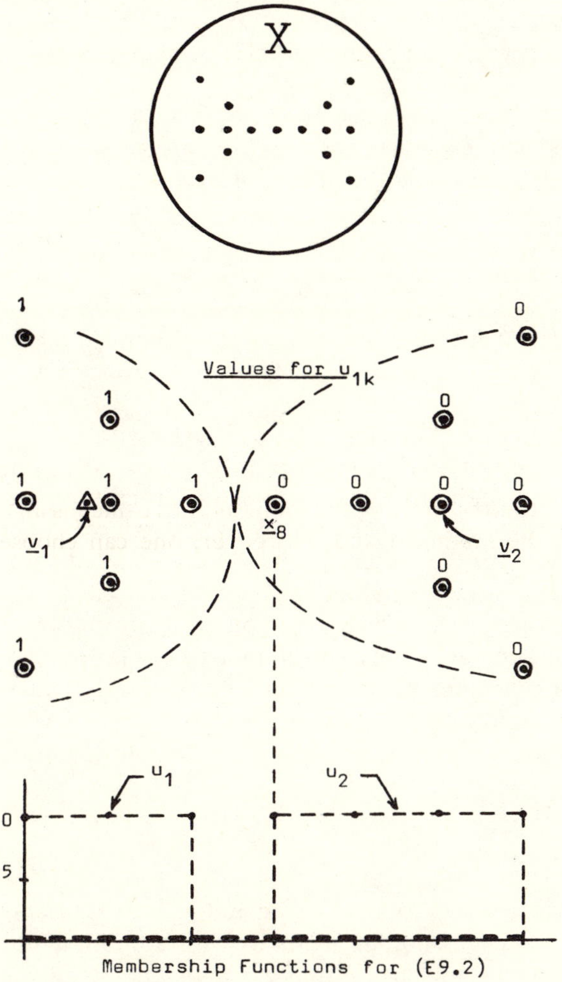

Figure 9.2. The butterfly, (E9.2): membership assignments using hard c-means algorithm (A9.2).

even though the data X are themselves perfectly symmetrical, This fact is mirrored by the terminal centroids $\mathbf{v}_1^{(4)} = (0.71, 2.00)$; $\mathbf{v}_2^{(4)} = (5, 2)$. If the neck had one more point ($n = 16$), the 2-means solution could be established entirely by symmetry. •

Comparing examples 9.1 and 9.2, it seems fair to assert that Ruspini's partition of the butterfly conveys more information about substructure in X than the 2-means partition does. Low memberships signal the investigator—here is a data point which bears a closer look: high membership in a single class usually points towards the prototypical individuals of distinct classes. If hard memberships are necessary, one way to realize them from a terminal fuzzy partition is via the maximum membership conversion.

(D9.2) *Definition 9.2* (*Nearest MM Hard Partition*). If $U \in M_{fco}$, the nearest hard c-partition of X in the sense of maximum membership (MM) is the partition $U_m \in M_{co}$, whose ikth element is

$$(U_m)_{ik} = \begin{cases} 1, & u_{ik} = \max_{1 \le j \le c} \{u_{jk}\} \\ 0, & \text{otherwise} \end{cases} \tag{9.11}$$

For each k such that $\max\{u_{jk}\}$ over j is nonunique, $\mathbf{x}_k$ is arbitrarily assigned to the first cluster achieving the maximum.

Applying (D9.2) to U from (E9.1) yields precisely the 2-means partition of (E9.2), up to arbitrary assignment for $\mathbf{x}_8$. Doing this, however, constitutes in our opinion a definite loss of information concerning substructure in X: the advantages gained by using fuzziness should be retained! Comparing the algorithms themselves, (A9.1) is a gradient method which operates sequentially on columns of $U^{(l)}$ within a Picard loop on l. Hard c-means is a Picard process which is computationally simpler, more efficient, and has the decided advantage of simultaneously generating c prototypes (the $\mathbf{v}_i$'s). On the other hand, the imbedding M_{fc} provides a means for defining and analyzing numerical convergence of (A9.1), and fuzzy memberships seem to decompose the butterfly into a more plausible substructure than hard ones can.

Remarks

Ruspini's early algorithms,[89–93] including (A9.1), seem to be the first well-defined fuzzy partitioning methods with a substantial mathematical basis. His criteria were hard to interpret and difficult to implement: computational efficiency was poor, and generalizations to more than $c = 2$ classes met with little success. Nonetheless, certain types of patterns are amenable to J_R; see (71) for a nice real data application of Ruspini's methods to sleep stage clustering in chimpanzees. Perhaps the greatest contribution made by these papers, however, was that they paved the way for other families of seemingly more effective fuzzy clustering criteria.

The sum of squared errors criterion J_W was apparently first discussed in 1953 by Thorndike.[102] There are literally hundreds of papers detailing the theory and applications of J_W and its variants (see Duda and Hart[34] for a nice survey of much of this literature). We emphasize again that a primary concern is whether or not solutions of (9.3) or (9.9) suggest *meaningful* data substructure. A discussion of the difficulties inherent in statistical cluster validity tests using J_W is given in (34): Hartigan[57] contains some excellent sections concerning this question; cluster validity tests for fuzzy partitions are taken up in Chapter 4. MacQueen[76] discusses *stochastic* convergence of the hard c-means algorithm to parametric estimates based on certain probabilistic assumptions concerning X.

Exercises

H9.1. Verify that equations (9.6) support C9.1.

H9.2. Let $U \in M_c$ be a hard c partition of n data points. If $n_i = |u_i|$, and $\mathbf{v}_i$ is the centroid of u_i, $1 \leq i \leq c$, as in (9.8a), show that $J_W(U, \mathbf{v})$ can be written as $J_W(U, \mathbf{v}) = \frac{1}{2}\sum_{i=1}^{c} n_i m_i$, where $m_i = \sum_{\mathbf{x}\in u_i}\sum_{\mathbf{y}\in u_i}(\|\mathbf{x}-\mathbf{y}\|/n_i)^2$.

H9.3. Find the hard 3-partition of the data set X of H5.1 that groups together (1, 3) and (10, 3) and minimizes $J_W(U, \mathbf{v})$.

H9.4. Establish by inspection the hard 3-partition of X in H5.1 that minimizes $J_W(U, \mathbf{v})$. [Answer $\{(1, 1), (1, 3)\} \cup \{(5, 2)\} \cup \{(10, 1), (10, 3)\}$.]

H9.5. Find the unique vector in $\mathbb{R}^p$ that minimizes $f(\mathbf{x}) = \sum_{k=1}^{n} \|\mathbf{x}-\mathbf{x}_k\|^2$, where $\{\mathbf{x}_1, \ldots, \mathbf{x}_n\}$ are fixed in $\mathbb{R}^p$. (Answer $\mathbf{x}^* = \sum_{k=1}^{n} \mathbf{x}_k/n$.)

H9.6. Verify that (9.10) is necessary to minimize $J_W(U, \mathbf{v})$ for fixed $\{\mathbf{v}_i | 1 \leq i \leq c\}$.[35]

H9.7. Prove that $J_W(U, \mathbf{v})$ decreases monotonically as c increases.

H9.8. $J_W(U, \mathbf{v}) = \mathrm{Tr}(S_W)$, S_W being the scatter matrix at (9.8c). Show that the optimal c-partition of X may *change* upon scaling the data (i.e., scaling the axes of $\mathbb{R}^p$). Let $J_d(U, \mathbf{v}) = \det(S_W)$. Show that if L is any invertible $p \times p$ matrix, the linear change of variables $\mathbf{v} = L\mathbf{x}$ scales J_d by the constant factor $[\det(L)]^2$ for *all* $U \in M_c$, so that optimal clusterings of X relative to J_d are *invariant* to linear transformations of the data. J_W is not invariant to all linear transformations of the data. Find a subset of the linear transformations on $\mathbb{R}^p \to \mathbb{R}^p$ for which J_W *is* invariant. [Answer: orthogonal linear maps (rotations).]

H9.9. Let $X \subset \mathbb{R}^2$ have $n = 17$ vectors, the x coordinates being 0, 1, 0, 1, 2, 1, 2, 3, 6, 7, 8, 9, 7, 8, 9, 8, 9; and the corresponding y coordinates—in order—being 0, 0, 1, 1, 1, 2, 2, 2, 6, 6, 6, 7, 7, 7, 8, 8, 9. For example, vector $\mathbf{x}_1^T = (0, 0)$; $\mathbf{x}_{17}^T = (9, 9)$.

(i) Sketch X.

(ii) Choose $\mathbf{v}_1^{(0)} = \mathbf{x}_1$; $\mathbf{v}_2^{(0)} = \mathbf{x}_2$ to initialize A9.2 at (A9.2b). Apply (A9.2) to X until $\mathbf{v}_i^{(l+1)} = \mathbf{v}_i^{(l)}$, $i = 1, 2$, with agreement to two decimal places. Sketch the final hard 2-partition and cluster centers for X. [Answer: $l = 3$; $X = \{\mathbf{x}_1, \ldots, \mathbf{x}_8\} \cup \{\mathbf{x}_9, \ldots, \mathbf{x}_{20}\}$; $\mathbf{v}_1^T = (1.25, 1.13)$; $\mathbf{v}_2^T = (8.87, 8.00)$.]

S10. Likelihood Functionals: Congenital Heart Disease

The material in this section is based on Woodbury and Clive.[118] Their functional and algorithm provide for us a first example of a *mixed* mathematical model. The functional employed involves products of fuzzy memberships and discrete probabilities, resulting in a composite fuzzy-statistical criterion (a case in point for the discussion in S2 concerning complementary models). The algorithm is appropriate only for data sets in $\mathbb{R}^p$, each of whose features is a categorical variable. Specifically, let $K_j = \{1, 2, \ldots, m_j\}$ for $j = 1, 2, \ldots, p$. We assume in this section that data set X is a subset of the Cartesian product of the K_j's over j:

$$X \subset (K_1 \times K_2 \times \cdots \times K_p) = K \subset \mathbb{R}^p \tag{10.1}$$

$|K_j| = m_j$, and as usual, $|X| = n$, the number of observations in sample X. If $m_j = 2$ for each j, X is a binary data set with p variables; more generally, $x_{kj} \in K_j$ is the jth indicant or feature (an integer) for individual $\mathbf{x}_k$. The algorithm described below utilizes frequency counts obtained from X as follows:

$$n = \text{No. of observations in } X \tag{10.2a}$$

$$n_{kj} = \text{No. of times feature } j \text{ of } \mathbf{x}_k \text{ appears} \tag{10.2b}$$

$$n_{kjt} = \text{No. of times } t \in K_j \text{ appears for } \mathbf{x}_k. \tag{10.2c}$$

It is further assumed that $n_{kj} = \sum_t n_{kjt} = 1$ or $0 \ \forall\, k, j$, so $n_{kjt} = 1$ or $0 \ \forall\, k, j$, and t.

Although prototypes (class paradigms) do not explicitly appear in the mathematics to follow, their hypothetical existence is assumed in order to provide a rational philosophical basis for the definitions used. Accordingly, assume c prototypes ["clinically pure types" in (118)] for the observations in X. To be physically consistent, these pure types would be c vectors $\{\mathbf{v}_i\} \subset K$;

the $\mathbf{v}_i$'s merely simplify the semantic necessities concerning the following probabilities: let

$$p_{ijt} = \text{Prob(``pure'' } \mathbf{v}_i \text{ manifests outcome } t \in K_j \text{ in feature } j) \tag{10.4a}$$

for $1 \leq i \leq c$; $1 \leq j \leq p$; $1 \leq t \leq m_j$. To be probabilities, it is necessary that

$$p_{ijt} \geq 0, \qquad \sum_{t=1}^{m_j} p_{ijt} = 1 \qquad \forall i, j, t \tag{10.4b}$$

Let $\sum_j m_j = m$. There are in all cpm probabilities $\{p_{ijt}\}$ to estimate. For notational ease, we denote these as a vector $\mathbf{p}$, and let

$$P_{wc} = \{\mathbf{p} \in \mathbb{R}^{cpm} | p_{ijt} \text{ satisfies (10.4) } \forall i, j, t\}$$

Now let $U = [u_{ik}] \in M_{fc}$. In the present context, $u_{ik} = u_i(\mathbf{x}_k)$—the membership of $\mathbf{x}_k$ in the ith fuzzy cluster of U—is interpreted as the extent to which the features of $\mathbf{x}_k$ agree with those of (hypothesized) "clinically pure" prototype $\mathbf{v}_i$. Finally, let

$$p_{kj(t)} = \left(\sum_{i=1}^{c} u_{ik}\, p_{ijt}\right) \Big/ \left[\sum_{i=1}^{c} u_{ik}\left(\sum_{t=1}^{m_j} p_{ijt}\right)\right] \tag{10.5}$$

$p_{kj(t)}$ is interpreted as the probability that data vector $\mathbf{x}_k$ will manifest outcome $t \in K_j$ in feature j. If the p features are statistically independent, the likelihood functions of the observed frequencies with $(U, \mathbf{p})$ as parameters is, using the multinomial theorem,

$$L(U, \mathbf{p}; X) = \prod_{k=1}^{n} \prod_{j=1}^{p} (n_{kj})! \prod_{t=1}^{m_j} (p_{kj(t)})^{n_{kjt}} / (n_{kjt})! \tag{10.6}$$

Taking logarithms of (10.6) and eliminating constants not involved in the estimation procedure yields the objective function of Woodbury and Clive: specifically, let $J_{wc} : M_{fc} \times P_{wc} \to \mathbb{R}$ be

$$J_{wc}(U, \mathbf{p}) = \sum_{k=1}^{n} \sum_{j=1}^{p} \left(\sum_{t=1}^{m_j} n_{kjt} \left\{ \log\left(\sum_{i=1}^{c} u_{ik}\, p_{kjt} \right) - \log\left[\sum_{i=1}^{c} u_{ik} \left(\sum_{t=1}^{m_j} p_{kjt} \right) \right] \right\} \right) \tag{10.7}$$

J_{wc} is a "fuzzy-statistical" criterion, in that it combines fuzzy memberships of observed data with probabilities of (hypothetical) pure prototypical features. Optimal c-partitions of X are part of optimal pairs $(\hat{U}, \hat{\mathbf{p}})$ which solve

$$\underset{M_{fc} \times P_{wc}}{\text{maximize}} \{J_{wc}(U, \mathbf{p})\} \tag{10.8}$$

and with the interpretation of (10.5) as outcome probabilities, solutions of (10.8) may be called maximum likelihood estimates for $(\hat{U}, \hat{\mathbf{p}})$. In an earlier paper (118), the authors attempt to optimize a *related* functional via classical Kuhn–Tucker theory. This approach is inadequate, as pointed out in (119), because the parameters $(U, \mathbf{p})$ to be estimated via (10.8) are not strictly identifiable in the statistical sense. Instead, the numerical example presented below is based on a gradient search method. The algorithm used in (E10.1) below is a sequential ascent method with line search control.

(A10.1) *Algorithm 10.1* (*Woodbury and Clive*[119])

(A10.1a) Fix c, $2 \leq c < n$. Initialize $U^{(0)} \in M_c$. Use (10.4) to compute $\mathbf{p}^{(0)}$ by counting. Then at step l, $l = 1, 2, \ldots$:

(A10.1b) Update $U^{(l)}$ as follows: for $1 \leq i \leq c$ and $1 \leq k \leq n$,

$$u_{ik}^{(l+1)} = u_{ik}^{(l)} + \alpha_k^{(l)} \frac{\partial J_{wc}}{\partial u_{ik}} (U^{(l)}, \mathbf{p}^{(l)}) \tag{10.9a}$$

where step length $\alpha_k^{(l)}$ is chosen for each k, $1 \leq k \leq n$, such that $u_{ik}^{(l+1)} \geq 0$ and $\sum_{i=1}^{c} u_{ik}^{(l+1)} = 1$.

(A10.1c) Update $\mathbf{p}^{(l)}$ as follows: for $1 \leq i \leq c$; $1 \leq j \leq p$; $1 \leq t \leq m_j$,

$$p_{ijt}^{(l+1)} = p_{ijt}^{(l)} + \beta_{ij}^{(l)} \frac{\partial J_{wc}}{\partial p_{ijt}} (U^{(l+1)}, \mathbf{p}^{(l)}) \tag{10.9b}$$

where step length $\beta_{ij}^{(l)}$ is chosen for each t, $1 \leq t \leq m_j$, so that $p_{ijt}^{(l+1)} \geq 0$ and $\sum_{t=1}^{m_j} p_{ijt}^{(l+1)} = 1$.

(A10.1d) Compute $\delta^{(l)} = \min_{i,j,k}\{\alpha_k^{(l)}, \beta_{ij}^{(l)}\}$. If $\delta^{(l)} < \varepsilon_L = 10^{-6}$, stop. Otherwise, compare $J_{wc}^{(l+1)}$ to $J_{wc}^{(l)}$. If $J_{wc}^{(l+1)} > J_{wc}^{(l)}$, put $l = l + 1$ and return to (A10.1b). If $J_{wc}^{(l+1)} \leq J_{wc}^{(l)}$, discard $(U^{(l+1)}, \mathbf{p}^{(l+1)})$, halve the step sizes $\{\alpha_k^{(l)}, \beta_{ij}^{(l)}\}$, and return to (A10.1b) without incrementing l.

Two important observations should be made here. First, the step size in (10.9a) and (10.9b) is not controlled by necessary conditions; this requires the "cut and try" criterion in (A10.1d). Woodbury and Clive have experimented with several methods for step size selection based on (the magnitudes of) ratios of (u_{ik}) to $(\partial J_{wc}/\partial u_{ik})$; and of $(\partial J_{wc}/\partial u_{ik})$ to $(\partial^2 J_{wc}/\partial u_{ik}^2)$.

Second, note that $U^{(0)}$ in (A10.1a) is *hard*. The authors recommend preprocessing of X with a hard clustering algorithm such as (A9.2) or the ditto algorithm of Hartigan[57] to secure an "educated" initial guess. The following example is a condensation of Section 6 of (119).

(E10.1) *Example 10.1* (*Tetralogy of Fallot*[119]). The data for (E10.1) consist of $n = 60$ congenital heart disease patients. Each patient $\mathbf{x}_k$ provided

$p = 108$ measured responses ranging from 2 to 13 in number (i.e., $2 \leqslant m_j \leqslant 13\ \forall j$). Of these, 91 features were chosen, the others being rejected as either incomplete or clinically irrelevant. An additional variable for (E10.1) is time t: a data set $X(t)$ of size $n \times p = 60 \times 91$ for each of 371 different patient dates was collected over a total time span of 10 years from birth. One of the major objectives was to trace the evolution *over time* of the membership of each patient "in" each of c clinically pure prototypes. The number of clusters c was a variable, $c = 3, 4, 5, 6$. Over this range, $c = 4$ (corresponding roughly to "asymptomatic," "moderate," "acyanotic severe," and "cyanotic severe" pure types) was chosen, using a statistical test of cluster validity based on χ^2 ratios discussed in (119).

Algorithm (A10.1) was applied to each $X(t)$, resulting in 371 optimal pairs $\{\hat{U}(t), \hat{\mathbf{p}}(t)\}$. Some typical membership assignments due to this procedure are listed for patient 53 in Table 10.1 as a function of time t = age of $\mathbf{x}_{53}$ in years from birth. Figure 10.1 plots cumulative memberships of $\mathbf{x}_{53}$ from Table 10.1 over time. These graphs portray the preoperative course of patient 53's membership in each of the four postulated pure types. The clinical trend suggested by these membership graphs is one of increasing sickness, as characterized by a sharp rise in $u_4(t)$, cyanotic severe, with age. The decision to operate on $\mathbf{x}_{53}$ at 9.28 years of age was made independently of the computer analysis described here; the point made by Woodbury and Clive is that the evolution of fuzzy memberships for $\mathbf{x}_{53}$ seems to accurately reflect the ongoing clinical facts, and could quite possibly have been an effective means of screening prior to final medical examination. Another advantage of this method is data reduction: the 108 measured features of each patient are reduced to $c = 4$ memberships which appear to summarize fairly effectively the relevant clinical information about each patient having a form of Fallot's heart disease.

Table 10.1. Tetralogy of Fallot (E10.1). Memberships of Patient 53 as a Function of Patient Age

	Membership of x_{53} in four clinically pure types: $\hat{u}_{i,53}(t)$			
Age (t) of $\mathbf{x}_{53}$ (yr)	Asymptomatic	Moderate	Acyanotic severe	Cyanotic severe
4.20	0.70	0.00	0.30	0.00
6.95	0.46	0.35	0.19	0.00
7.33	0.32	0.34	0.25	0.09
8.40	0.22	0.78	0.00	0.00
9.06	0.00	0.34	0.00	0.66
9.29	0.00	0.27	0.00	0.73

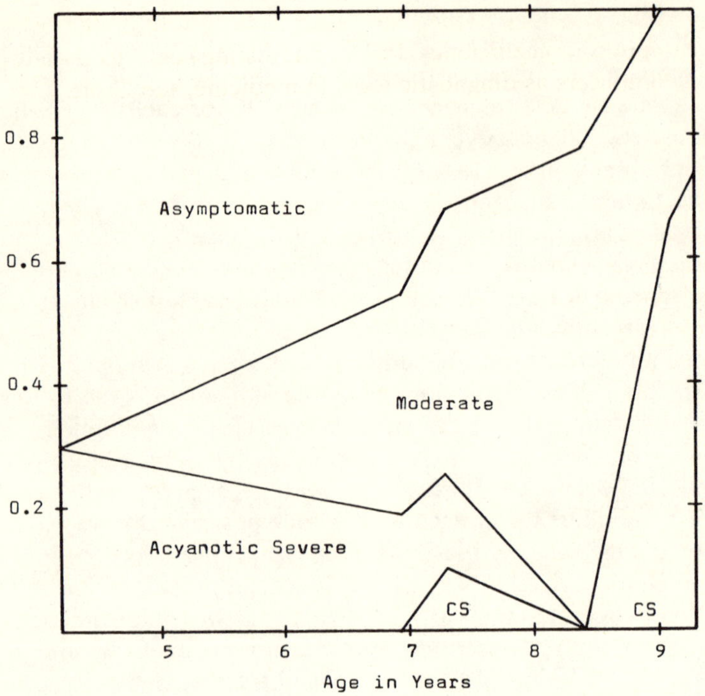

Figure 10.1. Cumulative memberships of patient 53: (E10.1). CS ≡ Cyanotic severe.

Remarks

The algorithms of Woodbury *et al.* reported in (118), (119), and (120) are all based on likelihood functionals similar to J_{wc}. The measure of similarity in (10.7) is somewhat obscured by the combination of memberships and probabilities. Interpreting J_{wc} as a maximum likelihood criterion depends upon (10.5). Further, one must recall that this functional applies only when data set X has the special discrete form specified in (10.1). Nonetheless, (E10.1) certainly suggests that (A10.1) or some variant thereof may provide substantial improvements in current methods for reduction and interpretation of categorical data. A point to which we return later concerns the hypothetical prototypes $\{\mathbf{v}_i\}$: (A10.1) does not generate them, but draws its rationale from their tacit existence; the fuzzy c-means algorithm described in S11 goes one step farther—the prototypes are actually *constructed* [but, in an interesting turn of events, lie outside data space K for data sets having the special construction in (10.1)]. Open mathematical questions concerning J_{wc} include existence and uniqueness for (10.8); and convergence, and rate of convergence for (A10.1). Some variant, theoretical or otherwise, of step size choice in the gradient search employed should improve computational efficiency. For an interesting pessimistic view on the efficacy of computers as diagnostic tools in medicine, see Croft.[30]

Exercises

H10.1. Suppose the null hypothesis regarding categorized data, as in (10.1), is that only $c = 1$ pure clinical prototype best explains the data. Assume the features to be statistically independent; and let $L_0(U, \mathbf{p}; X)$ be the likelihood function of X under the null hypothesis. By taking the logarithm of the likelihood ratio (L/L_0), L in (10.6), show that $(L - L_0)$ is distributed as $(\chi^2/2)$, with $[\sum_k \sum_j n_{kj} - (c-1)q - (c-1)(r-p) + c]$ degrees of freedom, where $c > 1$ is the number of pure prototypes under alternative L, and q is the number of "patient dates" as in E10.1, $r = |K|$, K, p as in (10.1).

S11. Least-Squares Functionals: Fuzzy c-Means

The next fuzzy clustering criterion we discuss generalizes the within-groups sum of square errors function J_W defined at (9.7a). This leads to many infinite families of fuzzy clustering algorithms which have been developed and used by a number of investigators. The initial generalization of J_W and an algorithm akin to hard c-means (A9.2) was reported by Dunn in (35). Shortly thereafter, Dunn's function and algorithm became a special case of the first infinite family of fuzzy clustering algorithms based on a least-squared errors criterion.[10]

(D11.1) *Definition 11.1 (Fuzzy c-Means Functionals).* Let $J_m : M_{fc} \times \mathbb{R}^{cp} \to \mathbb{R}^+$ be

$$J_m(U, \mathbf{v}) = \sum_{k=1}^{n} \sum_{i=1}^{c} (u_{ik})^m (d_{ik})^2 \tag{11.1a}$$

where

$$U \in M_{fc} \tag{11.1b}$$

is a fuzzy c-partition of X;

$$\mathbf{v} = (\mathbf{v}_1, \mathbf{v}_2, \ldots, \mathbf{v}_c) \in \mathbb{R}^{cp} \qquad \text{with } \mathbf{v}_i \in \mathbb{R}^p \tag{11.1c}$$

is the cluster center or prototype of u_i, $1 \le i \le c$;

$$(d_{ik})^2 = \|\mathbf{x}_k - \mathbf{v}_i\|^2 \qquad \text{and } \|\cdot\| \tag{11.1d}$$

is any inner product induced norm on $\mathbb{R}^p$; and

$$\text{Weighting exponent } m \in [1, \infty) \tag{11.1e}$$

Dunn's functional[35] is obtained by setting $m = 2$ in (11.1): the family $\{J_m | 1 \leq m < \infty\}$ was introduced in (10). Examination of J_m reveals that the measure of dissimilarity is $d_{ik} = \|\mathbf{x}_k - \mathbf{v}_i\|$, the distance between each data point $\mathbf{x}_k$ and a fuzzy prototype $\mathbf{v}_i$; the squared distance is then weighted by $(u_{ik})^m = (u_i(\mathbf{x}_k))^m$, the mth power of $\mathbf{x}_k$'s membership in fuzzy cluster u_i. Since each term of J_m is *proportional* to $(d_{ik})^2$, J_m is a squared error clustering criterion, and solutions of

$$\underset{M_{fc} \times \mathbb{R}^{cp}}{\text{minimize}} \{J_m(U, \mathbf{v})\} \tag{11.2}$$

are least-squared error stationary points of J_m. An infinite family of fuzzy clustering algorithms—one for each $m \in (1, \infty)$—is obtained via necessary conditions for solutions of (11.2). The basic theorem follows.

(T11.1) *Theorem 11.1.*[10] Assume $\|\cdot\|$ to be inner product induced: fix $m \in (1, \infty)$, let X have at least $c < n$ distinct points, and define $\forall k$ the sets

$$I_k = \{i | 1 \leq i \leq c; d_{ik} = \|\mathbf{x}_k - \mathbf{v}_i\| = 0\}$$

$$\tilde{I}_k = \{1, 2, \ldots, c\} - I_k$$

then $(U, \mathbf{v}) \in M_{fc} \times \mathbb{R}^{cp}$ may be globally minimal for J_m only if

$$I_k = \varnothing \Rightarrow u_{ik} = 1 \Big/ \left[\sum_{j=1}^{c} \left(\frac{d_{ik}}{d_{jk}} \right)^{2/(m-1)} \right] \tag{11.3a1}$$

or

$$I_k \neq \varnothing \Rightarrow u_{ik} = 0 \; \forall \, i \in \tilde{I}_k \quad \text{and} \quad \sum_{i \in I_k} u_{ik} = 1 \tag{11.3a2}$$

$$\mathbf{v}_i = \sum_{k=1}^{n} (u_{ik})^m \, \mathbf{x}_k \Big/ \sum_{k=1}^{n} (u_{ik})^m \; \forall i \tag{11.3b}$$

Analysis. (11.3a1) is derived by fixing $\mathbf{v} \in \mathbb{R}^{cp}$ and applying Lagrange multipliers to the variables $\{u_{ik}\}$. There is one technical trick: we relax U, allowing it to be in M_{fco}, so that minimization can be done term by term on the uncoupled columns of U. Ensuing solutions always turn out to be where we want them—in M_{fc}! (11.3a2) is the alternate necessary form for memberships of $\mathbf{x}_k$ when $\exists \, i$ so that $d_{ik} = 0$. We call this a "singularity," and whenever it occurs, $\mathbf{x}_k$ must have no membership in any cluster u_i where $d_{ik} > 0$; membership of $\mathbf{x}_k$ in clusters where $d_{ik} = 0$ is arbitrary up to column constraint (5.6b). This gives the only requisites for resolution of algorithmic singularity in step (A11.1c) below. We remark that singularity—$\mathbf{x}_k = \mathbf{v}_i$, $\exists i, k$—hardly ever occurs in practice, since machine roundoff usually precludes this eventuality. Equations (11.3b) are straightforward necessary conditions derived via unconstrained optimization of J_m with $\{\mathbf{v}_i\}$ as variables and U fixed.

Proof. First, fix $\mathbf{v} \in \mathbb{R}^{cp}$ and define $g_m(U) = J_m(U, \mathbf{v})$ for any $U \in M_{fco}$. Since U is degenerate, its columns are independent, and therefore

$$\min_{U \in M_{fco}} \{g_m(U)\} = \min_{U \in M_{fco}} \left\{ \sum_{k=1}^{n} \sum_{i=1}^{c} (u_{ik})^m \, (d_{ik})^2 \right\}$$

$$= \sum_{k=1}^{n} \left[\min_{\mathbf{u}_k \in \text{conv}(B_c)} \left\{ \sum_{i=1}^{c} (u_{ik})^m (d_{ik})^2 \right\} \right] \tag{11.4}$$

where

$$\text{conv}(B_c) = \left\{ \mathbf{u}_k \in \mathbb{R}^c \,\middle|\, \sum_{i=1}^{c} u_{ik} = 1; u_{ik} \geq 0 \right\}$$

as in Section 6. Solution of (11.4) is effected with Lagrange multipliers. For each term, let

$$g_{mk}(\mathbf{u}_k) = \sum_{i=1}^{c} (u_{ik})^m (d_{ik})^2$$

and let its LaGrangian be

$$F_k(\lambda, \mathbf{u}_k) = \sum_{i=1}^{c} (u_{ik})^m (d_{ik})^2 - \lambda\left(\sum_{i=1}^{c} u_{ik} - 1 \right)$$

$(\lambda, \mathbf{u}_k)$ is stationary for F_k only if $\nabla_{\lambda, \mathbf{u}_k} F_k(\lambda, \mathbf{u}_k) = (0, \mathbf{\theta} \in \mathbb{R}^c)$. Setting this gradient equal to zero yields

(A) $$\frac{\partial F_k}{\partial \lambda}(\lambda, \mathbf{u}_k) = \left(\sum_{i=1}^{c} u_{ik} - 1 \right) = 0$$

(B) $$\frac{\partial F_k}{\partial u_{st}}(\lambda, \mathbf{u}_k) = [m(u_{st})^{m-1} (d_{st})^2 - \lambda] = 0$$

From this,

(C) $$u_{st} = \left[\frac{\lambda}{m(d_{st})^2} \right]^{1/(m-1)}$$

Using (A),

$$\sum_{j=1}^{c} u_{jt} = \sum_{j=1}^{c} \left(\frac{\lambda}{m} \right)^{1/(m-1)} \left[\frac{1}{(d_{jt})^2} \right]^{1/(m-1)}$$

$$= \left(\frac{\lambda}{m} \right)^{1/(m-1)} \left\{ \sum_{j=1}^{c} \left[\frac{1}{(d_{jt})^2} \right]^{1/(m-1)} \right\} = 1$$

Thus,

$$\left(\frac{\lambda}{m}\right)^{1/(m-1)} = 1\Big/\sum_{j=1}^{c}\left(\frac{1}{(d_{jt})^2}\right)^{1/(m-1)}$$

Returning to (C),

$$u_{st} = \left\{1\Big/\sum_{j=1}^{c}\left[\frac{1}{(d_{jt})^2}\right]^{1/(m-1)}\right\}\left[\frac{1}{(d_{st})^2}\right]^{1/(m-1)}$$

$$= 1\Big/\sum_{j=1}^{c}\left(\frac{d_{st}}{d_{jt}}\right)^{2/(m-1)}$$

At this point, one of two possibilities exists: if $I_t = \varnothing$, then (11.3a1) follows for column t; if $I_t \neq \varnothing$, then choosing $\{u_{st}\}$ as in (11.3a2) results in $g_{mt}(\mathbf{u}_t) = 0$, because the non-zero weights are placed on zero distances, while positive distances will increase $g_{mt}(\mathbf{u}_t)$, contradicting minimality. Continuing in this way, one arrives at n vectors $\{\mathbf{u}_1, \mathbf{u}_2, \ldots, \mathbf{u}_n\}$, which, taken together as an array U, define a stationary point for g_m. If all of the $\mathbf{u}_k$'s are computed by (11.3a1), $u_{ik} \in (0, 1)\,\forall i, k$ and U is clearly in M_{fc} (*every* entry is fuzzy!). If singularities occur, necessitating the use of (11.3a2) for some columns, U is still nondegenerate. To see this, suppose, to the contrary, that for row r, $u_{rk} = 0\,\forall k$. Define

$$\mathbf{v}_i^* = \begin{Bmatrix} \mathbf{v}_i, & 1 \leq i \leq c; \quad i \neq r \\ \mathbf{x}_p, & i = r \end{Bmatrix}$$

where $\mathbf{x}_p \in X$, $\mathbf{x}_p \neq \mathbf{v}_i$ for $i \neq r$. Such a vector exists because X contains c distinct points. Now

$$0 = \sum_{k=1}^{n} (u_{rk})^m (d_{rk})^2 = \sum_{k=1}^{n} (u_{rk})^m (d_{rk}^*)^2$$

where $d_{rk}^* = \|\mathbf{x}_k - \mathbf{v}_r^*\|\,\forall r$. Then $(U, \mathbf{v}^*)$ is a *global* minimum of J_m. Since $\mathbf{v}_r^* = \mathbf{x}_p$, the set $I_p = \{r\}$ with $u_{rp} = 1$ from (11.3a2). This contradiction shows that $U \in M_{fc}$ whenever (11.3a) is used to construct it.

To establish (11.3b), fix $U \in M_{fc}$ and set $h_m(\mathbf{v}) = J_m(U, \mathbf{v})$. Minimization of h_m is unconstrained over $\mathbb{R}^{cp}$, so we have

$$h_m(\mathbf{v}) = \sum_{k=1}^{n}\sum_{i=1}^{c} (u_{ik})^m (d_{ik})^2$$

$$= \sum_{k=1}^{n}\sum_{i=1}^{c} (u_{ik})^m \langle \mathbf{x}_k - \mathbf{v}_i, \mathbf{x}_k - \mathbf{v}_i \rangle$$

where $\langle \cdot, \cdot \rangle$ is the norm-inducing inner product. Then for each i, it is necessary that the directional derivatives $h'_m(\mathbf{v}_i; \mathbf{w})$ vanish for all unit vectors $\mathbf{w} \in \mathbb{R}^c$:

$$h'_m(\mathbf{v}_i; \mathbf{w}) = \sum_{k=1}^{n} (u_{ik})^m \frac{d}{dt}(\langle \mathbf{x}_k - \mathbf{v}_i - t\mathbf{w}, \mathbf{x}_k - \mathbf{v}_i - t\mathbf{w}\rangle)|_{t=0}$$

$$= -2\left[\sum_{k=1}^{n} (u_{ik})^m \langle \mathbf{x}_k - \mathbf{v}_i, \mathbf{w}\rangle\right] = 0 \qquad \forall \mathbf{w}$$

$$\Leftrightarrow \left\langle \sum_{k=1}^{n} (u_{ik})^m (\mathbf{x}_k - \mathbf{v}_i), \mathbf{w}\right\rangle = 0 \qquad \forall \mathbf{w}$$

$$\Leftrightarrow \sum_{k=1}^{n} (u_{ik})^m (\mathbf{x}_k - \mathbf{v}_i) = \boldsymbol{\theta}$$

from which (11.3b) follows.

Note that conditions (11.3) hold for *any* inner-product-induced norm metric. In particular, any positive-definite matrix $A \in V_{pp}$ induces such a norm via the weighted inner product

$$\langle \mathbf{x}, \mathbf{y}\rangle_A = \mathbf{x}^T A \mathbf{y} = \sum_{i=1}^{p} \sum_{j=1}^{p} \mathbf{x}_i a_{ij} \mathbf{y}_j \tag{11.5}$$

$\forall \mathbf{x}, \mathbf{y} \in \mathbb{R}^p$. Relative to this special class of norms J_m can be written as

$$J_m(U, \mathbf{v}, A) = \sum_{k=1}^{n} \sum_{i=1}^{c} (u_{ik})^m \|\mathbf{x}_k - \mathbf{v}_i\|_A^2 \tag{11.6}$$

where

$$(d_{ik})^2 = \|\mathbf{x}_k - \mathbf{v}_i\|_A^2 = \langle \mathbf{x}_k - \mathbf{v}_i, \mathbf{x}_k - \mathbf{v}_i\rangle_A = (\mathbf{x}_k - \mathbf{v}_i)^T A (\mathbf{x}_k - \mathbf{v}_i)$$

This form emphasizes the dependence of J_m on the matrix A defining the norm for $\mathbb{R}^p$ via (11.5). There are two reasons to do so: under certain special conditions, A may be included as a *theoretical* variable for optimization, as in the modification of Gustafson and Kessel[54] discussed in S22; and in any case, A is an *algorithmic* variable for the fuzzy c-means methods based on (T11.1). Some specific cases are discussed in Chapter 4. Dependence of J_m on A will be suppressed hereafter unless the situation warrants it.

The fuzzy c-means clustering algorithms are simply Picard iteration through necessary conditions (11.3). In earlier literature, the term fuzzy ISODATA is often used to describe the following:

(A11.1) *Algorithm 11.1* [*Fuzzy c-Means* (*FCM*), *Bezdek*[10]]

(A11.1a) Fix c, $2 \leq c < n$; choose any inner product norm metric for $\mathbb{R}^p$; and fix m, $1 \leq m < \infty$. Initialize $U^{(0)} \in M_{fc}$. Then at step l, $l = 0, 1, 2, \ldots,$:

(A11.1b) Calculate the c fuzzy cluster centers $\{\mathbf{v}_i^{(l)}\}$ with (11.3b) and $U^{(l)}$.

(A11.1c) Update $U^{(l)}$ using (11.3a) and $\{\mathbf{v}_i^{(l)}\}$
(A11.1d) Compare $U^{(l)}$ to $U^{(l+1)}$ in a convenient matrix norm: if $\|U^{(l+1)} - U^{(l)}\| \leq \varepsilon_L$ stop: otherwise, return to (A11.1b).

Although it is not immediately obvious, (A11.1) reduces to (A9.2) in the nonsingular case when $m = 1$ and the norm on $\mathbb{R}^p$ is the Euclidean norm, because in this instance, $J_1(U, \mathbf{v}, A_E = I_p) = J_W(U, \mathbf{v})$, where I_p is the identity matrix for V_{pp}. (If U is constrained to lie in M_c, then of course, $J_m = J_1 \forall m$ automatically.) To see that local minima of $J_1(U, \mathbf{v}, I_p)$ over $M_{fc} \times \mathbb{R}^{cp}$ actually lie in $M_c \times \mathbb{R}^{cp}$ and satisfy necessary conditions for *hard* c-means solutions, first note that for $m = 1$, (11.2) is a *linear* programming problem in the $\{u_{ik}\}$, so solutions for fixed $\mathbf{v}_i$'s must lie in the extreme points of M_{fco}—namely, in M_{co} (cf. (6.11b), and Luenberger[75]). That the $\{u_{ik}\}$ necessarily satisfy (9.10) is proven in Dunn.[35] [We verify this analytically in (E11.2) as m approaches 1 from above.] Necessity for the $\{\mathbf{v}_i\}$ follows by letting $m \overset{+}{\to} 1$ from above, whereby (11.3b) converges to (9.8a).

Fuzzy c-means has a number of algorithmic parameters: c, m, $U^{(0)}$, $\|\cdot\|_A$, ε_L. For each A there is an infinite family parametrized by m: As $m \overset{+}{\to} 1$, fuzzy c-means converges in theory to a "generalized" hard c-means solution, where J_W has the added variable A. Conversely, as $m \to \infty$, it is easy to see that u_{ik} in (11.3a) $\to (1/c) \forall i, k$, so $\mathbf{v}_i \to \boldsymbol{\mu}$ the centroid of $X \forall i$. Thus as $m \to \infty$, the only optimal pair for J_m is $(\bar{U}, \boldsymbol{\mu})$ = (centroid of M_{fc}, centroid of X), and $J_m \to 0$. In general, the larger m is, the "fuzzier" are the membership assignments; and conversely, as $m \overset{+}{\to} 1$, fuzzy c-means solutions become hard. Weighting exponent m thus controls the extent of membership sharing between fuzzy clusters in X. This can be good [cf. (16)]: on the other hand, one must choose m to actually implement (A11.1)! Heuristic guidelines for m are available in examples to follow: no theoretical basis for an optimal choice for m has emerged to date. The following example provides for $m = 2$ a physical distinction of sorts, and a nice interpretation of the clustering criterion J_2 with any $\|\cdot\|_A$ in the nonsingular case.

(E11.1) *Example 11.1.* Let $(U, \mathbf{v})$ be stationary for $J_1 = J_W$: denote $d_{ik}^2 = \|\mathbf{x}_k - \mathbf{v}_i\|^2$ as r_{ik}, and interpret this distance as the "resistance" $\mathbf{x}_k$ offers to connection with $\mathbf{x}_{k+1}$ via the prototype $\mathbf{v}_i$. For each k, let $r_k = \min_j\{r_{jk}\} = \min_j\{\|\mathbf{x}_k - \mathbf{v}_j\|^2\}$. Then r_k is the minimum "resistance": $r_k = 0$ (a short circuit) in case $\mathbf{x}_k = \mathbf{v}_i \exists i$, the singular case for column k with J_1. Then at optimal pairs for J_1,

$$J_1(U, \mathbf{v}) = \sum_{k=1}^{n} \sum_{i=1}^{c} u_{ik}(d_{ik})^2 = \sum_{k=1}^{n} \sum_{i=1}^{c} u_{ik} r_{ik} = \sum_{k=1}^{n} r_k$$

which is exactly the total minimum "resistance" offered by a series connec-

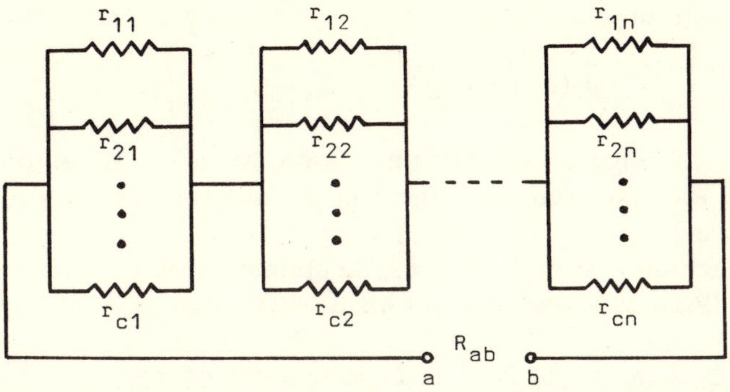

Figure 11.1. Electrical analogs for the hard and fuzzy c-means functionals J_1 and J_2. R_{ab} is the equivalent resistance from a to b through the data. $r_{ik} = (d_{ik})^2 = \|\mathbf{x}_k - \mathbf{v}_i\|^2 > 0 \; \forall \; i, k$. $m = 1$: $U \in m_c$ is hard, $(c-1)$ branches in each block are open, $r_k = \min_j\{r_{jk}\}$ is the MSE path $\forall k$, and $J_1(U, \mathbf{v}) = R_{ab} = \sum_k r_k$. $m > 1$: $U \in M_{fc}$ is fuzzy, all cn branches have positive resistance, $r_k = 1/\sum_j(1/r_{jk})$, and $J_m(U, \mathbf{v}) = R_{ab} = \sum_k r_k \Leftrightarrow m = 2$.

tion of the n data points via the links of minimum resistance [i.e., using nearest prototype rule (9.10)]. See Fig. 11.1 for a schematic of the electrical analog, where R_{ab} is the equivalent resistance of the n data points linked as shown. Now suppose $m > 1$ is fixed, and that $(U, \mathbf{v})$ is optimal for J_m. At such pairs,

$$J_m(U, \mathbf{v}) = \sum_{k=1}^{n} \sum_{i=1}^{c} (u_{ik})^m r_{ik}$$

$$= \sum_{k=1}^{n} \sum_{i=1}^{c} (u_{ik})(u_{ik})^{m-1} r_{ik}$$

Substitution of (11.3a1) for $(u_{ik})^{m-1}$ and a little algebra leads to

$$\text{(A)} \quad J_m(U, \mathbf{v}) = \sum_{k=1}^{n} \left\{ 1 \Big/ \left[\sum_{j=1}^{c} \left(\frac{1}{r_{jk}} \right)^{1/(m-1)} \right]^{(m-1)} \right\}$$

Using (11.3a1) supposes that $(d_{ik})^2 = r_{ik} > 0 \forall i, k$, i.e., that every $\mathbf{x}_k$ has positive membership in each of the c fuzzy clusters in X. In the electrical analog, this corresponds to c parallel paths of connection for each $\mathbf{x}_k$ (via the c prototypes $\{\mathbf{v}_i\}$). Since the equivalent resistance r_k for c parallel paths of resistance $\{r_{jk}\}$ is

$$\text{(B)} \quad r_k = \left(1 \Big/ \sum_{j=1}^{c} \frac{1}{r_{jk}} \right)$$

it follows by comparing (A) and (B) that J_m still corresponds to a path of (minimum) total equivalent resistance connecting the data for $m > 1$ *if and only if* $m = 2$, for then

$$J_2(U, v) = \sum_{k=1}^{n} \left(1 \Big/ \sum_{j=1}^{c} \frac{1}{r_{jk}}\right) = \sum_{k=1}^{n} r_k$$

as illustrated in Figure 11.1. For this reason, it seems natural to call J_2 the equivalent total squared error incurred by allowing all c $\mathbf{v}_i$'s to partially represent each $\mathbf{x}_k$. •

An algebraic interpretation of J_m at optimal pairs $(U, \mathbf{v})$ which connects (11.3a) to (9.10) is based on a factorization like that used in (E11.1).

(E11.2) *Example 11.2.* Let $(U, \mathbf{v})$ be optimal for J_m, and $r_{jk} = (d_{jk})^2 = \|\mathbf{x}_k - \mathbf{v}_j\|_A^2 > 0 \forall j, k$. Repeating equation (A) in (E11.1), we have for $m > 1$

$$J_m(U, \mathbf{v}) = \sum_{k=1}^{n} \left\{1 \Big/ \left[\sum_{j=1}^{c} \left(\frac{1}{r_{jk}}\right)^{1/(m-1)}\right]^{m-1}\right\}$$

$$= \sum_{k=1}^{n} \left[\sum_{j=1}^{c} (r_{jk})^{1/(1-m)}\right]^{1-m}$$

Now, let $\mathbf{r}_k^T = (r_{1k}, r_{2k}, \ldots, r_{ck})$ and $(\mathbf{1/c})^T = (1/c, 1/c, \ldots, 1/c)$. Since all the r_{jk}'s are positive, the weighted mean of order t of $\mathbf{r}_k$ relative to the weight vector $(\mathbf{1/c})$ is by definition, for all real t,

$$M_t(\mathbf{r}_k; (\mathbf{1/c})) = \left[\sum_{j=1}^{c} \left(\frac{1}{c}\right)(r_{jk})^t\right]^{1/t}$$

and the sum of order t is

$$S_t(\mathbf{r}_k) = \left[\sum_{j=1}^{c} (r_{jk})^t\right]^{1/t} = c^{1/t} \cdot M_t(\mathbf{r}_k; (\mathbf{1/c}))$$

Let $t = [1/(1 - m)]$ and $(1/t) = (1 - m)$, and substitute into the last expression for J_m

$$J_m(U, \mathbf{v}) = J_{(t-1)/t}(U, \mathbf{v}) = \sum_{k=1}^{n} c^{1/t} M_t(\mathbf{r}_k; (\mathbf{1/c})) = \sum_{k=1}^{n} S_t(\mathbf{r}_k)$$

$$= \sum_{k=1}^{n} S_{1/(1-m)}(\mathbf{r}_k)$$

This shows that at stationary points the fuzzy c-means criterion is just the sum of n sums of order $[1/(1 - m)]$ of squared error vectors (one for each data point $\mathbf{x}_k$). Since each $r_{jk} > 0$, $r_{jk} = |r_{jk}|$, and were it not for the fact that $m \in (1, \infty] \Rightarrow t = [1/(1 - m)] \in (-\infty, 0)$, one could write J_m as a sum of t norms. Using well-known facts concerning limiting properties of M_t (cf. Roberts and Varberg[84]), we can examine J_m as a function of m.

Case 1. $m \overset{+}{\to} 1 \Leftrightarrow t \to -\infty$. Then $M_t(\mathbf{r}_k; (\mathbf{1/c})) \to \min_j\{r_{jk}\} \forall k$, and since $(c^{(1-m)}) \to 1$, we have, provided the minimum over j is unique $\forall k$:

$$\lim_{m \overset{+}{\to} 1} \{J_m(U, \mathbf{v})\} = \sum_{k=1}^{n} \min_j\{r_{jk}\}$$

$$= \sum_{k=1}^{n} \sum_{i=1}^{c} u_{ik} r_{ik}$$

$$= \sum_{k=1}^{n} \sum_{i=1}^{c} u_{ik}(d_{ik})^2 = J_1(U, \mathbf{v})$$

if and only if $\{u_{ik}\}$ are computed via (9.10). This is an analytic proof of the necessity of using the nearest prototype rule for membership updating in the hard *c*-means algorithm.

Case 2. $m = 2 \Leftrightarrow t = -1$. Then $\forall k M_{-1}(\mathbf{r}_k; (\mathbf{1/c}))$ is the harmonic mean of the c distances $\|\mathbf{x}_k - \mathbf{v}_j\|^2$, so

$$J_2(U, \mathbf{v}) = \sum_{k=1}^{n} M_{-1}(\mathbf{r}_k; (\mathbf{1/c}))\Big/c$$

that is, J_2 averages the harmonic means of the n vectors of squared errors over the c fuzzy clusters in U.

Case 3. $m \to \infty \Leftrightarrow t \to 0$. Then $\forall k\ M_t(\mathbf{r}_k; (\mathbf{1/c})) \to \prod_{j=1}^{c} (r_{jk})^{1/c}$, the geometric mean of the c distances $\|\mathbf{x}_k - \mathbf{v}_j\|^2$, and since $(c^{(1-m)}) \to 0$, we have

$$\lim_{m \to \infty} \{J_m(U, \mathbf{v})\} = 0 \cdot \sum_{k=1}^{n} \left(\prod_{j=1}^{c} r_{jk} \right)^{1/c} = 0$$

The same result can be obtained by recalling that optimal u_{ik}'s $\to (1/c)$ as $m \to \infty$, so $(u_{ik})^m \to 0$ as $m \to \infty\ \forall i, k$. •

Since $J_m \to 0$ as $m \to \infty$, one might suspect that it decreases monotonically with m on optimal pairs. To see that this is indeed the case, differentiate J_m with respect to m. For $m > 1$

$$\frac{dJ_m}{dm}(U, \mathbf{v}) = \sum_{k=1}^{n} \sum_{i=1}^{c} (u_{ik})^m \log(u_{ik})(d_{ik})^2$$

$$= \sum_{k=1}^{n} \sum_{i=1}^{c} [u_{ik} \log(u_{ik})][(u_{ik})^{m-1}(d_{ik})^2]$$

With the usual convention that $x \log(x) = 0$ if $x = 0$, we have $[u_{ik} \log(u_{ik})] \leq 0$ and $[(u_{ik})^{m-1}(d_{ik})^2] \geq 0\ \forall i, k$, both inequalities being strict whenever $0 < u_{ik} < 1$. Since $0 < u_{ik} < 1$ at least twice for each $m > 1$, $(dJ_m/dm) < 0\ \forall m > 1$, so J_m is *strictly monotone decreasing* (evaluated at optimal pairs) on every finite interval of the form $[1, b]$ with $1 < b$.

The monotonic nature of J_m points up the fact that an adjunct measure of cluster validity would be highly desirable in choosing "optimal" clusterings of X, since J_m can be decreased simply by raising m, all other algorithmic parameters being fixed. Examples of fuzzy c-means clustering solutions using large real data sets abound in the literature. Since most of the interesting ones are inextricably tied to cluster validity functionals, we defer these to Chapter 4. As a first example, we discuss two fuzzy c-means clusterings of the Butterfly:

(E11.3) *Example 11.3.* The butterfly data of Examples 9.1 and 9.2 was processed with fuzzy 2-means, using for $U^{(0)}$ the matrix

$$U^{(0)} = \begin{bmatrix} 0.854 & 0.146 & 0.854 & 0.854 & \ldots & 0.854 \\ 0.146 & 0.854 & 0.146 & 0.146 & \ldots & 0.146 \end{bmatrix}_{2\times 15}$$

This certainly looks like a bizarre choice for $U^{(0)}$: we shall see in exercise H15.8 that it is not entirely capricious! $\varepsilon_L = 0.01$ with the max norm on $V_{2,15}$ as in (E9.2) was the cutoff criterion. We chose for $\|\cdot\|_A$ the Euclidean norm, and arbitrarily set $m = 1.25$ for this example. Termination in six iterations resulted in the memberships and cluster centers shown in Fig. 11.2 in the same fashion as those of Fig. 9.1. Comparing Figs. 9.1 and 9.2 with 11.2, observe that this partitioning of X seems to "combine" the two previous solutions, in the sense that membership assignments here are, for all practical purposes, identical with hard c-means except at $\mathbf{x}_8$, and with Ruspini's at $\mathbf{x}_8$. Membership functions u_1 and u_2 in Fig. 11.2 are nearly hard and symmetrical about $\mathbf{x}_8$. A finer threshold ε_L would force u_{18} and u_{28} closer to 0.50, and the $\mathbf{v}_i$'s would be driven to even closer symmetry. At $m = 1.25$ then, fuzzy c-means suggests that X has two quite distinct clusters centered at

$$\mathbf{v}_1 = (0.83, 2.00) \quad \text{and} \quad \mathbf{v}_2 = (5.14, 2.00)$$

and that these two "wings" are connected through $\mathbf{x}_8$, as indicated by its low membership in both clusters. Note that the MM partition U_m derived via (D9.2) using the fuzzy 2-partition on Fig. 11.2 is identical to the hard 2-means partition of (E9.2). •

(E11.4) *Example 11.4.* To study the effect of m in (A11.1), (E11.3) was replicated in every detail, except that weighting exponent $m = 2.00$. The results, shown in Fig. 11.3, are fairly predictable. Membership functions are symmetric with respect to $\mathbf{x}_8$ in both data coordinate directions; memberships are, in general, somewhat fuzzier; and the cluster centers are very nearly unchanged, now being

$$\mathbf{v}_1 = (0.85, 2.00) \quad \text{and} \quad \mathbf{v}_2 = (5.14, 2.00)$$

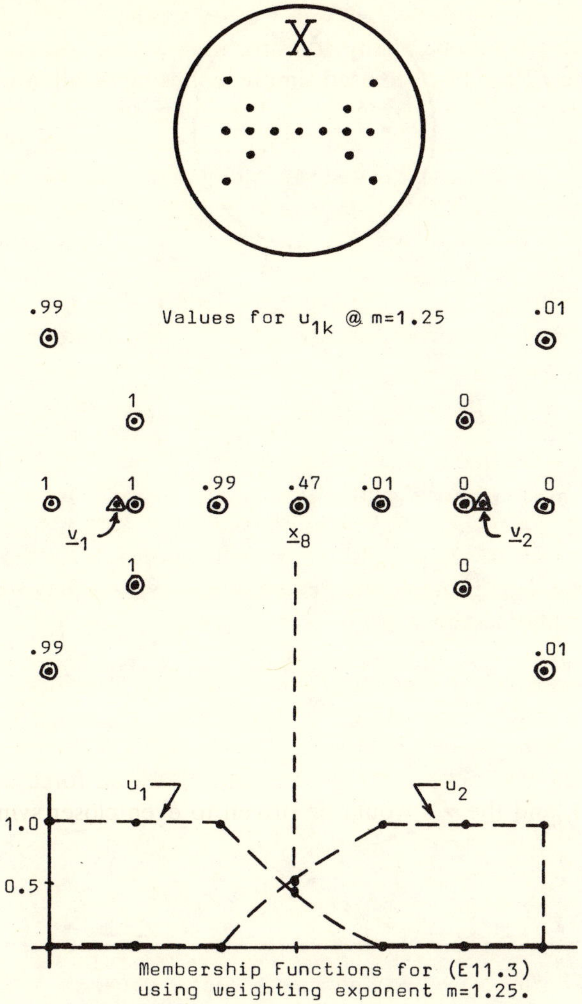

Figure 11.2. The butterfly, (E11.3): membership assignments using fuzzy c-means (A11.1).

This is the first partitioning of the butterfly where graphs of u_1 and u_2 are somewhat distorted when represented in $\mathbb{R}^2$ as in Fig. 11.3, rather than as points on surfaces in $\mathbb{R}^3$. The $\mathbb{R}^2$ graph in Fig. 11.3 is obtained by averaging memberships along the vertical variates. Note in particular that $\mathbf{x}_8$ now has exactly 0.50 membership in each cluster, and that peak memberships occur in a tighter zone around each $\mathbf{v}_i$. This partition suggests that $\mathbf{x}_8$ is a bridge connecting two compact clusters of four points each; in addition, each of these two zones has three "satellites" (the points with memberships in the

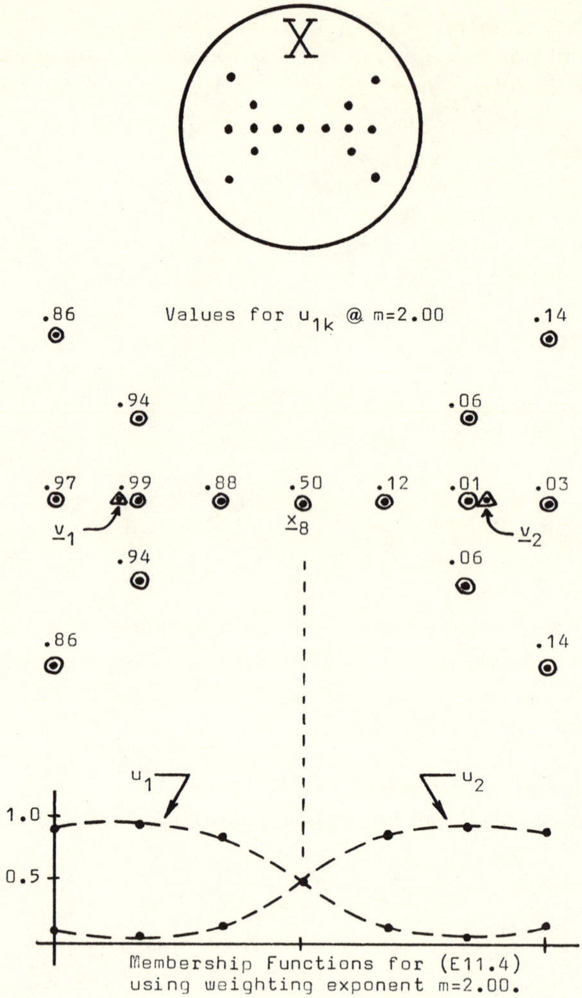

Figure 11.3. The butterfly, (E11.4): membership assignments using fuzzy c-means (A11.1).

0.8–0.9 range). This is different from the partitioning of (E9.1); Ruspini's algorithm suggests that the most distinct points (ones with highest memberships) are those furthest apart (in the outermost tips of the wings), whereas fuzzy 2-means regards part of each wingtip as outliers to its centers. Note, however, that hard MM partitions obtained from the fuzzy U's of (E9.1) and (E11.4) will be identical, even though suggested within-cluster relationships from the fuzzy 2-partitions are quite different. This is because unit balls are spherical in the Euclidean metric, so fuzzy c-means tends to assign equal memberships to points which are radially symmetric with respect to each cluster center. In other words, J_m with the Euclidean norm tries to force "circular" clusters in X; when this is not desirable, it may be altered by using a different norm. •

A popular idea often associated with fuzzy sets is their "core." The usual definition involves a threshold, say α, to identify the core.

(D11.2) *Definition 11.2* (*α-Level Set*). Let u_i be a fuzzy subset of X. The *α-core or α-level* set of X derived from u_i at each $\alpha \in [0, 1]$ is the hard set

$$C(u_i; \alpha) = \{x \in X | u_i(x) > \alpha\} \tag{11.7}$$

Note that u_i is its own core for every α in case u_i is hard.

The core of a fuzzy set provides a different way to compare c-partitions of data. As an example, we find the cores for the four partitions of Examples 9.1, 9.2, 11.3, and 11.4 using $\alpha = 0.90$.

(E11.5) *Example 11.5.* Figure 11.4 shows the cores for each cluster of the butterfly using $\alpha = 0.90$, with each of the four solutions previously discussed. Perhaps the most useful aspect of this type of comparison is its ability to depict how dramatically one's interpretation of substructure in X can change using different clustering strategies. For example, the cores at fixed α vary in size from 6 points to all 15 points in the data. •

Solutions of (11.2) can be regarded as solutions of generalized minimum variance partitioning problems by introducing a fuzzy extension of the hard scatter matrices of Wilks.

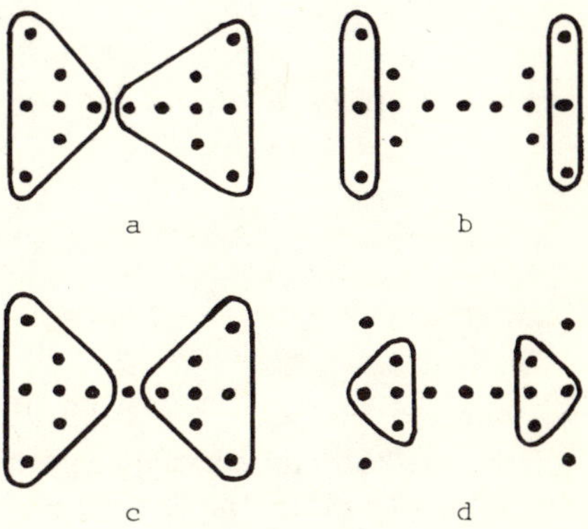

Figure 11.4. α-Cores for butterfly clusters: $\alpha = 0.90$. (a) Hard 2-means: (E9.2). (b) Ruspini's solution: (E9.1). (c) Fuzzy 2-means, $m = 1.25$: (E11.3). (d) Fuzzy 2-means, $m = 2.00$: (E11.4).

(D.11.3) *Definition 11.3* (*Fuzzy Scatter Matrix*). Assume $m \in (1, \infty)$; $X = \{\mathbf{x}_1, \mathbf{x}_2, \ldots, \mathbf{x}_n\} \subset \mathbb{R}^p$; and $(U, \mathbf{v}) \subset M_{fc} \times \mathbb{R}^{cp}$. We define the following:

Fuzzy centroid or prototype of cluster u_i:

$$\mathbf{v}_i = \sum_{k=1}^{n} (u_{ik})^m \mathbf{x}_k \Big/ \sum_{k=1}^{n} (u_{ik})^m \tag{11.8a}$$

Fuzzy scatter matrix of cluster u_i:

$$S_{fi} = \sum_{k=1}^{n} (u_{ik})^m (\mathbf{x}_k - \mathbf{v}_i)(\mathbf{x}_k - \mathbf{v}_i)^T \tag{11.8b}$$

Fuzzy within-cluster scatter matrix:

$$S_{fW} = \sum_{i=1}^{c} S_{fi} \tag{11.8c}$$

These definitions extend equations (9.8) just as J_m extends $J_W = J_1$, and in this sense may seem at first glance somewhat contrived. Such is not the case, however! We shall see in Chapter 5 that the fuzzy scatter matrices $\{S_{fi}\}$ at (11.8b) arise naturally in two generalizations of J_m: the fuzzy "covariance" algorithm of Gustafson and Kessel[54]; and the fuzzy c-varieties algorithms of Bezdek *et al.*[16] As an immediate application, we calculate the trace of the fuzzy within-cluster scatter matrix:

$$\begin{aligned}
\operatorname{Tr}(S_{fW}) &= \operatorname{Tr}\left(\sum_{i=1}^{c} S_{fi}\right) = \sum_{i=1}^{c} \operatorname{Tr}(S_{fi}) \\
&= \sum_{i=1}^{c} \operatorname{Tr}\left[\sum_{k=1}^{n} (u_{ik})^m (\mathbf{x}_k - \mathbf{v}_i)(\mathbf{x}_k - v_i)^T\right] \\
&= \sum_{k=1}^{n} \sum_{i=1}^{c} (u_{ik})^m \operatorname{Tr}[(\mathbf{x}_k - \mathbf{v}_i)(\mathbf{x}_k - \mathbf{v}_i)^T] \\
&= \sum_{k=1}^{n} \sum_{i=1}^{c} (u_{ik})^m \left[\sum_{t=1}^{p} (x_{kt} - v_{it})^2\right] \\
&= \sum_{k=1}^{n} \sum_{i=1}^{c} (u_{ik})^m \|\mathbf{x}_k - \mathbf{v}_i\|^2 \\
&= J_m(U, \mathbf{v})
\end{aligned} \tag{11.9}$$

as long as the norm for J_m is Euclidean. Evidently, then, (11.2) is a *generalized* minimum variance partitioning problem. Form (11.9) imparts a statistical flavor to the criterion J_m which is entirely analogous to that enjoyed by J_1 as long as the measure of dissimilarity is Euclidean. This result is cast in an even more general form in Chapter 5.

Remarks

The material in (E11.1) originally appeared in (13); (E11.2) is new. The amount of literature concerning the fuzzy c-means algorithms is quite extensive. There are probably several reasons for this: first, these functionals extend the classical WGSS criterion J_W, which is a very popular and well-studied basis for hard clustering; mathematically, J_m is intimately related to the Hilbert space structure of $\mathbb{R}^p$ (in particular, to orthogonal projection and least-squares approximation theory, which emerges in S23), and is thus tied to a far richer and in some sense more profound mathematical structure than other functionals; finally, the fuzzy c-means algorithms seem to work! Examples of this last point will be given throughout this book. Some references which contain detailed numerical examples for the convenience of applications-oriented readers interested in the fuzzy c-means procedure are, (by application): numerical taxonomy,[11] pharmacology,[87] medical diagnosis,[14,15,19,119] space guidance systems,[52] and sampling rate control.[53] Papers dealing with related topics—cluster validity for example—will be referenced in more appropriate places. Finally, an empirical observation: algorithmic singularities—nonunique minima in (A9.2), zero distances in (A11.1)—are not a serious problem for the c-means algorithms. The prevalence of this occurrence increases dramatically in the fuzzy c-varieties algorithms discussed in Chapter 5.

We should also, at this juncture, mention the "dynamic clusters" method of Diday, described in great detail in (33), which in turn alludes to many other papers concerning his algorithms. Diday's functionals are in some ways more general than J_m (the weights u_{ik} are simply real numbers with otherwise arbitrary values). This simplifies analysis, because the optimization problem in U is unconstrained, and hence (A11.1) is not a special case of the dynamic clusters method. All of the c-means algorithms can be cast nicely in the generalized setting of Picard iteration on prototypes. Diday does this for hard partitioning algorithms with various kinds of prototypes: points in $\mathbb{R}^p$, linear manifolds in $\mathbb{R}^p$, and probability density functions. Reference (33) has an excellent survey of many classical methods, and bears remarkable similarities (in the hard domain) to many of the ideas discussed here and later.

Exercises

H11.1. Let $U^{(0)} \in M_{fc}$ initialize (A11.1) or (A9.2). Prove that if any two or more rows of $U^{(0)}$ are identical, this will be the case for all subsequent iterations of either algorithm.

H11.2. Verify the result of H11.1 numerically, by calculating $U^{(1)}$ with A11.1 using X from H5.1, $m = 2$, $c = 3$, and

$$U^{(0)} = \begin{bmatrix} \frac{1}{3} & \frac{1}{4} & \frac{1}{5} & \frac{1}{6} & \frac{1}{7} \\ \frac{1}{3} & \frac{1}{4} & \frac{1}{5} & \frac{1}{6} & \frac{1}{7} \\ \frac{1}{3} & \frac{1}{2} & \frac{3}{5} & \frac{2}{3} & \frac{5}{7} \end{bmatrix}$$

H11.3. Problem H9.2 has an analog for J_m. Let $(U, \mathbf{v})$ be a local minimum of J_m; $m \in (1, \infty)$, $n = |X|$, and conditions (11.3a1), (11.3b) in force. Show that for $2 \leq c < n$,

$$J_m(U, \mathbf{v}) = \sum_{i=1}^{c} \left\{ \left[\sum_{k=1}^{n} (u_{ik})^m \right]^{-1} \sum_{\mathbf{x} \in X} \sum_{\mathbf{y} \in X} (u_i(\mathbf{x}) u_i(\mathbf{y}))^m \|\mathbf{x} - \mathbf{y}\|^2 / 2 \right\}$$

H11.4. Find the unique vector in $\mathbb{R}^p$ that minimizes $f(\mathbf{x}) = \sum_{k=1}^{n} \|\mathbf{x} - \mathbf{x}_k\|_A^2$, where $\{\mathbf{x}_1, \mathbf{x}_2, \ldots, \mathbf{x}_n\} \subset \mathbb{R}^p$ are fixed, and $\|\mathbf{x} - \mathbf{x}_k\|_A^2 = (\mathbf{x} - \mathbf{x}_k)^T A (\mathbf{x} - \mathbf{x}_k)$, with $A \in V_{pp}$, A positive-definite (cf. H9.5). [Answer: $\mathbf{x}^* = (1/n) \sum_{k=1}^{n} \mathbf{x}_k$.]

H11.5. Show that the result of H11.4 is valid for the function $g(\mathbf{x}) = \sum_{k=1}^{n} (\mathbf{x} - \mathbf{x}_k)^T B (\mathbf{x} - \mathbf{x}_k)$ whenever $B \in V_{pp}$ is invertible.

S12. Descent and Convergence of Fuzzy *c*-Means

Ruspini's algorithm converges (theoretically) to local minima of J_W because (A9.1) is a gradient descent technique. The situation for (A11.1) is a bit more complex. Fuzzy c-means is Picard iteration through first-order necessary conditions; as such, analysis of its convergence requires a different approach. For readers anxious to forge ahead, the gist of this section is that sequences $\{(U^{(l)}, \mathbf{v}^{(l)}) | l = 0, 1, 2, \ldots\}$ generated by (A11.1) at fixed $m > 1$ always converge (in theory) to strict local minima (not just stationary points) of J_m; or at worst, every convergent subsequence does, provided that singularity never necessitates the use of (11.3a2).

To begin, we specify the two parts of (A11.1) in functional form: let $X = \{\mathbf{x}_1, \mathbf{x}_2, \ldots, \mathbf{x}_n\} \subset \mathbb{R}^p$; $m > 1$, $2 \leq c < n$, $\|\cdot\|$ for J_m be fixed, and define

$$F: \mathbb{R}^{cp} \to M_{fc}, \qquad F(\mathbf{v}) = U = [u_{ik}] \tag{12.1a}$$

where each u_{ik} is calculated via (11.3a); and

$$G: M_{fc} \to \mathbb{R}^{cp}, \qquad G(U) = \mathbf{v} \tag{12.1b}$$

where each $\mathbf{v}_i$ is calculated via (11.3b). Indicating functional composition by a circle ($\circ$), (A11.1) can be specified as a Picard loop in U or in $\mathbf{v}$, depending on the choice for initialization:

$$U^{(l)} = (F \circ G)(U^{(l-1)}) = \cdots (F \circ G)^{(l)}(U^{(0)})$$

or

$$\mathbf{v}^{(l)} = (G \circ F)(\mathbf{v}^{(l-1)}) = \cdots (G \circ F)^{(l)}(\mathbf{v}^{(0)})$$

where $l = 1, 2, \ldots$. Either specifies the computational strategy outlined in (A11.1) to (hopefully) minimize $J_m(u, \mathbf{v})$. Our goal is to prove that this strategy is theoretically sound; in order to proceed, we must recast the algorithm in the form of a Picard sequence in both variables. To this end, let $T_m: (M_{fc} \times \mathbb{R}^{cp}) \to (M_{fc} \times \mathbb{R}^{cp})$ be the fuzzy c-means operator:

$$T_m = A_2 \circ A_1 \tag{12.2a}$$

where

$$A_1: (M_{fc} \times \mathbb{R}^{cp}) \to \mathbb{R}^{cp}; A_1(U, \mathbf{v}) = G(U) \tag{12.2b}$$

$$A_2: \mathbb{R}^{cp} \to (M_{fc} \times \mathbb{R}^{cp}); A_2(\mathbf{v}) = (F(\mathbf{v}), \mathbf{v}) \tag{12.2c}$$

Explicitly, the action of T_m on $(U^{(l)}, \mathbf{v}^{(l)})$ is

$$\begin{aligned}(U^{(l+1)}, \mathbf{v}^{(l+1)}) &= T_m(U^{(l)}, \mathbf{v}^{(l)})\\ &= (A_2 \circ A_1)(U^{(l)}, \mathbf{v}^{(l)})\\ &= A_2(A_1(U^{(l)}, \mathbf{v}^{(l)}))\\ &= A_2(G(U^{(l)}))\\ &= (F(G(U^{(l)}), G(U^{(l)})))\\ &= ((F \circ G)(U^{(l)}), G(U^{(l)}))\end{aligned}$$

From this it follows that properties of T_m ultimately reside in F and G defined at (12.1). We emphasize this, because the ensuing analysis depends upon precisely this fact.

Now suppose $\{T_m^{(l)}(U^{(0)}, \mathbf{v}^{(0)}) | l = 1, 2, \ldots\}$ to be an iterate sequence generated by the Picard operator T_m, and let S^* be the set of local minima of J_m. Our goal is to discover the circumstances whereby $\{T_m^{(l)}(U^{(0)}, \mathbf{v}^{(0)})\}$ converges to a point in S^*. The usual means for resolving questions of this type involve fixed point theorems (e.g., the contraction mapping theorem). However, the complexity of T_m renders this approach untenable for fuzzy c-means; instead, the convergence theorem of Zangwill[125] will be applied. The results are somewhat weaker than one might hope for, but may be the best available. In order to proceed, we repeat without proof the results of Zangwill essential for this section.

Let $f: \mathbb{R}^p \to \mathbb{R}$ be a real functional with domain D_f, and let

$$S^* = \{\mathbf{x}^* \in D_f | f(\mathbf{x}^*) < f(\mathbf{y}), \mathbf{y} \in B^0(\mathbf{x}^*, r)\}$$

where

$$B^0(\mathbf{x}^*, r) = \{\mathbf{x} \in \mathbb{R}^p | \|\mathbf{x} - \mathbf{x}^*\| < r, \|\cdot\| \text{ any norm}\}$$

is an open ball of radius r about the point $\mathbf{x}^*$. We may refer to S^* as the *solution set* of the optimization problem

$$\underset{D_f}{\text{minimize}}\{f(\mathbf{x})\} \tag{12.3}$$

Let $P(D_f)$ be the power set of D_f. Zangwill defined an iterative algorithm for solving (12.3) as any *point-to-set* map $A: D_f \to P(D_f)$. The case at hand, $T_m = A_2 \circ A_1$, is a *point-to-point* map, so our interest lies with algorithmic sequences of generic form

$$\mathbf{x}^{(l+1)} = A(\mathbf{x}^{(l)}) = \cdots A^{(l)}(\mathbf{x}^{(0)}), \qquad l = 1, 2, \ldots$$

We call $\{\mathbf{x}^{(l)}\}$ an iterative sequence generated by A.

Next we attach to A a descent functional g.

(D12.1) *Definition 12.1 (Descent Functional).* $g: D_f \to \mathbb{R}$ is a descent function for (A, S^*) if

$$g \text{ is continuous on } D_f \tag{12.4a}$$

$$\mathbf{x}^* \notin S^* \Rightarrow g(A(\mathbf{x}^*)) < g(\mathbf{x}^*) \tag{12.4b}$$

$$\mathbf{x}^* \in S^* \Rightarrow g(A(\mathbf{x}^*)) \leq g(\mathbf{x}^*) \tag{12.4c}$$

g monitors the progress of A at driving sequences $\{\mathbf{x}^{(l)}\}$ towards a point in S^*; g may or may not be f itself. Note the strict inequality in (12.4b); this will preclude a proof of convergence for fuzzy c-means in the singular case! With these preliminaries, we state the following theorem.

(T12.1) *Theorem 12.1 (Zangwill*[125]*).* Regarding (12.3), if

$$g \text{ is a descent function for } (A, S^*) \tag{12.5a}$$

$$A \text{ is continuous on } (D_f - S^*) \tag{12.5b}$$

$$\{\mathbf{x}^{(l)}\} \subset K \subseteq D_f \text{ and } K \text{ is compact} \tag{12.5c}$$

then for every sequence $\{\mathbf{x}^{(l)}\}$ generated by A, either

$$\{\mathbf{x}^{(l)}\} \to \mathbf{x}^* \in S^* \qquad \text{as } l \to \infty \tag{12.6a}$$

or the limit of *every* convergent subsequence $\{\mathbf{x}^{(l_k)}\}$ of

$$\{\mathbf{x}^{(l)}\} \text{ is a point } \mathbf{x}^* \in S^*, \qquad \text{for arbitrary } \mathbf{x}^{(0)} \in D_f, \tag{12.6b}$$

Proof. See (125). ∎

(T12.1) and its generalizations yield convergence proofs for most of the classical iterative algorithms, e.g., steepest descent, Newton's method, etc.

Note that $\mathbf{x}^* \in S^*$ are strict *local* minima of f, while (T12.1) is *global*, in the sense that convergence to $\mathbf{x}^* \in S^*$ is obtained from arbitrary $\mathbf{x}^{(0)}$ in D_f.

To apply these results to fuzzy c-means, we must assume that $d_{ik} = \|\mathbf{x}_k - \mathbf{v}_i\| > 0\ \forall i, k$. This is evident in the following proposition.

(P12.1) *Proposition 12.1.* Assume $(d_{ik}) > 0\ \forall i, k$, and $m > 1$. Let $\phi : M_{fc} \to \mathbb{R}$, $\phi(U) = J_m(U, \mathbf{v})$ with $\mathbf{v} \in \mathbb{R}^{cp}$ fixed. Then U^* is a *strict local minimum* of ϕ if and only if $U^* = F(\mathbf{v})$, where u^*_{ik} is calculated with (11.3a1) $\forall i, k$.

Analysis and Proof. (P12.1) says this: if $d_{ik} > 0\ \forall i, k$, calculation of U^* using nonsingular formula (11.3a1) is necessary and sufficient for strict local minima of J_m for fixed cluster centers $\{\mathbf{v}_i\}$. The proof is long, so it is left to (18). It applies the classical constrained second-order necessary and sufficient conditions to ϕ; the assumption $d_{ik} > 0\ \forall i, k$ is needed to ensure that the Hessian of the Lagrangian of ϕ at U^* is positive definite. •

The companion result for the "other half" of the fuzzy c-means loop is exhibited as follows.

(P12.2) *Proposition 12.2.* Let $\psi : \mathbb{R}^{cp} \to \mathbb{R}$, $\psi(U) = J_m(U, \mathbf{v})$ with $U \in M_{fc}$ fixed. Assume $m \geq 1$. Then $\mathbf{v}^*$ is a *strict local minimum* of ψ if and only if $\mathbf{v}^* = G(U)$, where $\mathbf{v}^*_i$ is calculated via (11.3b) $\forall i$.

Analysis and Proof. This proposition simply asserts that (11.3b) is both necessary and sufficient for strict local minima of J_m when regarded as a function of $\mathbf{v}$ for fixed U. The proof is quite straightforward, using classical unconstrained second-order conditions on the Hessian of ψ at $\mathbf{v}^*$, and is again left to (18). •

Propositions 12.1 and 12.2 are now combined to show that J_m descends *strictly* on iterates of T_m.

(T12.2) *Theorem 12.2.* Let $S^* = \{(U^*, \mathbf{v}^*) \in D_{J_m} | J_m(U^*, \mathbf{v}^*) < J_m(U, \mathbf{v})\ \forall (U, \mathbf{v}) \in B^0((U^*, \mathbf{v}^*), r)\}$. Then J_m is a descent functional [*à la* (D12.1)] for (T_m, S^*).

Analysis. (T12.2) says this: as long as no singularity occurs in the fuzzy c-means loop of (A11.1), each application of T_m will *strictly* reduce J_m unless it is at a local minimum; in this case, J_m will not go back up! Quite simply, T_m makes steady progress towards S^*, and once in S^*, will not leave it.

Proof. We must check the three conditions in (12.4). First, J_m is continuous in $(U, \mathbf{v})$, because $\{\mathbf{y} \to \|\mathbf{y}\|\}$, $\{z \to z^m\}$, and sums and products of continuous functions are all continuous. Next, if $(U, \mathbf{v}) \notin S^*$, then

$$\begin{aligned} J_m(T_m(U, \mathbf{v})) &= J_m((A_2 \circ A_1)(U, \mathbf{v})) \\ &= J_m(F(G(U)), G(U)) \\ &< J_m(U, G(U)) \qquad \text{by (P12.1)} \\ &< J_m(U, \mathbf{v}) \qquad \text{by (P12.2)} \end{aligned}$$

Finally, if $(U, \mathbf{v}) \in S^*$, equality prevails in the argument above. •

Theorem 12.2 says that J_m serves as its own descent functional, the first requirement of (T12.1). The second requirement is that T_m be continuous on $D_{J_m} - S^*$. T_m is actually continuous on all of $M_{fc} \times \mathbb{R}^{cp}$:

(T12.3) *Theorem 12.3.* T_m is continuous on $M_{fc} \times \mathbb{R}^{cp}$.

Analysis and Proof. T_m is built up with F and G at (12.1). It is fairly routine to show that F and G are continuous using standard facts about continuity; from these the continuity of T_m follows. See (18) for the proof. •

The last condition we need for (T12.1) is an affirmative for the question: does every fuzzy c-means sequence $\{(U^{(l)}, \mathbf{v}^{(l)})\}$ generated by T_m lie in a compact subset of $M_{fc} \times \mathbb{R}^{cp}$? The required answer is contained in the following theorem.

(T12.4) *Theorem 12.4.* $M_{fc} \times [\text{conv}(X)]^c$ is compact, and contains $\{(U^{(l)}, \mathbf{v}^{(l)}) | l = 1, 2, \ldots\}$ generated by T_m from any $(U^{(0)}, \mathbf{v}^{(0)}) \in M_{fc} \times \mathbb{R}^{cp}$.

Analysis. (T12.4) guarantees that Picard iterates never "wander off" so far from S^* that they will not come back sooner or later. (For the practitioner, of course, sooner is better than later, and usually comes in less than 100 iterations for $\varepsilon_L \approx 10^{-4}$ or larger.)

Proof. M_{fc} and conv(X) are both closed and bounded, and hence compact. From this follows the compactness of $M_{fc} \times [\text{conv}(X)]^c$. It remains to be seen that every iterate of T_m falls in this set. We have proven above that in the nonsingular case, $U^{(l)} \in M_{fc}$ at every $l \geq 1$. For $\mathbf{v}^{(l)}$, we have from (11.3b) for each i,

$$\mathbf{v}_i^{(l)} = \sum_{k=1}^{n} \rho_{ik}^{(l)} \mathbf{x}_k, \qquad \sum_{k=1}^{n} \rho_{ik}^{(l)} = 1, \qquad \rho_{ik}^{(l)} \geq 0$$

where

$$\rho_{ik}^{(l)} = (U_{ik}^{(l)})^m \Big/ \sum_{s=1}^{n} (U_{is}^{(l)})^m, \qquad l \geq 1.$$

Thus, $\mathbf{v}_i^{(l)} \in \text{conv}(X)\ \forall i$ and l; (T12.4) is done. •

Finally, we are in a position to establish the convergence result for fuzzy c-means, which will follow by applying (T12.1) for (T_m, S^*) with Theorems 12.2, 12.3, and 12.4.

(T.12.5) *Theorem 12.5.* Let $X = \{\mathbf{x}_1, \mathbf{x}_2, \ldots, \mathbf{x}_n\}$ be bounded in $\mathbb{R}^p$. If $d_{ik}^{(l)} = \|\mathbf{x}_k - \mathbf{v}_i^{(l)}\| > 0\ \forall i, k$, and l, then for any initial $(U^{(0)}, \mathbf{v}^{(0)}) \in M_{fc} \times [\text{conv}(X)]^c$ and $m > 1$, either

$\{T_m^{(l)}(U^{(0)}, \mathbf{v}^{(0)})\}$ converges to a strict local minimum $(U^*, \mathbf{v}^*)$ of J_m as $l \to \infty$ (12.7a)

or

the limit of every convergent subsequence of $\{T_m^{(l)}(U^{(0)}, \mathbf{v}^{(0)})\}$ is a strict local minimum of J_m (12.7b)

Analysis. This theorem assures one that sequences generated by fuzzy c-means (A11.1) are on the right track. Remember, however, that local minima of J_m need not suggest "good" clusters, and that convergence in (12.7) is *theoretical* (i.e., as $l \to \infty$). There is no guarantee that actual sequences stabilize or "settle down" near a theoretical limit for finite l—but whenever they do (in practice, almost always!), they are near a local solution of (11.2).

Proof. Take for g in Zangwill's Theorem (T12.1) J_m itself; (T12.2) shows that J_m is a descent function for (T_m, S^*), where S^* is the set of strict local minima of J_m. (T12.3) asserts continuity of T_m on $M_{fc} \times [\text{conv}(X)]^c$; and by (T12.4), iterate sequences generated by fuzzy c-means operator T_m always lie in a compact subset of the domain of J_m. Results (12.7) now follow as special cases of the corresponding equations (12.6) of (T12.1). •

Remarks

This section is based on (18). The main result, (T12.5), exhibits a nice interplay between conventional numerical analysis and fuzzy mathematics, and is a prelude to the same result for more generalized families of fuzzy objective functional algorithms. An analogous result for J_1 cannot hold, because "local" minima of J_1 are undefined (M_c is discrete). We remark that conditions (11.3) are jointly *necessary* for strict local minima of J_m, but joint *sufficiency* has yet to be established.[18] In view of the fact that fuzzy c-means prototypes $\{\mathbf{v}_i\}$ always lie in $[\text{conv}(X)]^c$ [as in the proof of (T12.4)], the optimization problem (11.2) could have been formulated over the convex domain $M_{fc} \times [\text{conv}(X)]^c$; in which case a unique global minimum for convex functional J_m is guaranteed.[125] (T12.5) does not ensure that (A11.1) will find the global minimum, and even if it does, this may be a poor clustering of X [(E14.1) illustrates this].

Exercises

H12.1. Let $A:\mathbb{R} \to \mathbb{R}$ be a point-to-point algorithm,

$$A(y) = \begin{Bmatrix} (y-1)/2+1, & y > 1 \\ (y/2), & y \leq 1 \end{Bmatrix} \qquad \text{and let } S^* = \{0\}$$

(i) Show that $g(y) = |y|$ is a descent function for (A, S^*).
(ii) Show that $\{y_{k+1} = A(y_k)\} \to 1$ for any $y_0 > 1$.
(iii) What condition of T12.1 is violated?

H12.2. Let $A:\mathbb{R} \to \mathbb{R}$, $A(w) = w + 3$. Take $S^* = \varnothing$ (the empty set), and $g(w) = e^{-w}$. Show that g is a descent function for (A, S^*). $\{w_{k+1}\} \to \infty$ regardless of w_0: what condition of (T12.1) is violated?

H12.3. The Hessian matrix of any scalar field $f:\mathbb{R}^p \to \mathbb{R}$ is the $p \times p$ matrix of second-order partials of f, $[D_{ij}f(\mathbf{x}) = H_f(\mathbf{x})]$. Assuming the hypotheses of (P12.1), let $H_\phi(U)$ be the Hessian of the Lagrangian of ϕ at any $U \in M_{fc}$. Show that $H_\phi(U^*)$ is a diagonal matrix, with n eigenvalues $\{\lambda_j\}$, each of multiplicity c. Find λ_j explicitly. {Answer: $\lambda_j = 4m(m-1)k_j$; $d_{ij}^* = \|\mathbf{x}_j - \mathbf{v}_i^*\|^2 \forall i, j$; $k_j = [\sum_{i=1}^c (d_{ij}^*)^{1/(m-1)}]$.}

H12.4. Assuming the hypotheses of (P12.2), let $H_\psi(\mathbf{v})$ be the Hessian of ψ at any $\mathbf{v} \in \mathbb{R}^{cp}$. Show that $H_\psi(\mathbf{v}^*)$ is a diagonal matrix, with c eigenvalues $\{\beta_i\}$, each of multiplicity p. Find β_i explicitly. {Answer: $\beta_i = 2[\sum_{k=1}^n (u_{ik}^*)^m]$.}

S13. Feature Selection for Binary Data: Important Medical Symptoms

In this section we develop a feature selection scheme for *binary data* based upon the fuzzy prototypes $\{\mathbf{v}_i\}$ derived from (11.3) of the fuzzy c-means algorithms. For S13, data set X must be binary valued in each feature; every K_j in (10.1) is the set $\{0, 1\}$, and for p features, we have

$$X \subset (\{0,1\} \times \{0,1\} \times \cdots \times \{0,1\}) = [\{0,1\}]^p \subset \mathbb{R}^p \qquad (13.1)$$

We assume $x_{kj} = 0$ or 1, respectively, according as "patient" $\mathbf{x}_k$ does not have or has symptom j (a special case of the medical context of S10). More generally, 0 (= absent) and 1 (= present) are observed features in many applications, and the method below will apply to them as well. Medical diagnosis offers a particularly nice context for feature selection, because the objective is quite intuitive:

> A doctor gathers p responses to clinical questions. Which questions (which measured features) of patient k enable the clinician to effect a correct diagnosis? Are there redundant features? Confusing symptoms? Too much data? Not enough data? In short, we seek a means for identifying the "best" features for medical diagnosis.

The basis for a method of feature selection based on the fuzzy c-means algorithms (A11.1) is contained in the following theorem.

(T13.1) *Theorem 13.1.* Let $X \subset [\{0, 1\}]^p$, let $\{\mathbf{v}_i\}$, $1 \leqslant i \leqslant c$ be optimal fuzzy c-means for J_m as in (11.3). Then for each $m > 1$, if $d_{ik} = \|\mathbf{x}_k - \mathbf{v}_i\| > 0 \ \forall i, k$, we have

$$0 \leqslant v_{ij} \leqslant 1 \qquad \forall i, j \tag{13.2a}$$

$$v_{ij} = 0 \Leftrightarrow x_{kj} = 0 \qquad \forall k \tag{13.2b}$$

$$v_{ij} = 1 \Leftrightarrow x_{kj} = 1 \qquad \forall k \tag{13.2c}$$

Analysis and Proof. If all of the data have entries 0 or 1, each prototype will have entries *between* 0 and 1 because $\mathbf{v}_i \in \text{conv}(X) \ \forall i$: the only time feature center v_{ij} achieves an extreme is when the corresponding feature has this value for all n data vectors. To prove the theorem, write the optimal cluster center of u_i as

$$\mathbf{v}_i = \sum_{k=1}^{n} \rho_{ik} \mathbf{x}_k$$

where

$$\rho_{ik} = (u_{ik})^m \Big/ \sum_{t=1}^{n} (u_{it})^m \qquad \forall i, k.$$

These numbers are all positive, and $\sum_{k=1}^{n} \rho_{ik} = 1$ so $\mathbf{v}_i \in \text{conv}(X) \subset \text{conv}([\{0, 1\}]^P) = [0, 1]^P$, which proves (13.2a). Now consider the jth component of $\mathbf{v}_i$:

$$v_{ij} = \sum_{k=1}^{n} \rho_{ik} x_{kj}$$

By (11.3) we know $\rho_{ik} > 0 \ \forall i, k$ in the nonsingular case of (A11.1). From this (13.2b) and (13.2c) are immediate. •

In the proof of (T13.1), there is no dependence on the index i. This observation is the basis for the more general results of the following corollary.

(C13.1) *Corollary 13.1.* With hypotheses as in (T13.1),

$$0 < v_{ij} < 1 \Leftrightarrow 0 < v_{kj} < 1, \qquad 1 \leqslant k \leqslant c \tag{13.3a}$$

$$v_{ij} = 0 \Leftrightarrow v_{kj} = 0, \qquad 1 \leqslant k \leqslant c \tag{13.3b}$$

$$v_{ij} = 1 \Leftrightarrow v_{kj} = 1, \qquad 1 \leqslant k \leqslant c \tag{13.3c}$$

The implications of (C13.1) are these: the jth element of all c fuzzy cluster centers can be 0 if and only if all n data vectors "lack" attribute j; and it can be 1 if and only if all n data vectors "possess" feature j. In either case, feature j contains no information concerning substructure in X, because either no individual exhibited this attribute, or all n of them did. On the other hand, if any of the c cluster centers $\{\mathbf{v}_i\}$ has a value in element j greater than 0 and less than 1, then all c v_{ij}'s do. In this case, feature center values $\{v_{ij}\}$ ostensibly rank the efficacy of feature j as a descriptor of the c subclasses identified by $U \in M_{fc}$.

At the extreme, any one of the prototypes $\mathbf{v}_i$ is *entirely* binary valued if and only if all c of them are, if and only if X contains n identical vectors and no cluster structure exists. This is interesting, because one would ordinarily assume that "prototypes" should be the same kind of objects as the data they represent: in particular, that the $\mathbf{v}_i \subset [\{0, 1\}]^p$, in which case $v_{ij} \in \{0, 1\}$ could be interpreted in the original system of data units as the characteristic response to question j of patients with disease i (the "clinically pure types" of Woodbury *et al.*, in S10). Instead, clusters in X via (A11.1) are possible when and only when at least some of the v_{ij}'s lie in (0, 1), and these values then seem to indicate (by their relative magnitude) the usefulness of feature j as an indicator of the c diseases.

(E13.1) *Example 13.1.* In (14) are listed 107 binary-valued data vectors having 11 features each. The features were symptoms adjudged clinically relevant for patients suffering from abdominal pain caused by either hiatal hernia or gallstones. For example, the data record for patient 104 was

$$x_{104} = (1, 1, 1, 0, 0, 1, 0, 1, 0, 0, 1)$$

The feature list is given in (14): feature 2 is epigastric pain; 3 is upper right quadrant pain; 10 is at least 20 pounds weight loss; and so forth. The first 57 patients had hiatal hernia; the remaining 50 had gallstones. Fuzzy c-means (A11.1) was applied to these data using $c = 2$; $m = 1.10$; $\|\cdot\| =$ Euclidean;

$\varepsilon_L = 0.01$; and the initialization of $U^{(0)}$ given at equation (12) of (14). Table 13.1 lists a portion of the membership function u_1 (row 1 of $U \in M_{fc}$) found by (A11.1) in ten iterations.

The values in Table 13.1 indicate which patients (A11.1) thinks have hernia, and which have gallstones. Patients assigned low memberships clearly cause trouble for (A11.1)—the membership of patient 25, (0.44), suggests that his symptoms are uncharacteristic of either disease. This is a nice example of how fuzziness can point to the atypical individuals in data sets: for the doctor, a signal to investigate patient 25 more carefully!

Table 13.2 lists the optimal cluster centers $\{\mathbf{v}_1, \mathbf{v}_2\}$ which correspond to $U \in M_{fc}$ exhibited in Table 13.1. Corollary 13.1 suggests that symptoms 2 and 3 are, respectively, the most characteristic of hernia and gallstones; whereas symptoms 10 and 9 are, respectively, least typical of these two diseases. For example, hiatal hernia patients are most likely to exhibit epigastric pain ($v_{12} = 0.98$); and rarely exhibit weight loss ($v_{1,10} = 0.02$). •

Table 13.1. Membership for Hiatal Hernia Cluster Generated by Fuzzy c-Means: (E13.1)

Patient	Hiatal hernia membership u_1	Patient	Hiatal hernia membership u_1
1	0.99	58	0.00
2	0.99	59	0.00
3	0.86	60	0.00
4	0.86	61	0.01
5	0.86	62	0.00
.	.	.	.
.	.	.	.
.	.	.	.
21	1.00	81	0.00
22	0.76	82	0.00
23	0.08	83	0.00
24	0.26	84	0.00
25	0.44	85	0.00
26	0.00	86	0.00
.	.	.	.
.	.	.	.
.	.	.	.
51	0.96	101	0.26
52	0.76	102	0.00
53	0.99	103	0.00
54	0.98	104	0.00
55	0.49	105	0.96
56	0.94	106	0.04
57	0.85	107	0.04

Table 13.2. Prototypes for the Membership Functions of Table 13.1: (E13.1)

	Feature centers		
Symptom	(Hernia) v_{1j}	(Gallstones) v_{2j}	Absolute differences $f_{12,j}$
1	0.57	0.27	0.30
2	0.98	0.67	0.31
3	0.06	0.93	→ 0.87
4	0.22	0.55	0.33
5	0.17	0.10	0.07
6	0.77	0.84	0.07
7	0.42	0.05	0.37
8	0.39	0.84	0.45
9	0.48	0.04	0.44
10	0.02	0.16	0.14
11	0.12	0.25	0.13

Example 13.1 shows how fuzzy c-means centers rank the significance of each feature as a subclass descriptor. These rankings do *not*, however, establish the features which possess optimal discriminatory power for interclass separation. A measure which appears to do this is the absolute difference between subclass feature means: for $i \neq j$ we define the *separation vector*

$$\mathbf{f}_{ij.} = (|v_{i1} - v_{j1}|, |v_{i2} - v_{j2}|, \ldots, |v_{ip} - v_{jp}|) \tag{13.4}$$

A direct application of (C13.1) yields for the components of $\mathbf{f}_{ij}$ the following properties.

(P13.1) *Proposition 13.1.* Let X be binary-valued data; if $\{v_{ij} | 1 \leq i \leq c; 1 \leq j \leq p\}$ are optimal fuzzy c-means derived from (11.3) in the nonsingular case with $m > 1$, then

$$0 \leq f_{ij,k} < 1 \qquad \forall \quad i, j, \text{ and } k \tag{13.5a}$$

$$f_{ij,k} = 0 \Leftrightarrow \left\{ \begin{array}{l} \text{(i) all vectors in } X \text{ have feature } k, \text{ or} \\ \text{(ii) no vector in } X \text{ has feature } k \end{array} \right\} \tag{13.5b}$$

Equation (13.5b) shows that feature k is useless as a measure of variability between classes i and j when $f_{ij,k} = 0$. At the opposite extreme, $f_{ij,k}$ cannot be 1 because v_{ik} and v_{jk} are simultaneously 0 or 1 by (C13.1); however, $f_{ij,k}$ *approaches* 1 as the number of members in one class possessing feature k

increases while the number of members in the other class possessing feature k decreases.

Based on the reasoning above, a plausible heuristic for ranking features as discriminators of classes i and j is to order them by the components of $\mathbf{f}_{ij}$. At $c = 2$ this procedure is direct; for $c > 2$, a reasonable procedure is to average $f_{ij,k}$ over the $c(c-1)/2$ pairs (i, j) with $i \neq j$ to secure an *overall average efficiency* for feature k, say

$$f_k = (2/(c)(c-1)) \sum_{j=i+1}^{c} \sum_{i=1}^{c-1} f_{ij,k} \tag{13.6}$$

f_k measures the relative ability of feature k for interclass separation over all the distinct pairwise fuzzy clusters in X. Optimal features are then selected by ordering the $\{f_k | 1 \leq k \leq p\}$.

(E13.2) *Example 13.2.* The vector $\mathbf{f}_{12}$ of differences between optimal feature centers associated with (E13.1) is shown as column 4 of Table 13.2. From this we infer that feature 3 (upper right quadrant pain) is, among the 11 measured responses, the most efficient indicant of whether a patient has hiatal hernia or gallstones: from Table 13.2 we find that the presence of feature 3 would indicate gallstones; its absence, hiatal hernia. Feature 8 and then 9 appear the next most useful features, and so on, down to features 5 and 6, which appear to be least useful. •

To test the validity of the assertions in (E13.2), a method for assessing algorithmic performance on reduced feature spaces is needed. Since $U \in M_{fc}$ is a *clustering* of X, classification error rate is not a meaningful measure of performance here. If X is labeled data, an analogous measure of performance for *hard* clustering algorithms is the clustering error rate.

(D13.1) *Definition 13.1* (*Cluster Error*). Let $\hat{U} \in M_c$ label X, $|X| = n$. The *cluster error* associated with any $U \in M_c$ is the number of vectors in X mislabeled by U: one way to calculate cluster error is

$$E(U) = \sum_{k=1}^{n} \sum_{i=1}^{c} (\hat{u}_{ik} - u_{ik})^2/2n \tag{13.7}$$

Note that $E(U)$, or any equivalent formula, is an absolute measure of cluster validity of U *when X is labeled*, and is undefined when X is not labeled. Equation (13.7) is well defined for any $U \in M_{fco}$, e.g., $E(U) = 2.36$, might be interpreted as more than two mistakes, but less than three. In this way one might compare the overall validity of different fuzzy partitionings of labeled data. In the present context, a clinician must ultimately decide what disease each patient has, so a more useful measure of error can be derived by conversion of $U \in M_{fco}$ to U_m in M_c via (D9.2)—that is, convert U to its

Table 13.3. Clustering Errors Using Reduced Feature Spaces: (E13.3)

Symptoms used	Cluster error $E(U)$
{1–11}	23
{3}	23
{3, 8}	23
{3, 9}	36
{3, 8, 9}	36

closest hard partition via the maximum membership rule, and then count mistakes. This method of counting errors is used in the following example.

(E13.3) *Example 13.3.* Using the rankings in Table 13.2 as a guide, data set X of (E13.1) was reprocessed with fuzzy c-means as above by processing feature 3, then {3, 8}, {3, 9}, and {3, 8, 9}, to see whether these feature subsets contained as much information about hernia and gallstones as the 11 symptom data. The results are shown in Table 13.3. Evidently (A11.1) correctly labels [by conversion to U_m via (D9.2)] exactly the same number of patients using either all 11 symptoms, or just symptom 3! This corroborates the supposition of (E13.2) concerning the validity of feature selection with $\mathbf{f}_{ij}$. Observing symptoms 3 and 8 gives no improvement; replacing {3, 8} by {3, 9} results in an *increase* of 13 mistakes, suggesting that symptom 9 confounds diagnosis, while symptom 8 is neutral. •

Remarks

The series of examples presented in this section are discussed in more detail in (14). Reference (15) exemplifies the use of (13.6) on a larger data set—300 patients with one of $c = 6$ stomach disorders; for this set, symptom 7 (relief induced by food ingestion) was chosen as the optimal feature, a result which agrees with a nonfuzzy feature selection method applied to the same data by R. C. T. Lee.[72] We emphasize again that this method of feature selection applies only to binary data, because (T13.1) fails when this is not the case.

Exercises

H13.1 What is the minimum number d of binary features needed, in principal, to discriminate between c subclasses? (Answer: $2^d \geq c$.)

H13.2. Show that the feature selection scheme of S13 can be extended to apply to categorical data of the type described in (10.1).

H13.3. Under what circumstances can the number v_{ij} in (13.3) be construed as the relative frequency of occurrence of feature j in class i data vectors?

H13.4. The *error rate* of a classifier is defined in S25. Briefly, it is the probability of misclassifying an unlabeled data vector drawn from one of c classes. The following problem shows that the *single* feature among a set of independent binary features which yields the minimum error rate need *not* be a member of the best pair of such features [Toussaint, G. T., "Note on Optimal Selection of Independent Binary-Valued Features for Pattern Recognition," *IEEE Trans. Comp.* (1971)]. Let $c = 2$, $p_1 = p_2 = 0.5$ be the prior probability of each class. Suppose each $\mathbf{x}_k \in X$ has three independent binary features, and that the class conditional probabilities of each feature are

$$\left\{\begin{matrix} \alpha_i = \Pr(x_{ki} = 1|\mathbf{x}_k \in \text{class 1}) \\ \beta_i = \Pr(x_{ki} = 1|\mathbf{x}_k \in \text{class 2}) \end{matrix}\right\}, \qquad 1 \leqslant i \leqslant 3, \quad 1 \leqslant k \leqslant n$$

Let e_i denote the error rate due to using only feature i, and e_{ij} denote the error rate due to using features (i, j) for the classifier design.

Suppose the parameters above satisfy

$$\text{(A)}\left\{\begin{matrix} e_1 < e_2 < e_3 \\ \alpha_i < \beta_i \quad \forall i \\ \beta_1 - \alpha_1 > \beta_2 - \alpha_2 > \beta_3 - \alpha_3 \end{matrix}\right\}$$

and let $l_i = \beta_i - \alpha_i$, $h_i = \frac{1}{2}(1 - \alpha_i - \beta_i)$, $d_{ij} = |h_i| - |h_j|$. If variables i and j are conditionally independent, then

$$\text{(B)}\left\{\begin{matrix} e_{ij} = \frac{1}{2}(e_i + e_j - l_i|h_j| - l_j|h_i|) \\ e_i = \frac{1}{2}1 - \beta_i + \alpha_i) \end{matrix}\right\}$$

(i) Show that $|h_1| > |h_3| \Rightarrow e_{12} < e_{23}$.
(ii) Show that $2d_{31} < (l_1 - l_3)(1 + 2|h_2|)/l_2 \Leftrightarrow e_{12} < e_{23}$.
(iii) Choose $\alpha_1 = 0.1$; $\alpha_2 = 0.05$; $\alpha_3 = 0.01$, $\beta_1 = 0.90$, $\beta_2 = 0.80$, $\beta_3 = 0.71$. Show that these probabilities satisfy (A) and violate (ii).
(iv) Calculate e_i, e_{ij} using (B) and the data in (iii). (Answer: $e_1 = 0.10$, $e_2 = 0.125$, $e_3 = 0.15$, $e_{12} = 0.0825$, $e_{13} = 0.069$; $e_{23} = 0.058$.)
(v) From (iv), conclude that feature 1 is the best *single* binary feature, but features (2, 3) are the best *pair* of binary features for classifier design.

4

Cluster Validity

Section 14 reviews the validity problem. The main difficulty is this: complex algorithms stand squarely between the data for which substructure is hypothesized and the solutions they generate; hence it is all but impossible to transfer a theoretical null hypothesis about X to $U \in M_{fc}$ which can be used to statistically substantiate or repudiate the validity of algorithmically suggested clusters. As a result, a number of scalar measures of partition fuzziness (which are interesting in their own right) have been used as heuristic validity indicants. Sections 15, 16, and 17 discuss three such measures: the Anderson iris data surfaces in S15 and S17. S18 contains several approaches aimed towards connecting a null hypothesis about X to $U \in M_{fc}$: this idea is currently being heavily studied, and S18 is transitory at best. Sections 19 and 20 discuss measures of hard cluster validity which have been related by their inventors to fuzzy algorithms in several ways. S19 contains a particularly interesting application to the design of interstellar navigational systems. S20 provides an additional insight into the geometric property that data which cluster well using FCM algorithm (A11.1) must have.

S14. Cluster Validity: An Overview

Equation (13.7) provides a unique and absolute measure of cluster validity for partitions of labeled data. This is a useful expedient for performance comparisons of different algorithms, but $E(U)$ is useless if X is not labeled. This focuses attention squarely on the central issue of Chapter 4: if X is unlabeled, just how many clusters (if any!) are present? Processing a fixed data set with, say, half of the currently available clustering methods, might result in perhaps two or three *hundred* different hard and fuzzy partitions of X at *each* value for c: which of these clusterings is valid? To complicate matters, it may be physically plausible to expect "good" clusters at more than one value of c, as the data are separated into successively finer substructures (e.g., $c = 2$, sick and healthy; $c = 3$, heart disease, stomach disease, and healthy; $c = 4$, heart disease, hernia, gallstones, and healthy). We try to measure validity mathematically using a validity functional, say $V(c)$. The obvious difficulty in this enterprise ultimately injects a subjective element into most final judgments—this is not bad: computers can be used for advice, but the investigators should call on their special knowledge of the data to ask: is this algorithmically suggested substructure reasonable for our data?

The different classes of clustering algorithms described in S8 generate different kinds of information upon which validity functionals can be based. In many hierarchical schemes, a validity function measures the overall "compactness" of the c-partition found by a particular method; it is then often presumed that large jumps in the function indicate levels at which natural groups exist in the data. Sneath and Sokal[99] discuss several popular approaches along these lines.

Graph-theoretical methods often produce measures of the degree of connectivity, strength of bonds, size of edge weights, length of chains, and the like, which can be used as a basis for both heuristic and statistical validity tests. Hartigan,[57] Ling,[74] and Matula[78] are all excellent sources on methods of this kind.

Objective function methods lend themselves to several approaches. An obvious strategy involves the objective function itself. One processes X at $c = 2, 3, \ldots, n-1$, and records the optimal values of the criterion as a function of c. The most valid clustering is then taken to be an extremum of this function or some derivative of it, such as a large change in slope or large jump at the supposedly optimal c. A significant problem with this strategy is that many objective functions are monotonic in c, which tends to obscure pathological behavior unless the change is quite radical. For this reason, some authors have tried a more formal approach by posing the validity question in the framework of statistical hypothesis testing. The major difficulty here is that the sampling distribution for $V(c+1)$ is not known, so the null hypothesis—that there are exactly c clusters present—cannot be tested very reliably. Nonetheless, goodness of fit statistics such as chi-square and Kolmogorov–Smirnov tests have been tried; see Wolfe[117] for a nice discussion in this direction. Duda and Hart[34] exemplify this idea with the objective function J_W of S9: their insightful analysis illustrates some of the difficulties inherent with this approach. A difficulty they do not mention, which has been alluded to several times above, concerns the fact that objective functions usually have multiple *local* stationary points *at fixed* c, and global extrema are not necessarily the "best" c-partitions of the data. To fix this idea, we illustrate this phenomenon with the classical WGSS error functional J_W.

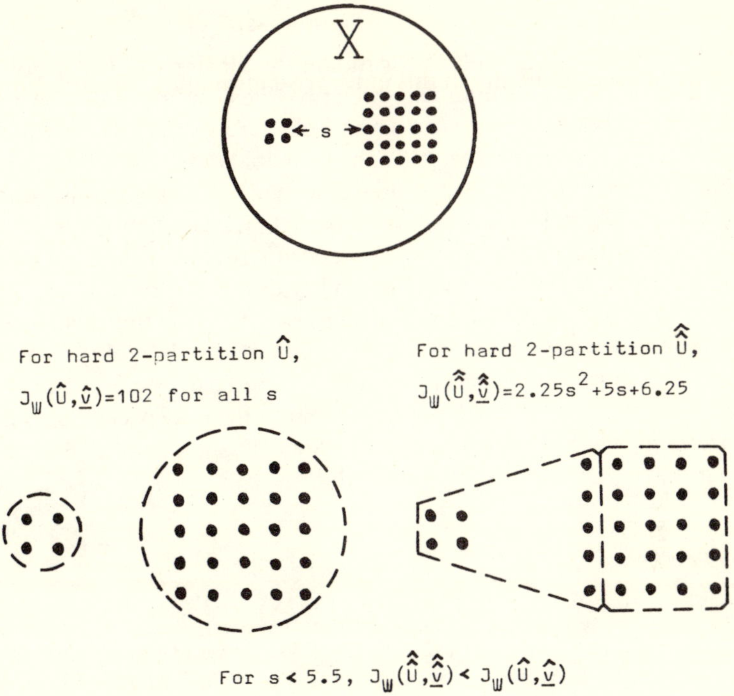

Figure 14.1. The global minimum of J_w may suggest the "wrong" 2-clusters: (E14.1).

(E14.1) *Example 14.1.* The inset of Fig. 14.1 shows $n = 29$ data vectors $\{\mathbf{x}_k\} \subset \mathbb{R}^2$. Interstitial spacing is unity in both directions within each cluster. The "correct" 2-partition of X, called $\hat{U}$ in Fig. 14.1, is shown on the left. Using J_W as the clustering criterion leads to the value $J_W(\hat{U}, \hat{\mathbf{v}}) = 102$ for any interfacial distance s between the two visually apparent hard clusters in X. On the right is a hard 2-partition $\hat{\hat{U}}$ of X which splits off the first column of the right cluster. One can check that $J_W(\hat{\hat{U}}, \hat{\hat{\mathbf{v}}}) = 2.25s^2 + 5s + 6.25$, that is, the value of J_W on this optimal pair is a function of interfacial distance s. For $s < 5.5$, $J_W(\hat{\hat{I}}, \hat{\hat{\mathbf{v}}}) < J_W(\hat{U}, \hat{\mathbf{v}})$, so the global minimum of J_W is hardly an attractive clustering solution! (The splitting tendency is due to the fact that a slight reduction in squared error achieved by moving to $\hat{\hat{U}}$ is multiplied by many terms in the sum.) This illustrates that objective functions can mislead one; the need for a validity functional that identifies $\hat{U}$ as more attractive than $\hat{\hat{U}}$ is quite clear. •

Remarks

It is exceedingly difficult to formulate the cluster validity problem in a mathematically tractable manner which provides useful results, because the basic question here—what is a "good" cluster?—rests on an even more delicate issue—what is a "cluster"? The principal difficulty is that the data X and every partition $U \in M_{fc}$ of X are separated by the algorithm generating U (and defining "cluster" in the process). Consequently, it is usually impossible to discern in U hypotheses about structure which are placed on X. It is, perhaps, precisely this intractability that is responsible for the large literature devoted to measures of fuzziness. Many of the techniques discussed in this chapter tacitly ignore the fundamental difficulty, and are thus heuristic in nature. Some heuristic methods are based on a rather amusing enigma: after eschewing hard methods because the data presumably contain real fuzziness, we then measure the amount of fuzziness in U, and presume the least fuzzy partitions to be most valid!

Exercises

H14.1. Verify that $J_W(U, \mathbf{v}) = 2.25s^2 + 5s + 6.25$ in (E14.1).

H14.2. $X = \{(1, 0), (1, 1), (1, 2), (2, 3)\}$. Let $c = 2$ and $J_1(U, \mathbf{v})$ be the criterion at (9.7a), with U, $\mathbf{v}$ as in (9.7c), (9.7d), respectively. Find the hard 2-partition of X that minimizes $J_1(U, \mathbf{v})$ if

(i) $\|\mathbf{x}_k - \mathbf{v}_i\| = [(x_{k1} - v_{i1})^2 + (x_{k2} - v_{i2})^2]^{1/2}$

(ii) $\|\mathbf{x}_k - \mathbf{v}_i\| = |x_{k1} - v_{i1}| + |x_{k2} - v_{i2}|$

(iii) $\|\mathbf{x}_k - \mathbf{v}_i\| = (x_{k1} - v_{i1}) \vee (x_{k2} - v_{i2})$

(iv) Which of these optimal 2-partitions satisfies the "generalized string property": if $\mathbf{x}_s, \mathbf{x}_t \in u_i$, then $\mathbf{x}_q \in u_i$ whenever $\|\mathbf{x}_q - \mathbf{x}_s\| \leq \|\mathbf{x}_s - \mathbf{x}_t\|$?

H14.3. Let X, $|X| = n$, be a fixed data set, and let $U \in M_c$, $W \in M_{c'}$ be two hard partitions of X, $1 \leq c, c' \leq n$. Define

$$c_{ij} = \left\{\begin{array}{lll} 1, & \exists k, k' \ni x_i \in (u_k \vee w_{k'}), & x_j \in (u_k \vee w_{k'}) \\ 1, & \exists k, k' \ni x_i \in (u_k \vee w_{k'}), & x_j \notin (u_k \wedge w_{k'}) \\ 0, & \text{otherwise} & \end{array}\right\}$$

and

$$R(U, W; c, c') = \sum_{j=i+1}^{n} \sum_{i=1}^{n-1} \left[c_{ij} \Big/ \binom{n}{2} \right].$$

This measure of hard cluster validity is defined in Rand, W. M., "Objective Criteria for the Evaluation of Clustering Methods", *J. Am. Stat. Assoc.* Vol. 66–336 (1971), pp. 846–850. Note that it applies to pairs of hard c-partitions having possibly different numbers of clusters. Show that

(i) if $n_{ij} = |(u_i \wedge w_j)|$, then

$$\binom{n}{2} R(U, W; c, c') = \binom{n}{2} - \left[\sum_i \left(\sum_j n_{ij}\right)^2 + \sum_j \left(\sum_i n_{ij}\right)^2\right] \Big/ 2 - \sum_i \sum_j (n_{ij})^2$$

(ii) $0 \leq R \leq 1 \forall U \in M_c, W \in M_{c'}$.
(iii) $R = 0 \Leftrightarrow c = 1$ and $c' = n$ or conversely.
(iv) $R = 1 \Leftrightarrow U = W$.
(v) $d(U, W) = 1 - R(U, W)$ is a metric on M_c when $c = c'$.

H14.4. Let $|X| = 6$, and suppose

$$U = \{x_1, x_2, x_3\} \cup \{x_4, x_5, x_6\}; \; W = \{x_1, x_2\} \cup \{x_3, x_4, x_5\} \cup \{x_6\}$$

Calculate $R(U, W; 2, 3)$. (Answer: $R = 0.6$.)

S15. The Partition Coefficient: Numerical Taxonomy

The first measure of cluster validity seems to have been the degree of separation between two fuzzy sets, defined by Zadeh[(122)] in the original 1965 paper.

(D15.1) *Definition 15.1* (*Degree of Separation*). Let $U \in M_{f2}$ be a fuzzy 2-partition of n data points. The degree of separation between fuzzy sets u_1 and u_2 is the scalar

$$\rho(U; 2) = 1 - \left[\bigvee_{k=1}^{n} (u_{1k} \wedge u_{2k})\right] \tag{15.1}$$

This number is well defined for any pair of fuzzy sets with finite and equal cardinalities. It interested Zadeh in connection with a generalization of the classical separating hyperplane theorem, rather than as a measure of partition quality. Allowing the minimum in (15.1) to range from 2 to c gives the obvious generalization

$$Z(U; c) = 1 - \left[\bigvee_{k=1}^{n} \left(\bigwedge_{i=1}^{c} u_{ik}\right)\right] \forall U \in M_{fc} \tag{15.2}$$

Note that $Z(U; c) = 1$ on every hard c-partition, so is not a potential validity functional over M_c alone. To see that Z is not very useful for fuzzy cluster validity, consider the following example.

(E15.1) *Example 15.1.* Suppose $n = 51, c = 2$, and we have two fuzzy 2-partitions of X:

$$U = \begin{bmatrix} 1 & \frac{1}{2} & \cdots & \frac{1}{2} & 0 \\ 0 & \frac{1}{2} & \cdots & \frac{1}{2} & 1 \end{bmatrix} \tag{15.3a}$$

and

$$V = \begin{bmatrix} 1 & \cdots & 1 & \frac{1}{2} & 0 & \cdots & 0 \\ 0 & \cdots & 0 & \frac{1}{2} & 1 & \cdots & 1 \end{bmatrix} \qquad (15.3b)$$

The value of Z is the same for U and V, viz., $Z(U;2) = Z(V;2) = 0.50$. U and V, however, are hardly equivalent, since U identifies two distinct points and 49 equimembership ones, while V indicates 50 distinct memberships and only 1 ambiguous one. The problem is simple: Z depends entirely on a single value, the largest minimum over the n columns of U; consequently, Z is too "coarse" to detect distinctions such as the one in this example. •

The first functional designed expressly as a cluster validity measure was the partition coefficient.[11]

(D15.2) *Definition 15.2 (Partition Coefficient)*. Let $U \in M_{fc}$ be a fuzzy c-partition of n data points. The partition coefficient of U is the scalar

$$F(U;c) = \sum_{k=1}^{n} \sum_{i=1}^{c} (u_{ik})^2/n \qquad (15.4)$$

There are several equivalent ways to write (15.4) which emphasize various properties of F. Specifically, the Euclidean inner product for two matrices A, $B \in V_{cn}$ is $\langle A, B\rangle = \mathrm{Tr}(AB^T)$. This induces the Euclidean matrix norm on V_{cn} in the usual way, and (15.4) has the alternate forms

$$F(U;c) = \frac{\mathrm{Tr}(UU^T)}{n} = \frac{\langle U, U\rangle}{n} = \frac{\|U\|^2}{n} \qquad (15.5)$$

Geometrically, F is a multiple of the squared Euclidean length of U when U is regarded as a vector in V_{cn}. Analytically, (15.5) shows that F is (uniformly) continuous and strictly convex on M_{fc}; since M_{fc} is convex, this guarantees a unique global minimum for F over M_{fc}.

(T15.1) *Theorem 15.1.* Let $U \in M_{fc}$ be a fuzzy c-partition of n data points. Then for $1 \leqslant c \leqslant n$,

$$(1/c) \leqslant F(U;c) \leqslant 1 \qquad (15.6a)$$

$$F(U;c) = 1 \Leftrightarrow U \in M_{co} \text{ is hard} \qquad (15.6b)$$

$$F(U;c) = 1/c \Leftrightarrow U = [1/c] \qquad (15.6c)$$

Analysis. (T15.1) brackets F in $[1/c, 1]$: F is constantly 1 on all hard c-partitions of X (i.e., at the extreme points of M_{fc}); conversely, F minimizes only at $[1/c]$, the centroid of M_{fc}. The theorem can be proved easily with Lagrange multipliers; see (10) for an algebraic proof.

Proof. Let $U \in M_{fc}$. Since $0 \leq u_{ik} \leq 1$, $(u_{ik})^2 \leq u_{ik} \forall i$ and k, equality prevailing when and only when $u_{ik} = 1$. In view of column constraint (5.6b), this means that

$$\sum_{i=1}^{c} (u_{ik})^2 \leq \sum_{i=1}^{c} u_{ik} = 1 \qquad \forall k$$

equality prevailing if and only if $u_{jk} = 1 \exists j$, $u_{tk} = 0 \forall t \neq j$, i.e., column k of U is hard. Summing over k and dividing by n gives

$$F(U; c) = \frac{1}{n} \sum_{k=1}^{n} \sum_{i=1}^{c} (u_{ik})^2 \leq \frac{1}{n} \sum_{k=1}^{n} \sum_{i=1}^{c} (u_{ik}) = 1$$

equality prevailing if and only if all n columns of U are hard.

Since F is the sum of n positive terms, F minimizes when each term does. Forming the Lagrangian of the kth term, we have, for $\mathbf{u}_k^T = (u_{1k}, u_{2k}, \ldots, u_{ck})$,

$$L(\lambda, \mathbf{u}_k) = \left(\frac{1}{n} \sum_{i=1}^{c} (u_{ik})^2\right) - \lambda\left(\sum_{i=1}^{c} u_{ik} - 1\right)$$

It is necessary for the gradient of L in $(\lambda, \mathbf{u}_k)$ to vanish at any minimum:

$$\text{(A)} \quad \frac{\partial L}{\partial \lambda}(\lambda, \mathbf{u}_k) = \left(\sum_{i=1}^{c} u_{ik} - 1\right) = 0$$

$$\text{(B)} \quad \frac{\partial L}{\partial u_{jk}}(\lambda, \mathbf{u}_k) = (2u_{jk}/n - \lambda) = 0, \qquad 1 \leq j \leq c$$

From (B) we find $u_{jk} = (n\lambda/2)$; summing this over j to satisfy (A) gives $\lambda = (2/nc)$, from which we find $u_{jk} = (1/c)$ for $j = 1, 2, \ldots, c$ and $k = 1, 2, \ldots, n$. Since F is strictly convex on convex M_{fc}, $[1/c]$ is its unique minimum. •

The value of $F(U; c)$ clearly depends on all (cn) elements of U, in sharp contradistinction to $Z(U; c)$, which depends on but one.

(E15.2) *Example 15.2.* We evaluate F on the partitions U and V of (E15.1). U has 49 equimembership columns, and two hard ones: it should yield an F close to the minimum (which is 0.50). Indeed, $F(U; 2) = 0.510$. On the other hard, V has 50 hard columns and 1 equimembership column, so we expect an F close to the maximum (which is 1.00): $F(V; 2) = 0.990$. In both instances, the value of F gives an accurate indication of the fuzziness of the partition measured relative to the most uncertain and certain states. •

There is an interesting way to interpret this idea in terms of the amount of "overlap" in pairwise intersections of the c fuzzy subsets in U. Towards this end, we make the following definition.

(D15.3) *Definition 15.3* (*Algebraic Product*). Let u_i and u_j be fuzzy subsets of X. Their algebraic product is the fuzzy set $u_{u_iu_j}$, whose membership function is

$$u_{u_iu_j}(x) \doteq u_i(x)u_j(x) \qquad \forall x \in X \tag{15.7}$$

If u_i and u_j are nonempty clusters i and j of $U \in M_{fc}$, since $0 \leq u_{ik} \leq 1 \forall i, k$, their algebraic product $(u_{u_iu_j})$ and intersection $(u_i \wedge u_j)$ are related as follows:

$$\mathbb{0} \leq u_{u_iu_j} \leq u_i \wedge u_j \leq \begin{Bmatrix} u_i \\ u_j \end{Bmatrix} \tag{15.8}$$

whenever $u_i \neq u_j$. The point of (15.8) is that the algebraic product is pinched between the empty set and pairwise intersections, so two fuzzy clusters must overlap if their algebraic product is not the empty set.

(D15.4) *Definition 15.4* (*Content and Coupling*). Let u_i and u_j be fuzzy subsets of X with $|X| = n$. We call

$$\left(\frac{1}{n} \sum_{k=1}^{n} u_{ik}u_{jk}\right) \text{ the average content of } u_{u_iu_j} \tag{15.9}$$

or the average coupling between u_i and u_j.

Returning to the definition of $F(U; c)$, we note that its value depends on the entries of the $c \times c$ *similarity matrix* $S(U) = (UU^T)/n$. The *ij*th entry of $S(U)$, say s_{ij}, is

$$s_{ij} = \sum_{k=1}^{n} u_{ik}u_{jk}/n \tag{15.10}$$

s_{ij}, the average coupling between u_i and u_j in U, actually defines a measure of similarity on pairs of fuzzy clusters: it gauges the extent to which u_i and u_j share memberships over the n data points in X. Note that for $i \neq j$

$$0 \leq s_{ij} < 1 \tag{15.11a}$$

$$0 = s_{ij} \Leftrightarrow \mathbb{0} = u_{u_iu_j} = u_i \wedge u_j \tag{15.11b}$$

Using (15.10), it is easy to rewrite $F(U; c)$ as

$$F(U; c) = \frac{\sum_{i=1}^{c} s_{ii}}{\sum_{i=1}^{c} s_{ii} + 2 \cdot \sum_{i=j+1}^{c} \sum_{j=1}^{c-1} s_{ij}} \tag{15.12}$$

Form (15.12) shows that $F = 1 \Leftrightarrow s_{ij} = 0 \forall i \neq j \Leftrightarrow$ there is no coupling (membership sharing) between any pairs of fuzzy clusters in $U \Leftrightarrow U$ contains no fuzzy clusters. In this form F is seen to be inversely proportional to the overall average coupling between pairs of fuzzy subsets in U. Maximizing F thus corresponds to minimizing the overall content of pairwise fuzzy intersections.

Since F maximizes at *every* hard c-partition, this seems to suggest that *any* hard partition is better (more valid) than all fuzzy ones: (E14.1) shows the fallacy in this logic. What does seem plausible, however, is this: if X really has distinct substructure (i.e., hard clusters), fuzzy partitioning algorithms should produce relatively hard U's, as measured by F. This is the heuristic rationale for the use of F as a cluster validity functional. The formal strategy is summarized as follows: using any fuzzy clustering algorithm, find at each $c = 2, 3, \ldots, n-1$, one or more "optimal" c-partitions of X; denote the optimality candidates at each c by Ω_c, and solve (by direct search)

$$\max_{c}\left\{\max_{\Omega_c}\{F(U; c)\}\right\} \tag{15.13}$$

If (U^*, c^*) solves (15.13), we presume U^* to be—among the candidates considered!—the most valid clustering of X. If $F(U^*, c^*)$ is not close to 1, we do not presume that X has no identifiable substructure; rather, that the algorithms used have not found any!

(E15.3) *Example 15.3.* Fuzzy c-means (A11.1) was applied to the data set of (E14.1) using $m = 2$, $\varepsilon_L = 0.001$, and the Euclidean norm. Algorithmic variables were $c = 2, 3, 4, 5, 6, 7$; and interface distance $s = 3, 4, 5, 6, 7$. Table 15.1 reports values of $F(U; c)$ as functions of U, c, and s. These values implicate $c^* = 2$ as the most plausible number of clusters at every s. Note that $F(U; 2)$ rapidly approaches 1 as s increases, indicating more and more algorithmic certainty concerning the choice $c^* = 2$; whereas

Table 15.1. Partition Coefficients $F(U; c)$ for (E15.3)

	Interfacial distance				
c	3	4	5	6	7
2	→ 0.74	→ 0.84	→ 0.88	→ 0.91	→ 0.92
3	0.65	0.67	0.68	0.69	0.70
4	0.62	0.63	0.64	0.64	0.65
5	0.61	0.62	0.63	0.60	0.60
6	0.56	0.58	0.58	0.59	0.60
7	0.53	0.53	0.53	0.54	0.54

$F(U; c) \forall c \geq 3$ remains uniformly lower, nearly constant, and independent of both c and s, indicating that only $c = 2$ should be considered. The hard MM 2-partition U_m closest to U^* via (D9.2) at each s was as follows: U_m coincided with $\hat{U}$ [the visually correct hard 2-partition of (E14.1)] for $s = 4$, 5, 6, and 7; U_m coincided with $\hat{\hat{U}}$ of (E14.1) at $s = 3$. Evidently, the range of reliability of J_W (hard c-means, $s > 5.5$) is extended by J_m to some s between 3 and 4 at $m = 2$. The decrease in algorithmic success suffered by (A11.1) as $s \to 3$ is mirrored by a corresponding decrease in $F(U^*; c^*)$ as a function of s. •

Example 15.3 illustrates one use of $F(U; c)$ and suggests some obvious questions:

(i) How low can $F(U^*; c^*)$ become before U^* is a poor interpretation of X?

(ii) Is F monotone increasing with c?

The range of F can be normalized, but the *expected value* of F over M_{fco} is, under certain statistical assumptions, still monotonic with c. Nonetheless, (15.13) provides a well-defined way to address the validity problem which has enjoyed some empirical success.

As a final example, we illustrate the use of F in connection with optimizing the performance of (A11.1) over choices for A, the matrix inducing the norm in (11.6). Of practical interest are the following choices for matrix A:

$$N_E \sim A_E = [\delta_{ij}] \tag{15.14a}$$

$$N_D \sim A_D = [\text{diag}(\sigma_j^2)]^{-1} \tag{15.14b}$$

$$N_M \sim A_M = [\text{cov}(X)]^{-1} \tag{15.14c}$$

δ_{ij} in (15.14a) is Kronecker's delta, A_E is the $p \times p$ identity, which induces the Euclidean norm (N_E for short). For (15.14b) and (15.14c), we need the sample means and variances of the p features: for $1 \leq j \leq p$

$$\bar{x}_j = \frac{1}{n} \sum_{k=1}^{n} x_{kj} \tag{15.15a}$$

$$\sigma_j^2 = \frac{1}{n} \sum_{k=1}^{n} (x_{kj} - \bar{x}_j)^2 \tag{15.15b}$$

With these, A_D is the $p \times p$ diagonal matrix

$$A_D = \begin{bmatrix} (1/\sigma_1)^2 & 0 & \cdots & 0 \\ 0 & (1/\sigma_2)^2 & \cdots & 0 \\ \vdots & & \ddots & \vdots \\ 0 & 0 & \cdots & (1/\sigma_p)^2 \end{bmatrix} \tag{15.14b$'$}$$

which induces the diagonal norm N_D on $\mathbb{R}^p$. A_M is the inverse of the $p \times p$ sample covariance matrix of X, the ijth entry of which is

$$[\text{cov}(X)]_{ij} = \frac{1}{n} \sum_{k=1}^{n} (x_{ki} - \bar{x}_i)(x_{kj} - \bar{x}_j) \tag{15.16}$$

A_M induces the Mahalonobis norm N_D on $\mathbb{R}^p$. These three norms endow J_m (and J_w) with different *invariance* properties discussed at length by Friedman and Rubin.[40] Perhaps more to the point, changes in the norm metric realized via (11.6) alter the clustering criterion by changing the measure of dissimilarity. Geometrically, each norm induces a topology for $\mathbb{R}^p$ which has a differently *shaped* unit ball. The Euclidean norm generates (hyper-)spherical balls, so is appropriate when clusters in X have the general shape of "spherical" clouds. The diagonal norm compensates for distortions of this basic shape due to large disparities among sample feature variances in the coordinate axes directions. Statistical independence of the p features is implicit in both N_E and N_D. N_M accounts for distortions in the same way as N_D does, and, additionally, tries to mitigate the effects of statistical dependence between pairs of measured features.

By varying the norm in any distance-based clustering criterion, one may infer geometric and statistical properties of the data from those of the "best" norm used. In particular, the norms above have "compatibilities" as follows:

N_E ~ statistically independent, equally variable features for hyperspherical clusters

N_D ~ statistically independent, unequally variable features for hyperellipsoidal clusters

N_M ~ statistically dependent, unequally variable features for hyperellipsoidal clusters

In the next example, $F(U; c)$ is used in this way on a classical set of botanical data.

(E15.4) *Example 15.4.* This example concerns the Iris data of Anderson,[2] a perennial favorite of clusterers since 1936, when R. A. Fisher[38] first used it to exemplify linear discriminant analysis. The data are a set of $n = 150$ four-dimensional vectors, each of which gives the sepal length, sepal width, petal length, and petal width (all in cm) of one of three subspecies of irises, viz., *Sestosa* (SE), *Versicolor* (VC), and *Virginica* (VG), Anderson measured 50 plants from each subspecies. It was his conjecture that (VC) was an (SE)–(VG) hybrid: Fisher confirmed this, even suggesting that (VG) exerted roughly twice as much influence on (VC) as (SE) did.

Figure 15.1 depicts X in its most discriminatory pair of dimensions: only distinct points are plotted; solid lines are convex hulls of the three

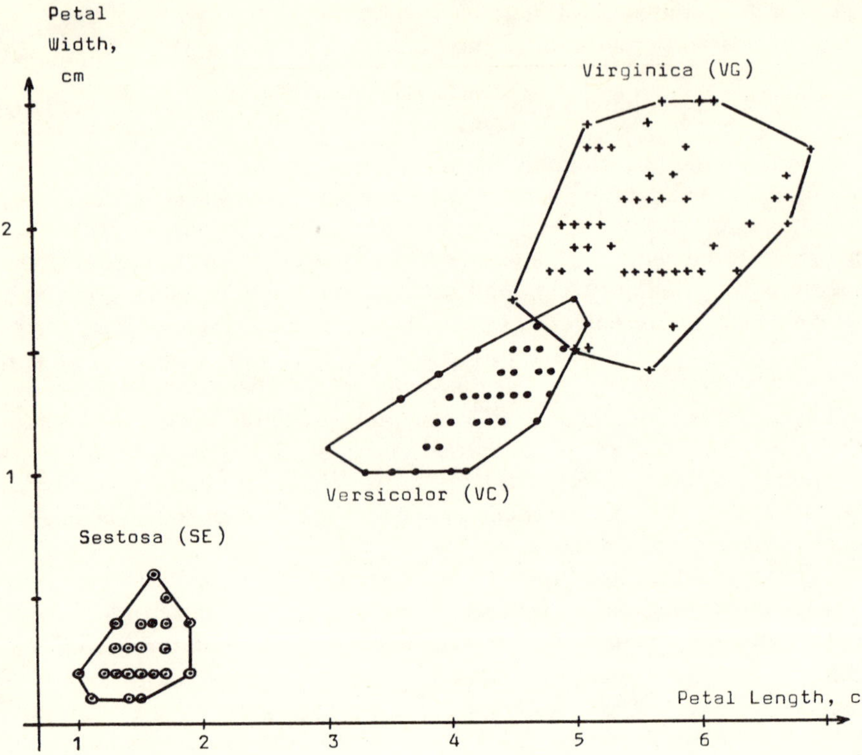

Figure 15.1. Anderson's iris data (petal features): (E15.1).

subspecies. Note that the geometric structure of Fig. 15.1 agrees quite well with the hybridization theory of Anderson and Fisher. Algorithm (A11.1) was applied to X with $m = 2.00$ and $\varepsilon_L = 0.01$. Algorithmic variables were $c = 2, 3, 4, 5, 6$; and the norm $\|\cdot\|_A$ in (11.6)—specifically, the three norms N_E, N_D, and N_M. Table 15.2 lists $F(U; c)$ for these alternatives. The solution of (15.13) indicates $c^* = 2$ with the diagonal norm; however, the Euclidean and diagonal norms yield practically the same validity indicators, which are uniformly higher at every c than that for N_M. From this one might infer that the iris data consist of four essentially statistically independent features with roughly hyperspherically shaped clusters in $\mathbb{R}^4$. Observe that $c^* = 2$ is indicated in *every* case (i.e., all three norms) as the most valid clustering of X. An anomaly? Figure 15.1 suggests not: the (visually apparent) *primary* substructure *is* two clusters: {SE} ∪ {VC–VG}. This seems to agree with Fisher's assertions, that the VC hybrid is "twice as similar" to VG as it is to SE, and is thus *first* lumped together with VG by virtue of its similar features. The MM partition U_m derived from U^* via (D9.2) at $c^* = 2$ with the

Table 15.2 Partition Coefficients $F(U;c)$ for (E15.4)

	Norm for (11.6) from (15.4)		
c	N_E	N_D	N_M
2	0.88	→ 0.89	0.56
3	0.78	0.80	0.46
4	0.71	0.71	0.36
5	0.66	0.68	0.31
6	0.60	0.62	0.37

diagonal norm, consisted of two clusters: all 50 (SE) with three (VC); and the remaining 97 (VC)–(VG) plants are grouped together. Note that the *next best* choice indicated by $F(U;c)$ is 3, the botanist's choice for c^*. In this example then, F first identifies primary substructure at $c^* = 2$; and then secondary structure at $c = 3$. •

Remarks

Disadvantages of the partition coefficient are its monotonic tendency and lack of direct connection to some property of the data themselves. Dunn[36] defined a separation index for hard c-partitions which identifies compact, well-separated clusters with a clear geometrical property: the connection between Dunn's index and F is discussed in S19. The rationale underlying strategy (15.13) is heuristic—why should fuzzy clustering algorithms necessarily progress towards "*good*" hard clusterings? This depends to some extent on the algorithm used: S19 provides an answer in the case of fuzzy c-means. These objections notwithstanding, the partition coefficient has enjoyed success in many (but not all) of its empirical trials. (E15.3) and others like it are discussed at greater length in (10). Reference (11) amplifies many details omitted from (E15.4). The Iris data have been used as a test set by at least a dozen authors, including Fisher,[38] Kendall,[64] Friedman and Rubin,[40] Wolfe,[117] Scott and Symons,[95] and Backer.[4] Backer[4] has modified both the form and interpretation of F, and uses it directly as an objective functional (cf. Chapter 5 below).

Exercises

H15.1. Let $U, W \in V_{cn}$, and define $N(U) = [\mathrm{Tr}(UU^T)]^{1/2}$. Show that

(i) $N(U) \geqslant 0$, equality $\Leftrightarrow U = [0]$, the zero matrix;

(ii) $N(\lambda U) = |\lambda| N(U) \forall \lambda \in \mathbb{R}$;

(iii) $N(U + W) \leqslant N(U) + N(W)$.

This proves that N is a norm on V_{cn}.

(iv) Show that M_{fco} is a closed set in V_{cn} relative to the topology induced on $\mathbb{R}^p$ by N.

H15.2. Let $U, W, V \in V_{cn}$, and define $S(U, W) = [\mathrm{Tr}(UW^T)]$. Show that
(i) $S(U, U) \geq 0$, equality $\Leftrightarrow U = [0]$, the zero matrix.
(ii) $S(\lambda U, W) = \lambda S(U, W) \forall \lambda \in \mathbb{R}$
(iii) $S(U, W) = S(W, U)$
(iv) $S(U + W, V) = S(U, V) + S(W, V)$
This proves that S is an inner product on V_{cn}.

H15.3. Let $F: M_{fc} \to [1/c, 1], F(U) = \mathrm{Tr}(UU^T)/n$ as in (15.5).
(i) Prove that M_{fco} is bounded, by finding a ball $B(\boldsymbol{\theta}, r)$ centered at $\boldsymbol{\theta}$, the zero matrix of V_{cn}, which entirely contains M_{fco}. (Answer: $r = n^{1/2}$.)
(ii) Calculate the diameter of M_{fc} relative to the norm in H15.1. [Answer: $(2n)^{1/2}$.]
(iii) Calculate the distance between the sets $\{[1/c]\}$ and M_c in M_{fc} relative to the norm in H15.1. {Answer: $[n - (n/c)]^{1/2}$.}
(iv) Show that the distance between adjacent vertices of M_{fc} relative to the norm in H15.1 is equal to $2^{1/2}$.
(v) Prove that $[1/c]$ is the geometric centroid of M_{fc}.

H15.4. With Z and F defined as in (15.2) and (15.4), respectively, show that for $U \in M_{f2}$,
(i) $Z(U; 2) = 1 \Leftrightarrow F(U; 2) = 1$
(ii) $F(U; 2) = \frac{1}{2} \Rightarrow D(U; 2) = \frac{1}{2}$
(iii) $\frac{1}{2} \leq Z(U; 2) \leq 1 \; \forall U \in M_{f2}$

H15.5.

$$U = \begin{bmatrix} 1 & 0.5 & \cdots & 0.5 \\ 0 & 0.5 & \cdots & 0.5 \end{bmatrix}, \qquad V = \begin{bmatrix} 0.5 & 1 & 1 & \cdots & 1 \\ 0.5 & 0 & 0 & \cdots & 0 \end{bmatrix}$$

both matrices having 101 columns. Compute Z and F on U and V. [Answer: $Z(U) = Z(V) = \frac{1}{2}$; $F(U) = 51/101$; $F(V) = 100.5/101$.]

H15.6. Show that $F(U; c)$ is not one to one into $[1/c, 1]$.

H15.7. Let $W \in M_c \subset V_{cn}$ be hard, $U \in M_{fc} \subset V_{cn}$ be fuzzy, and $\delta \in (0, \infty)$. Prove that
(i) $\|W - U\| \leq \delta \Rightarrow 1 - (\delta/n)(2n^{1/2} - \delta) \leq F(U; c) \leq 1$.
(ii) $F(U; c) = K \Rightarrow \|W - U\| \geq n^{1/2}(1 - K)$.
Assume the norm of H15.1 for these problems. These inequalities relate the partition coefficient of U to the distance between U and hard c-partitions of X.[10]

H15.8. Let $U = [1/c] \in V_{cn}$ be the centroid of M_{fc}, and $W \in M_c$ be any hard c-partition in V_{cn}. For $\alpha, \beta \in \mathbb{R}$,
(i) Show that $F(\alpha U + \beta W; c) = (\alpha^2 + 2\alpha\beta + c\beta^2)/c$.
(ii) Find $\alpha, \beta = 1 - \alpha \in (0, 1)$ so that $F(\alpha U + \beta W; c) = (c + 1)/(2c)$, the midpoint of $[1/c, 1]$, the range of F. Such a matrix is in some sense midway between the centroid and extreme points of M_{fc}, and is often

used as a starting guess ($U^{(0)}$) for fuzzy clustering algorithms. The $U^{(0)}$ in (E11.3) was chosen this way. [Answer: $1 - (2^{1/2}/2)$, $2^{1/2}/2$.]

H15.9. Find a scalar multiple of $F(U; c)$ that has range $[0, 1]$ for every c, $2 \leq c < n$.[4]

H15.10. Suggest several symmetric, positive-definite matrices for (11.5) other than those at (15.14) which might induce useful measures of distance in (11.6).

H15.11. $U, V \in M_{fc} \subset V_{cn}$. Suppose $F(U; c) > F(V; c)$. Let $W \in M_c$. Is $\|W - U\| < \|W - V\|$? (Hint: see H15.7.)

H15.12. Let $\{U_n\} \subset M_{fc} \subset V_{cn}$ be a sequence of fuzzy c-partitions, and suppose $\{U_n\} \to W \in M_c$. Show that $\{F(U_n; c)\} \to 1$. Is the converse true?

H15.13. Let u, w, and y be convex fuzzy subsets of $X \subset \mathbb{R}^p$ as in H6.14. The convex combination of u and w by y is [see (122)] the fuzzy set $(u, w; y) = yu + \tilde{y}w$, where juxtaposition indicates algebraic product as in (15.7)—i.e., $(u, w; y)(x) = y(x)u(x) + \tilde{y}(x)w(x)$. Show that

(i) $(u \wedge w) \subset (u, w; y) \subset (u \vee w) \forall y$

(ii) given fuzzy set z such that $(u \wedge w) \subset z \subset (u \vee w)$, $\exists$ a fuzzy set y for which $z = (u, w; y)$. {Answer: $y(x) = [z(x) - w(x)][u(x) - w(x)]^{-1}$.}

H15.14. Another definition of $(u, w; y)$ given in [Brown, J., "A Note on Fuzzy Sets", *Inf. Control.*, Vol. 18 (1971), pp. 32–39] is $(u, w; y) = (y \wedge u) \vee (\tilde{y} \wedge w)$. Do the results of H15.13 hold for this definition?

S16. Classification Entropy

Since its introduction by Shannon in (98), the information-theoretic concept of entropy has played an important role in many pattern recognition contexts. It was suggested in Chapter 1 that an essential difference between fuzzy and statistical models was the type and amount of information borne by them: in this section we discuss several measures of "fuzzy entropy," so called in deference to their formal (mathematical) similarity to Shannon's measure. Towards this end, a brief review of statistical entropy seems desirable.

Let $\{A_i | 1 \leq i \leq c\}$ denote a hard c-partition of events of any sample space connected with an experiment; and let $\mathbf{p}^T = (p_1, p_2, \ldots, p_c)$ denote a probability vector associated with the $\{A_i\}$. The pair $(\{A_i\}, \mathbf{p})$ is called a finite probability scheme for the experiment. The ith component of $\mathbf{p}$ is the probability of event A_i: c is here called the length of the scheme. Each scheme of this sort describes a state of statistical uncertainty: our concern is to devise a measure $h(\mathbf{p})$ of the amount of uncertainty attached to each state.

Shannon reasoned that the unique vector $(\mathbf{1/c})$ represented the state of maximum uncertainty concerning the observation of one of the A_i's, hence suggesting that h should maximize at $(\mathbf{1/c})$. Conversely, if $\mathbf{p} \in B_c$ is hard, no uncertainty exists concerning the scheme $(\{A_i\}, \mathbf{p})$, so h should be minimum for all hard (in this context, statistically certain) finite schemes. Further, h should be continuous in all c arguments; and the addition of impossible events to $\{A_i\}$ should not alter the original amount of uncertainty. Finally, Shannon required that the measure of two dependent schemes be additive under conditioning of one scheme's measure by dependence on the other. These properties determine Shannon's measure of uncertainty uniquely up to a positive constant. The measure is called the entropy of the scheme $(\{A_i\}, \mathbf{p})$ because its functional form is similar to a like-named quantity in statistical mechanics. Shannon's definition[98] is as follows.

(D16.1) *Definition 16.1* (*Statistical Entropy*). Let $(\{A_i\}, \mathbf{p})$ denote a finite probability scheme of length c. The entropy of this scheme is

$$h(\mathbf{p}) = -\sum_{i=1}^{c} p_i \log_a(p_i) \tag{16.1}$$

where logarithmic base $a \in (1, \infty)$ and $p_i \log_a(p_i) \doteq 0$ whenever $p_i = 0$.

Shannon actually proved that $\lambda h(\mathbf{p})$, $\lambda \in \mathbb{R}^+$, defines the class of functionals satisfying the requirements above; λ is almost always then taken as 1. The part of Shannon's theorem repeated here lists properties of h which prove useful in the sequel.

(T16.1) *Theorem 16.1.* Let $\mathbf{p} = (p_1, p_2, \ldots, p_c)$ with $\sum_{i=1}^{c} p_i = 1$ and $p_i \geqslant 0$ $\forall i = 1, 2, \ldots, c$. The entropy of $\mathbf{p}$ satisfies

$$0 \leqslant h(\mathbf{p}) \leqslant \log_a(c) \tag{16.2a}$$

$$h(\mathbf{p}) = 0 \Leftrightarrow \mathbf{p} \text{ is "hard" } (p_i = 1 \exists i) \tag{16.2b}$$

$$h(\mathbf{p}) = \log_a(c) \Leftrightarrow \mathbf{p} = (\mathbf{1/c}) \tag{16.2c}$$

Analysis and Proof. (T16.1) Shows that Shannon's measure of statistical uncertainty behaves in a very plausible way. A nice proof of (16.2) and the other properties of h alluded to above is given in Khinchin.[65] •

The scalar $h(\mathbf{p})$ is interpreted as a measure of "information" as follows: if we perform a trial of the scheme $(\{A_i\}, \mathbf{p})$, we find out which event *actually* occurs. Removal of this uncertainty concerning the scheme constitutes a gain in information. The more uncertainty we remove, the larger is the information gained. Hence, we gain zero information if no uncertainty is removed (16.2b); and maximum information by removing the most uncertain state [at (16.2c)].

Because fuzzy sets represent uncertainty, it seems natural to try to adapt the idea of information content to this context. Apparently Ruspini[89] first suggested this, even supposing that an entropylike scalar measure of partition quality would be useful for the cluster validity question, but he never implemented the idea. Shortly thereafter, Deluca and Termini[32] gave the following definition for the normalized entropy of a finite fuzzy set.

(D16.2) *Definition 16.2* (*Normalized Entropy*). Let u_i be a fuzzy subset of X, let $\tilde{u}_i$ be its complement, $|X| = n$, $a \in (1, \infty)$, $u_{ik} \log_a(u_{ik}) \doteq 0$ if $u_{ik} = 0$, and let $\mathbf{u}_i^T = (u_{i1}, u_{i2}, \ldots, u_{in})$; then $h_f(\mathbf{u})$, or

$$h_f(u_i, \tilde{u}_i) = -\left\{ \sum_{k=1}^{n} [u_{ik} \log_a(u_{ik}) + (1 - u_{ik}) \log_a(1 - u_{ik})]/n \right\} \tag{16.3}$$

is called the normalized entropy of u_i (or of $\mathbf{u}_i$).

The reason terms involving the complement $(1 - u_i) = \tilde{u}_i$ of u_i appear in (16.3) is that the memberships $\{u_{ik}\}$ of a general fuzzy set do not sum to unity, so the first term alone does not maximize at $(\mathbf{1/n})$: adding the complementary terms remedies this. Deluca and Termini verify properties (16.2) for $n \cdot h_f(\mathbf{u})$, give an additional monotonicity result, and interpret $h_f(\mathbf{u})$ in the context of fuzzy information.

A complete formal analogy to Shannon's entropy *is* possible for fuzzy c-partitions, because constraint (5.6b) on the columns of $U \in M_{fc}$ renders each of them *mathematically* identical to probability vector $\mathbf{p}$ of (D16.1). Following (12), we make the following definition.

(D16.3) *Definition 16.3* (*Partition Entropy*). The partition entropy of any fuzzy c-partition $U \in M_{fc}$ of X, where $|X| = n$, is, for $1 \leq c \leq n$,

$$H(U; c) = -\sum_{k=1}^{n} \sum_{i=1}^{c} u_{ik} \log_a(u_{ik})/n \tag{16.4}$$

where logarithmic base $a \in (1, \infty)$ and $u_{ik} \log_a(u_{ik}) \doteq 0$ whenever $u_{ik} = 0$.

Because $\sum_{i=1}^{c} u_{ik} = 1 \forall k$, (16.4) has the alternate form

$$H(U; c) = \sum_{k=1}^{n} h(\mathbf{u}_k)/n \tag{16.5}$$

where h is Shannon's entropy as in (D16.1), and $\mathbf{u}_k^T = (u_{1k}, u_{2k}, \ldots, u_{ck})$ is the kth column of $U \in M_{fc}$. If $c = 2$, then for $U \in M_{f2}$, $u_{2k} = (1 - u_{1k}) \forall k$, and (16.4) reduces to (16.3). In other words, the normalized entropy of

Deluca and Termini for any fuzzy set u_i can be regarded as a special case of $H(U; c)$ by arraying the values of u_i and u_i as a fuzzy 2-partition of X. In this way, $H(U; 2) = h_f(u_i, \tilde{u}_i)$. The following theorem shows that $H(U; c)$ is the appropriate formal generalization for fuzzy c-partitions analogous to Shannon's entropy for finite probability schemes.

(T16.2) *Theorem 16.2.* Let $U \in M_{fc}$ be a fuzzy c-partition of n data points. Then for $1 \leq c \leq n$ and $a \in (1, \infty)$,

$$0 \leq H(U; c) \leq \log_a(c) \tag{16.6a}$$

$$H(U; c) = 0 \Leftrightarrow U \in M_{co} \text{ is hard} \tag{16.6b}$$

$$H(U; c) = \log_a(c) \Leftrightarrow U = [1/c] \tag{16.6c}$$

Analysis and Proof. (T16.2) is for H what (T15.1) is for the partition coefficient $F(U; c)$. Here H is bracketed in $[0, \log_a(c)]$; it is zero on all hard c-partitions of X, and maximizes only at the equimembership (most uncertain) partition $[1/c]$. The proof is obvious: Results (16.6) are all immediate consequences of the corresponding equations (16.2) of (T16.1), using form (16.5) for H. •

(E16.1) *Example 16.1.* We calculate H for the 2-partitions U and V of (E15.1) and (E15.2). Recalling that U has two hard and 49 $(1/c)$ columns,

$$H(U; c) = 49 \log_e(2)/51 = 0.665$$

compared to a possible maximum of $\log_e(2) = 0.693$, indicating that U is a very uncertain partition. For V, with 50 hard and 1 equimembership columns,

$$H(V; c) = \log_e(2)/51 = 0.013$$

compared to an absolute minimum of 0.000 at hard 2-partitions. •

It might be well to observe at this point that (D16.2) could be generalized to $U \in M_{fc}$. Indeed, letting U' denote the matrix $[1 - u_{ik}]$ and summing (16.3) over c, we would have the normalized entropy of U in the sense of Deluca and Termini:

$$h_f(U; c) = H(U; c) + H'(U'; c)$$

Since columns of U' sum to $c - 1$, H and H' are dependent; accordingly, either term of this expression determines the other, so it suffices to utilize only one.

Now let $\mathbf{p}_k$ and $\mathbf{u}_k$ denote, respectively, probability and membership vectors of length c, $\mathbf{u}_k$ having components that sum to unity, as is the case for columns of $U \in M_{fc}$. The distinction between $\mathbf{p}_k$ and $\mathbf{u}_k$ made in S3 affords a similar distinction for the scalars $h(\mathbf{p}_k)$ and $h(\mathbf{u}_k)$. $h(\mathbf{p}_k)$ is the amount of statistical information gained upon removing (by observation) the probabilistic uncertainty due to $\mathbf{p}_k$. On the other hand, $h(\mathbf{u}_k)$ is interpreted in the present context as the amount of fuzzy information about the membership of $\mathbf{x}_k$ in c subclasses that is *retained* by column $\mathbf{u}_k$. At $U = [1/c]$, the most information is withheld because membership assignments are maximally uncertain: conversely, hard c-partitions cluster the data with total certainty (no information is retained by hard clusters). For cluster validity, we thus presume that minimization of $H(U; c)$ corresponds to maximizing the amount of information about substructural memberships an algorithm extracts from X, and suggests that $H(U; c)$ can be used for cluster validity as follows: let Ω_c denote any finite set of "optimal" U's $\in M_{fc}$, and let $c = 2, 3, \ldots, n-1$. Solve (by direct search)

$$\min_{c}\left\{\min_{\Omega_c}\{H(U; c)\}\right\} \tag{16.7}$$

Exactly as in (15.13), if (U^*, c^*) solves (16.7), U^* is presumed to be the most valid partitioning of X among those in the set Ω_c. Even though H is formally analogous to a measure of statistical uncertainty, one must be cautioned that this validity strategy is heuristic—its rationale being exactly the same as for $F(U; c)$—viz., if distinct substructure really exists in X, fuzzy partitioning algorithms should produce relatively hard U's.

Although H has an information-theoretic flavor, it is ultimately just a scalar measure of the amount of fuzziness in a given $U \in M_{fc}$. Comparing (15.6) and (16.6) shows that F and H extremize together, and in this sense they are equivalent. An even more direct connection between these two validity functionals is exhibited in the following theorem.

(T16.3) *Theorem 16.3.* Let $U \in M_{fc}$, $a \in (1, \infty)$, and $e = 2.718\ldots$ be the natural logarithmic base. Then

$$0 \leq 1 - F(U; c) \leq H(U; c)/\log_a(e) \tag{16.8}$$

with equality whenever $U \in M_{co}$ is hard; and strict inequality otherwise.

Analysis. $(1 - F)$ is strictly less than H unless U is hard, so it yields a tighter lower bound for entropy than zero. Since (16.7) asks us to minimize H, the utility of such a bound is obvious. The proof itself is quite geometric: note that the line tangent to $\log_a(x)$ lies above the graph of $\log_a(x)$ for positive x, and carry this result to U. Interested readers can find an upper bound {viz., $[c - F(U; c)]/2$} for H in (12).

Proof. For $x \in (0, \infty)$, $\log_a(x)$ is strictly concave, so the line tangent to it at $x = 1$ is entirely above it;

(A) $(x - 1)\log_a(e) \geq \log_a(x)$

with strict inequality unless $x = 1$. Let $x = u_{ik}$, and multiply (A) by $(-u_{ik}/n)$, obtaining

(B) $$\frac{1}{n}[u_{ik} - (u_{ik})^2] \leq -\frac{u_{ik}\log_a(u_{ik})}{n\log_a(e)}$$

Inequality (B) holds for $u_{ik} \in (0, \infty)$, and is strict unless $u_{ik} = 1$. If $u_{ik} = 0$, the left side of (B) is 0, and by the convention of (D16.3), the right side of (B) also vanishes. Accordingly, (B) holds for $u_{ik} \in [0, 1]$, equality at $u_{ik} = 0, 1$. Now sum (B) over i and k:

(C) $$0 \leq \frac{1}{n}\sum_{k=1}^{n}\sum_{i=1}^{c}[u_{ik} - (u_{ik})^2] = \frac{1}{n}\sum_{k=1}^{n}\sum_{i=1}^{c}u_{ik} - \frac{1}{n}\sum_{k=1}^{n}\sum_{i=1}^{c}(u_{ik})^2$$

$$= 1 - F(U; c) \leq -\sum_{k=1}^{n}\sum_{i=1}^{c}\frac{u_{ik}\log_a(u_{ik})}{n\log_a(e)}$$

$$= H(U; c)/\log_a(e)$$

If $U \in M_{co}$ is hard, $0 = 1 - F(U; c) = H(U; c)$. Otherwise, there are at least two elements of U in (0, 1); for both of these, (B) is strict, $1 - F(U; c)$ is strictly greater than zero, and (C) is strict throughout, $0 < 1 - F(U; c) < H(U; c)/\log_e(e)$.

On taking $a = e$ as the logarithmic base for entropy functional H, we have the following corollary.

(C16.1) *Corollary 16.1.* If $a = e$ in (T16.3), then

$$0 \leq 1 - F(U; c) \leq H(U; c) \tag{16.9}$$

with strict inequality unless $U \in M_{co}$ is hard.

(E16.2) *Example 16.2.* We calculate the bound $(1 - F)$ on H for the 2-partitions U and V from (E15.1). From (E15.2) we have

$$1 - F(U; 2) = 1 - 0.510 = 0.490 < H(U; 2) = 0.665$$

$$1 - F(V; 2) = 1 - 0.990 = 0.010 < H(V; 2) = 0.013$$

Next, we describe briefly the *unweighted pair-group method using arithmetic averaging* (UPGMA for short). This method of hierarchical clustering is quite popular with numerical taxonomists. It generates nested hard clusters in X by merging the two clusters at each step, minimizing the measure of hard cluster similarity due to Sokal and Michener[100] which we call J_{SM}. Let $U \in M_c$, and $\mathbf{u}_j$, $\mathbf{u}_k$ denote the jth and kth rows of U. Define

$$J_{SM}(\mathbf{u}_j, \mathbf{u}_k) = \frac{\sum_{i=t+1}^{n} \sum_{t=1}^{n-1} u_{ji}u_{kt}d(\mathbf{x}_i, \mathbf{x}_t)}{\left(\sum_{i=1}^{n} u_{ji}\right)\left(\sum_{t=1}^{n} u_{kt}\right)} \tag{16.10}$$

where $d : X \times X \to \mathbb{R}^+$ is any measure of dissimilarity on X, $|X| = n$. J_{SM} simply averages the dissimilarities between points in u_j and points in u_k. The UPGMA merges, at each step, a pair of clusters to minimize J_{SM}. In the following description, U_c denotes the (unique) hard c-partition of X formed by UPGMA at each step:

(A16.1.) *The UPGMA Algorithm* (*Sokal and Michener*[100])

(A16.1a) Given $X \subset \mathbb{R}^p$, $|X| = n$. Put $U_n = [\delta_{ij}]_{n\times n}$ [each $\mathbf{x}_k \in X$ is a singleton cluster at $c = n$; $V(n) = 0$].

(A16.1b) At step k, $k = 1, 2, \ldots, n-1$, put $c = n - k + 1$. Using $U_c \in M_c$, solve (directly)

$$\underset{j,k}{\text{minimize}} \{J_{SM}(\mathbf{u}_j, \mathbf{u}_k)\} \tag{16.11}$$

Let $(\mathbf{u}_r, \mathbf{u}_s)_c$ solve (16.11). Merge u_r and u_s, thus constructing from U_c the updated partition $U_{c-1} \in M_{c-1}$: record $V(c - 1) = J_{SM}((\mathbf{u}_r, \mathbf{u}_s)_c)$.

(A16.1c) If $k < n - 1$, go to (A16.1b); if $k = n - 1$, $c = 2$. Merge the two remaining clusters, set $U_1 = [1, 1, \ldots, 1]$, compute $V(c - 1) = J_{SM}(\mathbf{u}_1, \mathbf{u}_2)$, and stop.

This algorithm is a popular member of a family of linear combinatorial strategies; the graph-theoretic characterization of the entire family is discussed at length by Lance and Williams.[69] A validity strategy often employed with methods of this kind is to maximize the *jump* in $V(c)$; presumably, the largest jump corresponds to leaving the "most natural" grouping of X. Formally, let $\Delta V(c) = |V(c) - V(c-1)|$, and solve

$$\max_{1 \leq c \leq n-1} \{\Delta V(c)\} \tag{16.12}$$

for $\Delta V(c^*)$. Then take c^* as the most valid clustering of X. The clustering criterion in UPGMA, J_{SM}, has no easily discernible geometric interpretation: the measure of (dis)similarity is, of course, the function d in (16.10). Ordinarily, d is a metric for $\mathbb{R}^p$; more generally, any similarity measure can be used. UPGMA has performed well on many taxonomic data sets, and seems reliable when one expects to find treelike structure in the data on physical grounds. UPGMA solutions are often quite similar to complete linkage ones.(99)

(E16.3) *Example 16.3.* Figure 16.1 depicts an artificial data set X of $h = 16$ points in $\mathbb{R}^2$. These data appear as Table 5.1 in Sneath and Sokal(99): they applied (A16.1) to X, obtaining a phenogram [cf. Fig. 5-7 of (99)] which illustrates clustering with UPGMA. The measure d for J_{SM} was the Euclidean norm metric. Although cluster validity was not discussed as a part of Sneath and Sokal's example, the information needed to solve (16.12) is available therein, and is recorded as column 2 of Table 16.1. Evidently $\Delta V(2) = |V(1) - V(2)| = 1.65$ is more than twice the jump $\Delta V(4) = 0.71$, and indicates, via heuristic (16.12), that $c^* = 2$ exhibits the most natural groupings in X. The UPGMA clusters are $\{\mathbf{x}_1, \ldots, \mathbf{x}_8\}$ and $\{\mathbf{x}_9, \ldots, \mathbf{x}_{16}\}$ as listed by the UPGMA membership function u_1 in column 3 of Table 16.2. These clusters are illustrated graphically in Fig. 16.1: readers may enjoy arguing whether or not these are "visually apparent" clusters in X. Applying FCM algorithm (A11.1) to X using $\varepsilon_L = 0.001$, the Euclidean norm, $m = 2.00$, and c variable, $c = 2$, 3, 4, 5, and 6, resulted in the validity

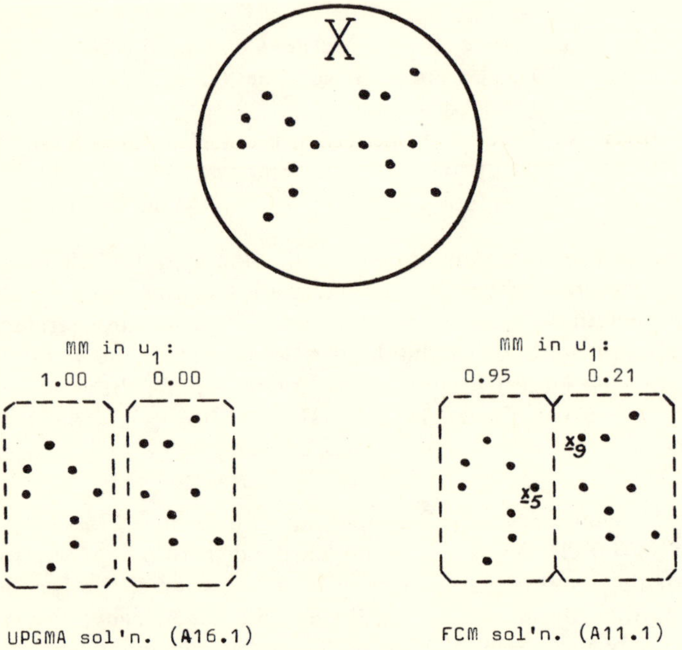

Figure 16.1. UPGMA and FCM 2-clusters for (E16.3).

Table 16.1. Validity Indicators for Clusterings of X: (E16.3)

c	UPGMA $\Delta V(c)$	Fuzzy c-means (A11.1): $m = 2$ $F(U;c)$	$1\text{-}F(U;c)$	$H(U;c)$
2	1.65 ←	0.79 ←	0.21	0.35 ←
3	0.35	0.68	0.32	0.57
4	0.71	0.70	0.30	0.60
5	0.47	0.66	0.34	0.70
6	0.00	0.65	0.35	0.76

indicants F, $(1 - F)$, and H listed in Table 16.1. The solutions of (15.13) and (16.7) were identical: $c^* = 2$ with U^* the last column of Table 16.2. The MM 2-partition U_m derived from this U^* via (D9.2) is the UPGMA partition. Note, however, that hybrids (e.g., $\mathbf{x}_5$) and outliers (e.g., $\mathbf{x}_9$) are clearly identified by low memberships in U^*; conversion to U_m loses this information. Observe that the sensitivity of H does seem to be greater than that of F: $H(U^*; 2)$ is 0.22 less than its nearest competitor; this difference for F is only 0.09. Note too, that ΔV and F both indicate $c = 4$ as the second best choice for c (albeit only slightly in the case of F), while H indicates $c = 3$ as the secondary choice. The reason H is perhaps more sensitive than

Table 16.2. Membership Assignments at $c = 2$: (E16.3)

Data k	X $\mathbf{x}_k$	Membership in cluster 1 (left side of X) u_{1k} from UPGMA	u_{1k} from (A11.1)
1	(0, 4)	1.00	0.92
2	(0, 3)	1.00	0.95
3	(1, 5)	1.00	0.86
4	(2, 4)	1.00	0.91
5	(3, 3)	1.00	0.80
6	(2, 2)	1.00	0.95
7	(2, 1)	1.00	0.86
8	(1, 0)	1.00	0.82
9	(5, 5)	0.00	0.21
10	(6, 5)	0.00	0.12
11	(7, 6)	0.00	0.18
12	(5, 3)	0.00	0.10
13	(7, 3)	0.00	0.02
14	(6, 2)	0.00	0.06
15	(6, 1)	0.00	0.16
16	(8, 1)	0.00	0.15

F lies with the geometric content of (16.9). That inequality simply asserts that the slope of $(x - x^2)$ is always less than that of $[-x \log_e(x)]$ for $x \in (0, 1)$; thus, H usually experiences a more rapid change than F as U varies.

Remarks

Partition entropy has some drawbacks: apparent monotonicity; lack of a suitable benchmark against which values of H can be judged as "acceptable," and the heuristic nature of the rationale underlying (16.7). The formal similarity of H to Shannon's measure should not lead one to attach the statistical connotations of information theory to it. After all is said and done, H (and h_f at $c = 2$) is just another way to define "amount of fuzziness"—it does seem, however, to be a useful concept. See (12) for a more extensive discussion of (E16.3).

Other authors have used Deluca and Termini's special form of $H(U; c)$ for applications quite apart from the present context. For example, Kokawa, Nakamura, and Oda[(68)] use $h_f(\cdot) \equiv H((\cdot); 2)$ to quantify certain aspects of memory and inference; (68) contains a numerical example using $H((\cdot); 2)$ to characterize these concepts with real data experiments using playing card sequences. Another scalar measure of fuzziness (of fuzzy *relations* in X) which assumes the mathematical *form* of Shannon's measure has been proposed by Ragade[(82)] for use in the context of group decision theory. Ragade's measure is not formally analogous to that of Shannon because the relations considered are not constrained like (5.6b). Nevertheless, this variant of fuzzy entropy provides yet another interesting application of the idea of measuring uncertainty in quite a different area.

A number of other measures of fuzziness have been proposed and investigated in the general context of fuzzy sets. While not specifically applicable to the cluster validity problem, it seems appropriate to mention here the works of Knopfmacher,[(67)] who discusses several scalar measures of uncertainty; Halpern,[(56)] who attaches to fuzzy graphs a measure of fuzziness having a strong connectivity flavor; and Trillas and Riera,[(108)] who have proposed a number of "algebraic entropies" in connection with the fuzzy integral of Sugeno.[(101)]

Exercises

H16.1. Find a scalar multiple of $H(U; c)$ that has range [0, 1] for every c, $2 \leq c < n$.

H16.2. Calculate $H(U; 2)$ and $H(V; 2)$ for the partitions U and V in H15.5. Use the results of H15.5 to check the bound in (16.9). [Answer: $H(U) = 100 \log(2)/101$; $H(V) = \log(2)/101$.]

H16.3. $U, V \in M_{fc} \subset V_{cn}$. Suppose $H(U;c) > H(V;c)$. Does this imply that $F(U;c) < F(V;c)$? What about the converse?

H16.4. $U \in M_{fc} \subset V_{cn}$, $u_{ik} \in (0,1)\forall i, k$. For any $a \in (1, \infty)$, show that[12]

$$H(U;c) < (\log_a e^{1/2})[c - F(U;c)]$$

H16.5. Show that H is not one to one into $[0, \log_a c]$.

H16.6. Calculate $H(\alpha U + \beta W; c)$, where $\alpha U + \beta W$ is the partition in H15.8 (ii). Is this the midpoint of the range of H?

H16.7. *Let* $(\{\mathbf{x}_k\}, \{p_k = \Pr(\mathbf{x}_k)\} | 1 \leq k \leq n)$ be a finite probability scheme, where $X = \{\mathbf{x}_k\}$ are the n values of a discrete p-variate random variable; and let u_i be a fuzzy subset of X. Zadeh proposed in (123) that, for $a \in (1, \infty)$,

$$h_p(u_i) = -\sum_{k=1}^{n} u_{ik} p_k \log_a(p_k)$$

be called the entropy of u_i *relative to the discrete probability* measure induced by the $\{p_k\}$.

(i) Show that h_p does not satisfy Shannon's theorem.

(ii) If $(\{\mathbf{x}_k\}, \{p_k\})$ and $(\{\mathbf{y}_j\}, \{q_j\})$ are independent schemes of arbitrary but finite lengths, and u_i, w_i are fuzzy subsets of $X = \{\mathbf{x}_k\}$, $Y = \{\mathbf{y}_j\}$, respectively, prove that

$$h_{pq}(u_i w_i) = \Pr(u) h_p(u_i) + \Pr(w) h_q(w_i)$$

where $(u_i w_i)$ is the algebraic product of u_i and w_i; $pq = \{p_k q_j\}$; and $\Pr(u)$, $\Pr(w)$ are probabilities of fuzzy sets u_i, w_i relative to $\{p_k\}$, $\{q_j\}$ (cf. H14.13); $\Pr(u_i) = \sum_{k=1}^{n} u_{ik} p_k$; $\Pr(w_i) = \sum_{j=1}^{m} w_{ij} q_j$.

S17. The Proportion Exponent: Numerical Taxonomy Revisited

Disadvantages of F and H provoked Windham[115] to seek yet another measure of partition quality. His functional is defined as follows.

(D17.1) *Definition 17.1* (*Proportion Exponent*). Let $U \in (M_{fc} - M_{co})$ be a fuzzy c-partition of X; $|X| = n$; $2 \leq c < n$. For column k of U, $1 \leq k \leq n$, let

$$\mu_k = \max_{1 \leq i \leq c} \{u_{ik}\} = \bigvee_{i=1}^{c} u_{ik} \tag{17.1a}$$

$$[\mu_k^{-1}] = \text{greatest integer} \leq (1/\mu_k) \tag{17.1b}$$

Then the proportion exponent of U is the scalar

$$P(U;c) = -\log_e\left\{\prod_{k=1}^{n}\left[\sum_{j=1}^{[\mu_k^{-1}]}(-1)^{j+1}\binom{c}{j}\left(1 - j \cdot \bigvee_{i=1}^{c} u_{ik}\right)^{c-1}\right]\right\} \tag{17.1c}$$

The calculations required by (17.1c) are rather awe-inspiring. However, this definition has an imminently logical basis: to understand its rationale, we unravel the stages in the construction of P. Note first that P utilizes the n maximum entries from the columns of U, the addresses of which indicate the maximum membership hard c-partition of X via (D9.2). In this sense, P uses more information than the single maximum entry in the degree of separation $Z(U; c)$; but less than all cn elements used for both $F(U; c)$ and $H(U; c)$. Windham asserts that it is natural to focus on the n column maxima, because large values of μ_k indicate a more distinct (i.e., harder) identification of substructure in X. Windham's analysis of P has the combinatorial flavor of random variables for discrete probability distributions.

Having settled on the n maxima $\{\mu_k\}$ as the building blocks for P, Windham noted that the μ_k values themselves are too dependent on c to be used directly. For example, at $c = 2$, 20% of the possible column vectors for each column of U have a maximum of at least 0.90, whereas only 0.4% can achieve this maximum at $c = 4$. There is a simple geometric interpretation of this: the measure of $\text{conv}(B_c)$ grows "faster" with c than the measure of its subset of points with a fixed maximum entry [recall that Fig. 6.1 illustrates $\text{conv}(B_3)$]. The sum in (17.1c) mitigates the dependence of μ_k on c: to prove this claim, some definitions from advanced calculus are needed.

In (T6.4) we saw that $\text{conv}(B_c)$—the source of columns for $U \in M_{fc}$, was a piece of the hyperplane $\text{HP}(\mathbf{e}_1; \bar{\mathbf{u}})$ and, as such, it is a $(c-1)$-dimensional surface in $\mathbb{R}^c$. The "size" of $\text{conv}(B_c)$ and its subsets can be defined via surface integrals in $\mathbb{R}^c$. If $c = 2$, these are lengths; for $c = 3$, areas; and so on. For convenience, we call this measure of size "area" for every $c \geq 2$. To calculate the area of any subset of $\text{conv}(B_c)$, parametrize it as follows: let

$$D_\phi = \left\{\mathbf{x} \in \mathbb{R}^{c-1} \middle| x_i \geq 0 \forall i; \sum_{i=1}^{c-1} x_i = 1\right\} \tag{17.2a}$$

and define the mapping $\phi: D_\phi \to \text{conv}(B_c)$ as

$$\phi(x_1, x_2, \ldots, x_{c-1}) = \left(x_1, x_2, \ldots, x_{c-1}, 1 - \sum_{i=1}^{c-1} x_i\right) \tag{17.2b}$$

Let $A(S)$ denote the area of any subset $S \subset \text{conv}(B_c)$, and let $\phi^{-1}[S]$ be the inverse image of S in D_ϕ. Then

$$A(S) = \int\int_{\phi^{-1}[S]} \cdots \int c^{1/2}\, dx_1\, dx_2 \cdots dx_{c-1} \tag{17.2c}$$

gives the area of S as a surface integral via the parametrization ϕ of S. In particular, it can be checked that

$$A(\operatorname{conv}(B_c)) = c^{1/2}/(c-1)! \tag{17.2d}$$

Finally, since $0 \leq A(S)/A(\operatorname{conv}(B_c)) \leq 1$, this quotient may be called the *proportion* of [possible column vectors in $\operatorname{conv}(B_c)$] subset S; say

$$p(S) = A(S)/A(\operatorname{conv}(B_c)) \tag{17.2e}$$

The basis for Windham's validity functional is contained in the following theorem.

(T17.1) *Theorem 17.1* (*Windham*[115]). Let $U \in M_{fc} - M_{co}$ be a fuzzy c-partition of X, $|X| = n$, $2 \leq c < n$. Let

$$1/c \leq \mu \leq 1 \tag{17.3a}$$

$$[\mu^{-1}] = \text{greatest integer} \leq (1/\mu) \tag{17.3b}$$

$$S_\mu = \left\{ \mathbf{u} \in \operatorname{conv}(B_c) \middle| \mu \leq \bigvee_{i=1}^{c} u_i \right\} \tag{17.3c}$$

Then the proportion of S_μ is

$$p(S_\mu) = \sum_{j=1}^{[\mu^{-1}]} (-1)^{j+1} \binom{c}{j} (1 - j\mu)^{c-1} \tag{17.4}$$

Analysis. Result (17.4) says this: if $p(S_\mu)$ is, say, 0.16, then 16% of all possible column vectors from $\operatorname{conv}(B_c)$ have a maximum entry at least as large as μ. Note especially that this interpretation is independent of c. The argument itself depends on constraints (5.6a) and (5.6b): these allow a formal analogy between column vectors in $\operatorname{conv}(B_c)$ and random vectors uniformly distributed over $\operatorname{conv}(B_c)$. Borrowing some elementary probability theory then enables one to do the surface integrations which culminate in formula (17.4).

Proof. For $1 \leq i \leq c$, let $E_i = \{\mathbf{u} \in \operatorname{conv}(B_c) | u_i \geq \mu\}$, so that

$$S_\mu = \bigcup_{i=1}^{c} E_i$$

We write the proportion of S_μ as

$$p(S_\mu) = p\left(\bigcup_{i=1}^{c} E_i\right)$$

$$= \sum_{i=1}^{c} p(E_i) + \sum_{j=2}^{c+1} (-1)^{j+1} \sum_{i_1 < \cdots < i_j} p(E_{i_1} \cap \cdots \cap E_{i_j}) \tag{17.5}$$

These equalities follow by regarding $\mathbf{u} \in \text{conv}(B_c)$ as one observation of a c-dimensional random variable uniformly distributed over $\text{conv}(B_c)$: then S_μ is an event, $p(S_\mu)$ its probability, in the probability space $\text{conv}(B_c)$.

Since $\text{conv}(B_c)$ is radially symmetric about $(\mathbf{1/c})$, $p(E_i) = p(E_j)\ \forall i \neq j$, and further,

$$p(E_{i_1} \cap \cdots \cap E_{i_j}) = p(E_1 \cap \cdots \cap E_{i_j})$$

for all $\binom{c}{j}$ choices of the indices $i_1 < \cdots < i_j$; hence, the proof can be completed by calculating the $c + 1$ integrals

$$p(E_1) = \underset{\phi^{-1}[E_1]}{\iint \cdots \int} dx_1\, dx_2 \cdots dx_{c-1} \tag{17.6a}$$

$$p\left(\bigcap_{i=1}^{j} E_i\right) = \underset{\phi^{-1}[\bigcap_{i=1}^{j} E_i]}{\iint \cdots \int} dx_1\, dx_2 \cdots dx_{c-1}, \qquad 2 \leqslant j \leqslant c+1 \tag{17.6b}$$

and substituting into (17.5).

To determine the limits of integration for these integrals, suppose

$$\mathbf{x} \in \phi^{-1}\left[\bigcap_{i=1}^{j} E_i\right]$$

Then

$$\mu \leqslant x_i \leqslant 1 - (j - i)\mu - x_1 - \cdots - x_{i-1}, \qquad 1 \leqslant i \leqslant j \tag{17.7a}$$

$$0 \leqslant x_i \leqslant 1 - x_1 - \cdots - x_{i-1}, \qquad j + 1 \leqslant i \leqslant c - 1 \tag{17.7b}$$

In particular, $\mu \leqslant x_1 \leqslant 1 - (i - 1)\mu$. Therefore, $\phi^{-1}[\bigcap_{i=1}^{j} E_i]$ is empty and (17.6b) is zero unless $\mu \leqslant 1 - (i - 1)\mu$, i.e., $i \leqslant [\mu^{-1}]$. If $\phi^{-1}[\bigcap_{i=1}^{c} E_i]$ is not empty, the inequalities at (17.7) provide the limits of integration for each of the $c + 1$ integrals in (17.6). For example, if $\mathbf{x} \in E_1$, then $x_1 \geqslant \mu$; $x_2 \leqslant 1 - x_1$; $x_3 \leqslant 1 - x_1 - x_2$; etc., and

$$p(E_1) = \int_\mu^1 dx_1 \int_0^{1-x_1} dx_2 \cdots \int_0^{1-\sum_{i=1}^{c-2} x_i} dx_{c-1}$$

Although the actual calculation of these integrals is admittedly rather arduous, they are otherwise routine. Their values, when substituted into (17.5), yield (17.4) for $p(S_\mu)$. •

Of particular interest are the extremes of $p(S_\mu)$ in (T17.1). If $\mu = 1$, $S_\mu = B_c$ are the extreme points of $\text{conv}(B_c)$—the c possible hard columns—and $p(S_\mu) = 0$: that is, the proportion of columns whose maximum exceeds unity is zero (this is because the area of B_c is zero). On the other hand, if $\mu = 1/c$, the sum in (17.4) runs from 1 to c, and equals 1. This proves the following corollary.

(C17.1) *Corollary 17.1.* With hypotheses as in (T17.1),

$$p(S_\mu) = 0 \Leftrightarrow \mu = 1 \tag{17.8a}$$

$$p(S_\mu) = 1 \Leftrightarrow \mu = 1/c \tag{17.8b}$$

Next, the product over k in (17.1c) yields a single scalar for $U \in M_{fc}$. This product is interpreted as the proportion of membership matrices in M_{fc} which have larger maxima in *every* column than those of the U being measured. This underlies the use of P for cluster validity: the number

$$\eta(U; c) = \prod_{k=1}^{n} \left[\sum_{j=1}^{[\mu_k^{-1}]} (-1)^{j+1} \binom{c}{j} \left(1 - j \bigvee_{i=1}^{c} u_{ik}\right)^{c-1} \right] \tag{17.9}$$

is interpreted as the number of c-partitions which would more clearly identify clusters in X than U does. (17.8a) shows that η is zero if *any* of the n columns in U is hard; (17.8b) implies that η is one if and only if all n columns are identically the centroid of the convex hull of B_c, i.e., $U = [1/c]$. Thus, η maximizes only at the centroid of M_{fc} [just as H and $(1 - F)$ do]: however, η attains its minimum as soon as *a single column* of U becomes hard, while H and $(1 - F)$ extremize only if all n columns of U are hard. This is an important distinction because it means that η will indicate optimal clusters as soon as a single $\mathbf{x}_k$ becomes unequivocally clustered, whereas all $n\mathbf{x}_k$'s must be algorithmically distinct before H and $1 - F$ indicate optimal validity.

The final step to obtain P is to take the negative of the natural logarithm of η. This explodes $[0, 1]$ to $[0, \infty)$, and reverses the extremes of η and P. Windham's reasoning here is that values of η can be *very* close to zero (on the order of e^{-100}) for partitions that have relatively fuzzy entries, so P is more suitable for detection of changes than η is. He also gives a heuristic interpretation of this step based on Shannon's notion of information: $-\log_e(\eta)$ quantifies the "surprise" one would feel upon observing an "event" (the algorithm producing a better partition in the sense of η) with "probability" η. This usage of the term "surprise" follows that in Ross.(86) The transformation $-\log_e(\eta) = P$ means that

$$0 \leq P(U; c) < \infty \tag{17.10}$$

$\forall U \in (M_{fc} - M_{co})$. (17.8b) shows that $P(U; c) = 0$ when and only when $U = [1/c]$: conversely, P approaches infinity when any column of U approaches some $e_i \in B_c$; hence P is undefined on *all* hard c-partitions of X. Based on the interpretation of η given above, the validity strategy of Windham is to solve the problem

$$\max_{2 \leq c < n} \left\{ \max_{U \in \Omega_c} \{P(U; c)\} \right\} \tag{17.11}$$

where, as in previous sections, Ω_c is the set of optimality candidates for U at each c found by any method whatsoever. If $(U^*; c^*)$ solves (17.11), we presume that U^* is the most valid clustering of X in Ω_c. One difficulty this strategy shares with validity tests based on F and H is the benchmark problem: how large must $P(U^*; c^*)$ be to instill in one confidence about the reliability of U^*? Put another way—what value of $P(U^*; c^*)$ is a "statistically significant" maximum? Analysis of this question is still in its infancy: suffice it to say here that the problem exists, and again necessitates an element of subjective judgment! For investigators who prefer unique solutions, an even more discomforting situation eventuates in our next example.

(E17.1) *Example 17.1* (*Windham*[115]). Windham processed Anderson's Iris data—the data set X of (E15.1) displayed in (Fig. 15.1)—with the FCM algorithm (A11.1) using $\varepsilon_L = 0.05$, $\|\cdot\|$ = Euclidean; $m = 4.00$, and variable $c = 2, 3, 4, 5$, and 6. The stopping criterion ε_L was applied with the max norm on successive iterates of the cp feature centers $\{v_{ij}^{(l)}\}$. Only one initial guess was used for $U^{(0)}$: consequently, Ω_c contained one optimal pair $(U, \mathbf{v})$ for J_4 at each c. Values of the three validity functionals are listed in Table 17.1. From this table we find that (15.13) and (16.7) have the identical solution $c^* = 2$, which agrees with the analysis of (E15.1) (although a different algorithm, $m = 2.00$, was used). The proportion exponent of Windham, however, indicates a slight preference for $c^* = 3$. Note that F and H are, respectively, monotone decreasing and increasing functions of c; S18 contains an analysis of their expected values, which have this property. Further observe that P is here *not* monotonic. When validity indicators are at odds with each other, one is forced to turn elsewhere, namely, to the data themselves. Windham accepts $c^* = 3$; indeed, X does contain representatives of three subspecies of irises. This is, of course, the botanically correct number of clusters [in fact, (E15.1) is the only analysis of the iris data that suggests *otherwise* for the primary or "coarse" substructure in X!]. It is left to the reader to decide whether $c^* = 2$ or $c^* = 3$ is a better choice (cf. Fig.

Table 17.1. Validity Indicators for (E17.1)

c	Partition coefficient $F(U; c)$	Partition entropy $H(U; c)$	Proportion exponent $P(U; c)$
2	0.63 ←	0.55 ←	109
3	0.45	0.92	113 ←
4	0.35	1.21	111
5	0.27	1.43	62
6	0.23	1.60	68

15.1). Our contention, based on examples 15.1 and 17.1, is that there is more than one *useful* interpretation of substructure in this data. There are *algorithmic* clusters at $c^* = 2$—the "coarse" substructure of (E15.1); and *apparent* clusters at $c^* = 3$—the finer (botanical) substructure identified here. Another example of this type is presented in S19.

For interested readers, the MM hard 3-partition U_m of X obtained from U^* by Windham clustered all 50 (SE) plants correctly; mislabeled three (VC); and made nine mistakes among the (VG). This yields via (13.7) the clustering error $E(U_m) = 0.08$ or 8%, a very typical value compared to various other clustering studies of the iris data.[11] The highest membership in U^* was 0.84; typical values for most of the (VG) and (VC) vectors were on the order of 0.50. In other words, U^* was a relatively fuzzy 3-partition of X [as a result of using $m = 4.00$ in (A11.1)]. •

Remarks

(E17.1) shows that P behaves somewhat differently from F and H. The reason is clear: P derives its behavior from n entries in U, whereas F and H use all cn elements. It would be philosophically inconsistent to expect any of these functionals to be most reliable on a regular basis. F and H have slightly larger domains than P, and in this sense are more general. F and H seem more sensitive to small changes in a few columns as U "approaches" hardness. Since a single column of U can drive P to infinity, it is perhaps least dependable when c-partitions have a lot of large column maxima (as happens, for example, as $m \overset{+}{\to} 1$ in (A11.1)). On the other hand, P seems more adept at detecting structural variations in U as its distance from M_{co} increases, that is, as U becomes "fuzzier."

Under what circumstances F, H, and P suggest *different* solutions remains to be seen. Since all of these mappings optimize at every hard c-partition of X—no matter how "incorrect"—it seems clear that something more than their values is needed. Dependence on both c and n should be eliminated. A more basic issue concerns the lack of a suitable way to threshold F, H, or P. In other words, when should the cluster hypothesis be rejected altogether? All of these functions do provide tests based on *order* and an arbitrary *threshold*, so they can be accorded the same analytical status as statistical hypothesis tests. Both normalization and statistical standardization of F and H (discussed in Section 18) have been proposed as ways to overcome their monotonic tendencies.

Perhaps the most fundamental objection to the validity strategies based on F, H, or P lies with a fact we have tacitly avoided so far: none of these methods explicitly involves the *data* or the *algorithm* used to partition it. Sections 19 and 20 describe two measures of cluster validity that identify well-defined geometric properties of the identified cluster structure: both of these measures have been used in the fuzzy sets context by their inventors.

Exercises

H17.1. Calculate $P(U;2)$ and $P(V;2)$ for the partitions U and V in H15.5. Compare with H15.5 and H16.2.

H17.2. $U, V \in M_{fc} \subset V_{cn}$. Suppose $P(U;c) > P(V;c)$. Does this imply $H(U;c) < H(V;c)$? That $F(U;c) > F(V;c)$? What about the converses?

H17.3. Derive (17.2d).

H17.4. Calculate the integral for $p(E_1)$ in the proof of (T17.1):

$$p(E_1) = \int_{\mu}^{1} dx_1 \int_{0}^{1-x_1} dx_2 \cdots \int_{0}^{1-\sum_{i=1}^{c-2} x_i} dx_{c-1}$$

S18. Normalization and Standardization of Validity Functionals

The validity strategies discussed so far all require $2 \leq c < n$. Dunn[37] observed that the endpoints of the range of H fail to establish suitable benchmarks: at $c = 1$, $H = 0$ because $M_{f1} = M_1$; at $c = n$, the $n \times n$ identity matrix is the n-partition of X which optimizes most fuzzy clustering criteria, and again H is zero. These facts led Dunn to suggest that as $c \to n$, H could be expected to decrease towards zero, thus negating the efficacy of strategy (16.7). Dunn suggested in (37) a normalization of H that might counter this tendency.

Let X_o be a set of $n = \rho^p$ vectors uniformly distributed over a hypercubical lattice in $\mathbb{R}^p$ (ρ is the number of points in each of p directions). X_0 has cluster structure *only* at $c = 1$ or $c = n$, so if one knew the expected behavior of H_o—the entropy of c-partitions of X_o—for $2 \leq c < n$, a much more cogent strategy than that of (16.7) would be to compare $H(U;c)$ to $H_o(U;c)$, if a significant deviation of H from the expected H_o (which represents the null hypothesis that X is cluster-free) occurs, the likelihood that X has cluster substructure increases. In other words, H_o provides a null hypothesis against which H can be tested. Unfortunately, the theoretical distribution of H_o seems hopelessly entangled with the algorithm generating the U's, so that this idea has not yet led to a genuine statistical validity test. However, Dunn conjectured, based on empirical experiments using X_o in $\mathbb{R}^2$ and $\mathbb{R}^3$ and the FCM algorithm (A11.1), that H_o was approximately linear, following the equation

$$H_o(c) = 1 - (c/n) \tag{18.1}$$

Formula (18.1) was taken by Dunn as the (approximate) solution of

$$\min_{U} \{H(U; c)\} \tag{18.2}$$

at $c = 2, 3, \ldots, n - 1$, where the minimum at fixed c in (18.2) is taken over a number of fuzzy c-partitions of X_o generated by (A11.1). Dunn's normalization of H is based on the conjecture that H_o at (18.1) might approximate the expected solution of (18.2)—and thus of (16.7) as well—for each c, and all reference hypercubes of ρ^p vectors in $\mathbb{R}^p$.

(D18.1) *Definition 18.1* (*Normalized Partition Entropy*). Let $U \in M_{fc}$ be a fuzzy c-partition of X, $|X| = n$. The normalized entropy of U is, for $2 \leq c < n$

$$\hat{H}(U; c) = H(U; c)/[1 - (c/n)] \tag{18.3}$$

The quasistatistical validity strategy of Dunn based on H is to solve (16.7) using $\hat{H}$ instead of H. Since $\hat{H} = (H/H_o)$ has the form of an inverse likelihood ratio test, minimizing $\hat{H}$ ostensibly corresponds to maximizing the approximate likelihood that X contains cluster substructure. In other words, if $(U^*; c^*)$ solves

$$\min_{2 \leq c < n} \left\{ \min_{\Omega_c} \{\hat{H}(U; c)\} \right\} \tag{18.4}$$

c^* is the most likely number of clusters in X, because $H(U^*; c^*)$ is smallest compared to the conjectured null value $H_o(c^*)$ which would be observed if the data were uniformly distributed like X_o.

A number of observations concerning (18.4) should be made. First, H is invariant to Euclidean similarity transformations, so the size, position, and orientation of X relative to X_o is unimportant. Next, note that Ω_c in (18.4) is implicitly restricted to approximate fixed points of J_m, because (18.1) was generated by FCM algorithm (A11.1). Furthermore, $H_o(n) = 0$, as anticipated by Dunn, but $H_o(1) = (n - 1)/n$, not zero, as it must be. The reason for this is simple: Dunn extrapolated the observed solutions of (18.2) to $c = 0$ [cf. Figs. 1 and 2 of (37)] to obtain (18.1), whereas, H_o clearly must be zero at $c = 1$. Further, $\hat{H}$ is undefined at $c = n$. Taken together, these facts again necessitate the restriction $2 \leq c < n$ for $\hat{H}$, which leaves $\hat{H}$ in the same "benchmark" predicament as F, H, and P concerning rejection of $c = 1$ or $c = n$. Moreover, even though $\hat{H} = H/H_o \to 0$ means that H_o is large compared to H, one must still select an arbitrary threshold at which to reject H_o.

From a practical standpoint, $\hat{H}$ and H will have nearly identical values when c is relatively small compared to n. For example, with the iris data of

(E17.1), the values of $\hat{H}$ and H listed in Table 17.1 will differ *at most* by the factor $1/[1-(6/150)] = 1.04$. Clearly, the ratio (c/n) determines the observed difference between $\hat{H}$ and H: for $(c/n) \ll 1$, $\hat{H}$ and H will behave nearly identically; as $(c/n) \to 1$, their behaviors will diverge. In applications, it is almost always the case that $c \ll n$: in this instance, either functional will usually suggest the same solutions. In applications where (c/n) is large, however, $\hat{H}$ yields substantially different results.

(E18.1) *Example 18.1* (*Dunn*[37]). Figure 18.1 depicts (to scale) an artificial data set of $n = 9$ data points in $\mathbb{R}^2$. Whether or not X contains $c^* = 3$ visually apparent clusters is a matter for debate. In any case, Dunn applied the FCM algorithm (A11.1) to X, using $m = 2.00$, the Euclidean norm, $\varepsilon_L = 0.01$ applied to the max norm on successive cluster center iterates, and three starting guesses for $\{\mathbf{v}_i^{(0)}\}$ at each value of $c = 2, 3, \ldots, 8$. Ω_c at each c contained but one $U \in M_{fc}$, generated as part of an optimal pair for J_2, independent of the initialization of (A11.1). Figure 18.1 illustrates graphs of F, H, and $\hat{H}$ as (if they were continuous) functions of c on the unique points in Ω_c. In this example, both F and H have "local" extrema at $c^* = 3$—i.e., "something funny" occurs at $c^* = 3$—but continue, as $c \to n = 9$, to trend in the manner expected by Dunn, towards "global" extrema that would choose $c^* = 9$ when using strategies (15.13) or (16.7). The

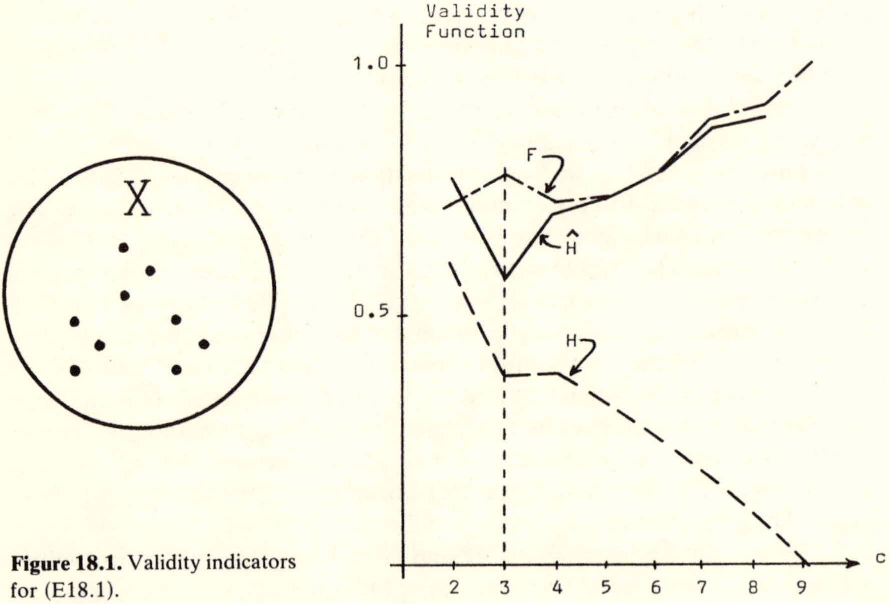

Figure 18.1. Validity indicators for (E18.1).

Table 18.1. The Optimal 3-Partition of X: (E18.1)

Data	Membership functions ~ (A11.1)		
$\mathbf{x}_k$	u_{1k}	u_{2k}	u_{3k}
1	0.05	0.04	0.91
2	0.05	0.10	→ 0.85
3	0.03	0.03	0.94
4	0.89	0.05	0.06
5	→ 0.78	0.14	0.08
6	0.96	0.02	0.02
7	0.10	0.75	0.15
8	0.04	→ 0.91	0.05
9	0.03	0.95	0.02

normalized entropy $\hat{H}$, however, provides via (18.4) the "global" minimum at $c^* = 3$. Table 18.1 lists the 3-partition U^* of X selected. Certainly these membership functions suggest the most visually appealing partition of X into three clusters. Letting

$$\mu(c) = [(n - c)\log_a(c)]/c$$

denote the maximum for $\hat{H}$ at fixed c, the number

$$\frac{\mu(c^*) - \hat{H}(U; c^*)}{\mu(c^*)}$$

is in some sense the "level of significance" at which one rejects the null hypothesis (accepts c^*): in the present case, this number is 0.22. The point of the present example, however, is this: when (c/n) is large, $\hat{H}$ and H do exhibit markedly different behavior. •

Another modification of H having a statistical flavor has been recently discussed in (23). Dunn's null hypothesis—that $X = X_o$ is uniformly distributed in $\mathbb{R}^p$—is replaced by an assumption concerning the distribution of U over M_{fco}: in particular, that U is uniformly distributed thereon. In other words, each column of U is viewed as one observation of a c-variate random variable uniformly distributed over conv(B_c). In this context, H (and F) become statistics on the n columns of U—namely, sample means—and if $U \in M_{fco}$, these n columns correspond to statistically independent observations. Our goal is to find the expected value and variance of H (and F) over M_{fc}. Towards this end, we prove the following theorem.

(T18.1) *Theorem 18.1.* Let $\mathbf{Y} = (Y_1, Y_2, \ldots, Y_c)$ be a random vector valued in $\text{conv}(B_c)$. For $2 \leq c < n$ define $h(\mathbf{y}) = -\sum_{i=1}^{c} y_i \log y_i$, using logarithms to base e. If $\mathbf{Y}$ is uniformly distributed over $\text{conv}(B_c)$, the mean and variance of statistic h are

$$E[h(\mathbf{y})] = \sum_{k=2}^{c} (1/k) \tag{18.5a}$$

$$\text{var}\,[h(\mathbf{y})] = \left(\sum_{k=2}^{c} (1/k)^2\right) - \frac{(c-1)(\pi^2/6 - 1)}{c+1} \tag{18.5b}$$

Analysis. (T18.1) shows that the expected value of h (and therefore H) is monotonic with c. But, it is monotone *increasing* with c. From this we infer that H itself can be expected to increase monotonically with c in the absence of structure in X that forces the distribution of U away from uniformity. The proof of Theorem 18.1 is long and rather tricky; the essential ingredient is surface integration similar to that required in (T17.1).

Proof. $\text{conv}(B_c)$ is the range of $\mathbf{Y}$, which is uniformly distributed over it. Accordingly, the probability measure dP is the surface area measure of $\text{conv}(B_c)$ divided by its area. Let $\phi : \mathbb{R}^{c-1} \to \mathbb{R}^c$ parametrize $\text{conv}(B_c)$ as in (T17.1):

$$\phi(x_1, x_2, \ldots, x_{c-1}) = \left(x_1, x_2, \ldots, x_{c-1}, 1 - \sum_{i=1}^{c-1} x_i\right)$$

The domain of ϕ is

$$D_\phi = \left\{\mathbf{x} \in \mathbb{R}^{c-1} \,\middle|\, 0 \leq x_i \leq 1;\ \sum_{i=1}^{c-1} x_i \leq 1\right\}$$

The area of $\text{conv}(B_c)$ is

$$A(\text{conv}(B_c)) = \int_{D_\phi} c^{1/2}\, dx_1\, dx_2 \cdots dx_{c-1} = \frac{c^{1/2}}{(c-1)!}$$

and so probability measure dP is

$$dP = \left[\frac{(c-1)!}{c^{1/2}}\right] dS$$

where dS is the surface area measure on $\text{conv}(B_c)$ which, in the parametrization of $\text{conv}(B_c)$ by ϕ, is

$$dS = c^{1/2}\, dx_1\, dx_2 \cdots dx_{c-1}$$

Consequently, for any statistic $g(\mathbf{y})$ on $\operatorname{conv}(B_c)$, we have

$$E[g(\mathbf{y})] = \int_{\operatorname{conv}(B_c)} g(\mathbf{y})\, dP = (c-1)! \int_{D_\phi} g(\phi(\mathbf{x}))\, dx_1\, dx_2 \cdots dx_{c-1}$$

In particular, for $g = h$ we have for the mean

$$E[h(\mathbf{y})] = \int_{\operatorname{conv}(B_c)} -\left(\sum_{i=1}^{c} y_i \log y_i\right) dP = -\sum_{i=1}^{c} \int_{\operatorname{conv}(B_c)} (y_i \log y_i)\, dP$$

$$= -c \int_{\operatorname{conv}(B_c)} y_1 \log y_1\, dP = -c! \int_{D_\phi} x_1 \log x_1\, dx_{c-1} \cdots dx_2\, dx_1 \tag{18.6}$$

For any integrable $G : D_\phi \to \mathbb{R}$, we have

$$\int_{D_\phi} G(\mathbf{x})\, dx_1\, dx_2 \cdots dx_{c-1}$$

$$= \int_0^1 dx_1 \int_0^{1-x} dx_2 \int_0^{1-x_1-x_2} dx_3 \cdots \int_0^{1-\sum_{i=1}^{c-2} x_i} G(\mathbf{x})\, dx_{c-1}$$

so the integral at (18.6) becomes

$$E[h(\mathbf{y})] = -c! \int_0^1 x_1 \log x_1\, dx_1 \int_0^{1-x_1} dx_2 \cdots \int_0^{1-\sum_{i=1}^{c-2} x_i} dx_{c-1}$$

$$E[h(\mathbf{y})] = \frac{-c!}{(c-2)!} \int_0^1 x \log x (1-x)^{c-2}\, dx \tag{18.7}$$

For $c = 2$, (18.7) reduces to

$$E[h(\mathbf{y})] = -2 \int_0^1 x \log x\, dx = -2\left(-\frac{1}{4}\right) = \frac{1}{2}$$

which is, for $c = 2$, (18.5a). For $c > 2$, expand $(1-x)^{c-2}$ by the binomial theorem; then from (18.7) there follows

$$E[h(\mathbf{y})] = \frac{-c!}{(c-2)!} \int_0^1 x \log x \left[\sum_{k=0}^{c-2} \binom{c-2}{k} (-1)^k x^k\right] dx$$

$$= \frac{-c!}{(c-2)!} \sum_{k=0}^{c-2} (-1)^k \binom{c-2}{k} \int_0^1 x^{k+1} \log x\, dx$$

$$E[h(\mathbf{y})] = \left[\frac{c!}{(c-2)!}\right] \sum_{k=0}^{c-2} \frac{(-1)^k}{(k+2)^2} \binom{c-2}{k} \tag{18.8}$$

Equation (18.8) can be simplified as follows: let

$$S_{n,m} = \sum_{k=0}^{n} \binom{n}{k} \frac{(-1)^k}{(k+2)^m}$$

Then $S_{n,0} = 0$ for $n \geqslant 1$;

$$S_{0,m} = \frac{1}{2^m} \text{ for } m \geqslant 1$$

and by rearrangement,

$$S_{n,m} = n\left[\sum_{k=0}^{n-1} \binom{n-1}{k} \frac{(-1)^k}{(n-k)(k+2)^m}\right] + \frac{(-1)^n}{(n+2)^m}$$

For $m = 1, 2$, we have for each $n \geqslant 1$ with partial fractions

$$\frac{1}{(n-k)(k+2)} = \left(\frac{1}{n+2}\right)\left(\frac{1}{n-k} + \frac{1}{k+2}\right)$$

$$\frac{1}{(n-k)(k+2)^2} = \frac{1}{(n+2)^2(n-k)} + \frac{1}{(n+2)^2(k+2)} + \frac{1}{(n+2)(k+2)^2}$$

With these equalities we find

$$S_{n,1} = \frac{1}{n+2}(S_{n,0}) + \frac{n}{n+2} S_{n-1,1} = \left(\frac{n}{n+2}\right) S_{n-1,1} \tag{18.9a}$$

$$S_{n,2} = \frac{1}{(n+2)^2} S_{n,0} + \frac{n}{(n+2)^2} S_{n-1,1} + \frac{n}{(n+2)} S_{n-1,2}$$

$$= \left(\frac{n}{(n+2)^2}\right) S_{n-1,1} + \frac{n}{n+2} S_{n-1,2} \tag{18.9b}$$

Solving this system of difference equations gives $\forall n \geqslant 1$

$$S_{n,1} = \frac{n!}{(n+2)!}$$

$$S_{n,2} = S_{n,1}\left(\sum_{k=2}^{n+2} \frac{1}{k}\right) = \frac{1}{(n+1)(n+2)}\left(\sum_{k=2}^{n+2} \frac{1}{k}\right)$$

Returning to (18.8), we have for $n = c - 2$ using $S_{n,2}$ above, that

$$E[h(\mathbf{y})] = \left[\frac{c!}{(c-2)!}\right] S_{c-2,2}$$

$$= \left[\frac{c!}{(c-2)!}\right]\left(\frac{1}{(c-1)(c)}\right) \sum_{k=2}^{c} \frac{1}{k}$$

$$= \sum_{k=2}^{c} \frac{1}{k}$$

which proves (18.5a). To find the variance, (18.5b), we follow essentially the same argument (admittedly more tedious, but with no major departures from the above) to find $E(h^2(\mathbf{y}))$. It is convenient to proceed with the simplified notation

$$E_c = E[h(\mathbf{y})]\,\forall c = 1, 2, \ldots, n-1$$

where we define $E_1 = 0$. Then at each fixed $c > 1$,

$$\operatorname{var}[h(\mathbf{y})] = E[h^2(\mathbf{y})] - E_c^2$$

$$E[h^2(\mathbf{y})] = \left(\int_{\operatorname{conv}(B_c)} \sum_{i=1}^{c} (y_i \log y_i)\,dP\right)^2 - E_c^2$$

$$= \left\{\int_{\operatorname{conv}(B_c)} \left[\sum_{i=1}^{c} (y_i \log y_i)^2 + 2\sum_{i=1}^{c-1}\sum_{j=i+1}^{c} y_i \log y_i\, y_j \log y_j\right] dP\right\} - E_c^2$$

$$= c\int_{\operatorname{conv}(B_c)} y_i^2 (\log y_i)^2\, dP$$

$$+ (c)(c-1)\int_{\operatorname{conv}(B_c)} (y_i \log y_i)(y_j \log y_j)\, dP - E_c^2$$

$$= \frac{c!}{(c-2)!}\int_0^1 x^2(\log x)^2(1-x)^{c-2}\, dx$$

$$+ \frac{(c-1)c!}{(c-3)!}\int_0^1\int_0^{1-x} (x \log x)(y \log y)(1-x-y)^{c-3}\, dy\, dx - E_c^2$$

$$= \frac{c!}{(c-2)!}\int_0^1 x^2(\log x)^2(1-x)^{c-2}\, dx$$

$$+ \frac{c!}{(c-2)!}\int_0^1 (x \log x)(1-x)^{c-1}\log(1-x)\, dx$$

$$+ \frac{c-1}{c+1} E_{c+1}E_{c-1} - E_c^2 \qquad (18.10)$$

The integrals in (18.10) have values as follows: expansion of $(1-x)^{c-2}$ by the binomial theorem leads to

$$\int_0^1 x^2(\log x)^2(1-x)^{c-2}\, dx = 2\left[\sum_{k=0}^{c-2}\binom{c-2}{k}\frac{(-1)^k}{(k+3)^3}\right]$$

$$= 4\frac{(c-2)!}{(c-1)!}\left(\sum_{k=3}^{c+1}\sum_{j=3}^{k}\frac{1}{jk}\right) \qquad (18.11a)$$

$$\int_0^1 (x \log x)(1-x)^{c-1} \log(1-x)\,dx = \int_0^1 x^{c-1} \log x \log(1-x)\,dx - \int_0^1 x^c \log x \log(1-x)\,dx \tag{18.11b}$$

Using the series expansion $\log(1-x) = -\sum_{k=1}^{\infty}(x^k/k)$, we have

$$\int_0^1 x^n \log x \log(1-x)\,dx = \sum_{k=1}^{\infty} \frac{1}{k(n+k+1)^2}$$

$$= \frac{1}{(n+1)^2}\left(\sum_{1}^{n+1}\frac{1}{k}\right) - \frac{1}{n+1}\left(\sum_{n+2}^{\infty}\frac{1}{k^2}\right) \tag{18.12}$$

Applying (18.12) to each of the terms in (18.11b), and then combining (18.10), (18.11a), and (18.11b) now yields (18.5b), and completes the proof of Theorem 18.1. •

Theorem 18.1 yields immediately the mean and variance of functional H, which is just the average or sample mean of h over n independent columns of $U \in M_{fco}$.

(C18.1) *Corollary 18.1.* Let $U \in M_{fco}$; $H(U;c)$ as in (16.4) with logarithmic base e. Then for each integer c, $2 \leq c < n$,

$$\mu_H(U;c) = E[H(U;c)] = \sum_{k=2}^{c}\frac{1}{k} \tag{18.13a}$$

$$\sigma_H^2(U;c) = \text{var}[H(U;c)] = \frac{1}{n}\left[\left(\sum_{k=2}^{c}\frac{1}{k^2}\right) - \left(\frac{c-1}{c+1}\right)\left(\frac{\pi^2}{6}-1\right)\right] \tag{18.13b}$$

Proof. $H(U;c) = (1/n)\sum_{i=1}^{n} h(\mathbf{U}^{(i)})$, where $\mathbf{U}^{(i)} \in \text{conv}(B_c)$ is the ith column of $U \in M_{fco}$. Because the rows of U are uncoupled via degeneracy, the $\{h(\mathbf{U}^{(i)})\}$ are n independent and identically distributed (i.i.d.) observations of $h(\mathbf{Y})$ in Theorem 18.1. Since $H(U;c)$ is just the sample mean of these n i.i.d. observations, results (18.13) follow. •

Now suppose $U \in M_{fc}$ is a *nondegenerate* fuzzy c-partition of X. Then $\sum_{k=1}^{n} u_{ik} > 0 \forall i$ produces dependency among the n columns of U. Since fuzzy clustering algorithms usually produce U's in M_{fc}, the behavior of H on this subset of M_{fco} is of interest. But $M_{fco} - M_{fc}$ has (probability) measure zero, i.e.,

$$\int_{M_{fco}-M_{fc}} dP = 0$$

Accordingly, the expected value of any function over M_{fco} agrees with its expected value over M_{fc}, so the hypothesis $U \in M_{fco}$ in Corollary 18.1 can be replaced by $U \in M_{fc}$.

(C18.2) *Corollary 18.2.* The mean and variance of $H(U; c)$ for $U \in M_{fc}$, holding the other hypotheses of Corollary 18.1, are given by (18.13a) and (18.13b), respectively.

Result (18.13a) supports the conjecture that observed values of H_c tend to increase monotonically with c; $E[H(U; c)]$ is just the cth partial sum of the divergent harmonic series (minus one).

As an aside, an interesting corollary of (T18.1) that relates $E[H(U; c)]$ to the maximum entropy ($\log_e c$) follows by recalling that

$$\lim_{c \to \infty} \left\{1 + \sum_{k=2}^{c} \frac{1}{k} - \log_e c\right\} = \gamma_E = 0.5772\ldots$$

γ_E being Euler's constant. This fact and (18.13a) yield the following corollary.

(C18.3) *Corollary 18.3.* Under the hypotheses of Corollary 18.2,

$$\lim_{c \to \infty} \{\log_e c - E[H(U; c)]\} = 1 - \gamma_E = 0.4227\ldots \qquad (18.14)$$

Since $c < n$, (18.14) requires $n \to \infty$. Asymptotically then, the distance between $E[H]$ and $\max\{H\}$ over M_{fc} is fixed at $1 - \gamma_E$. To three digits this distance is 0.400 at $c = 22$; 0.411 at $c = 41$; and 0.422 at $c = 390$.

The phrase "expected value of H over M_{fc}" is a particularly seductive one. Although Theorem 18.1 and its corollaries are interesting in their own right, they do not seem to bear directly on the cluster validity question. The difficulty lies with the hypothesis that U is uniformly distributed over M_{fc}, which gives substance to the mathematical content of (T18.1), but does not provide any null hypothesis about structure in X. At first glance, one is tempted to reason that if $X = X_o$ (i.e., that X is uniformly distributed as above), then a particular algorithm is "equally likely" to produce any $U \in M_{fc}$, hence $|H - \mu_H|$ might be a useful validity measure. Unfortunately, just the reverse appears true: fuzzy clustering algorithms yield quite predictable results on uniformly distributed data. For example, if X consists of n independent observations from the uniform distribution on [0, 1], and the FCM algorithm is applied to X with $c = 2$, the cluster centers will approach 1/4 and 3/4, and each MM cluster will have $(n/2)$ points as n grows large; in other words, for this cluster-free structure, there are only a few of the infinitely many possible fuzzy 2-partitions which are very probable.

Remarks

The two approaches discussed in this section raise many interesting questions. Concerning Dunn's H_0, one wonders if a theoretical analysis affirming (18.1) is possible. Further, would H_0 have the form (18.1) if the U's in (18.2) came from a different algorithm? If (c/n) is large—say 0.10 or more—$\hat{H}$ will apparently behave quite differently from H, and is probably a better validity indicator [especially since $\hat{H}$ references $(U^*; c^*)$ to a null hypothesis about substructure in X]. On the other hand, for $c \ll n$, $[1 - (c/n)] \approx 1$ and $\hat{H} \approx H$, so solutions of (16.7) and (18.4) will be the same. This is borne out by several examples in Dunn.[37]

The behavior of $F(U; c)$ under the hypotheses of (T18.1) is delineated in (23): in particular, $\mu_F(U; c) = 2/(c+1)$; and

$$\sigma_F^2(U; c) = 4(c-1)/[n(c+1)^2(c+2)(c+3)]$$

These results bear the same disclaimer as those of (T18.1): they do not involve a null hypothesis about the data. What they do provide is a possible way to attack the validity question based on a combination of the two approaches presented here: first, a null hypothesis (Dunn's $X = X_0$ seems perfectly reasonable) about cluster structure is made; then, find—here is the hard step!—theoretically or statistically, *the distribution of U generated by a particular algorithm* applied to X_0; finally, find the distribution of a statistic—H, F, or P, for example—on M_{fc}, using the distribution of U and methods similar to those in (T18.1). The resultant statistical validity test should be effective for the algorithm under consideration.

Other transformations of H, F, and P have been proposed. None of these have theoretical underpinnings similar to the ones above. Backer's[4] normalization of F, however, has been used effectively as an objective function, and will be discussed separately in Chapter 5. Data sets amenable to Woodbury's algorithm (A10.1) lend themselves well to a statistical validity test, which takes the form of a likelihood ratio test of the null hypothesis $(c = 1)$ against the alternative $(c > 1)$. This is a standard goodness-of-fit type of test based on a χ^2 distribution having the number of degrees of freedom specified in (119) (cf. H10.1).

Exercises

H18.1. Calculate $\hat{H}(U; 2)$ and $\hat{H}(V; 2)$ for the partitions U and V in H15.5. Compare with H15.5, H16.2, and H17.1.

H18.2. Under the hypotheses of C18.1, show that (23)
(i) $\mu_F(U; c) = 2/(c+1)$;
(ii) $\sigma_F^2(U; c) = 4(c-1)/(n)(c+2)(c+3)(c+1)^2$.

H18.3. Calculate the mean and variance of F, H, and $\hat{H}$ for $c = 2$. Compare with the values of F, H, and $\hat{H}$ on U and V in H15.5, H16.2, and H18.1.

S19. The Separation Coefficient: Star-Tracker Systems

In retrospect, validity functionals F, H, and P all measure a property (the amount of fuzziness) of $U \in M_{fc}$—not of X. Since the clusters themselves reside in the data, the properties that "good" clusters have when identified by any of these functions is obscured by the algorithm connecting X to U (it is precisely this fact that renders statistical tests of cluster validity so intractable). The separation coefficient of Gunderson[52] imbues "good" hard clusterings of X with a property derived from the topological structure induced on $\mathbb{R}^p$ by an arbitrary metric, say $d : \mathbb{R}^p \times \mathbb{R}^p \to \mathbb{R}^+$. The notion of "validity" is then defined relative to the geometric configuration that optimizes this coefficient. This validity measure is then applied to fuzzy clustering outputs by first converting them to hard ones—the usual (but not only) means being via (D9.2).

Gunderson's coefficient depends not only on X and d; but also on prototypes, or cluster centers, for its geometrical substance. The centers involved need *not* be centroids of the hard u_i's; however, (P12.2) shows that (9.8a) yields centers which minimize the overall sum of within cluster squared error between the n data points and c prototypes, so one might ordinarily take the prototypes to be these centroids. In the following definition, which is somewhat more general than that of Gunderson, it is merely required that the c prototypes lie in the convex hulls of the hard clusters they represent.

(D19.1) *Definition 19.1* (*Separation Coefficient*). Let $U \in M_c$ be a hard c-partition of X; let $\mathbf{v}_i \in \operatorname{conv}(u_i) \forall i = 1, 2, \ldots, c$ be a set of c cluster centers for the sets $\{u_i\}$. Let r_i be the radius of the smallest closed ball $\bar{B}(\mathbf{v}_i, r_i)$ centered at $\mathbf{v}_i$ which contains hard cluster u_i; for $1 \leq i \leq c$,

$$r_i = \max_{1 \leq k \leq n} \{u_{ik}\, d(\mathbf{x}_k, \mathbf{v}_i)\} \tag{19.1a}$$

Let c_{ij} denote the distance of separation between cluster centers $\mathbf{v}_i$ and $\mathbf{v}_j$; for $1 \leq i \neq j \leq c$,

$$c_{ij} = d(\mathbf{v}_i, \mathbf{v}_j) \tag{19.1b}$$

The separation coefficient of U is the scalar

$$G(U; \mathbf{v}; c; X, d) = 1 - \max_{i+1 \leq j \leq c} \{ \max_{1 \leq i \leq c-1} \{(r_i + r_j)/c_{ij}\}\} \tag{19.1c}$$

The variable list in (19.1c) exhibits the dependence of G not only on U and c (as do F, H, and P); but also on the centers of the clusters, the data themselves, and the measure of distance being used. If U and $\mathbf{v}$ are related via (9.8a) or (9.10) and d is the Euclidean metric, (19.1c) reduces to the coefficient originally proposed in (52). Using the triangle inequality for d, it's easy to check that $(r_i + r_j)/c_{ij} > 1$ for $i \neq j$, so G lies in the interval $(-\infty, 1)$. G is readily interpreted geometrically by observing that its basic element is the sum of two cluster radii divided by the distance of separation of their centers. Consequently, G has the following properties:

$$0 < G < 1 \Leftrightarrow \left\{\begin{matrix}\text{no pair of closed balls } \bar{B}(\mathbf{v}_i, r_i) \\ \text{intersect one another}\end{matrix}\right\} \tag{19.2a}$$

$$G = 0 \Leftrightarrow \left\{\begin{matrix}\text{the closest pairs of } \bar{B}(\mathbf{v}_i, r_i)\text{'s} \\ \text{are exactly tangent}\end{matrix}\right\} \tag{19.2b}$$

$$G < 0 \Leftrightarrow \left\{\begin{matrix}\text{at least one pair of } \bar{B}(\mathbf{v}_i, r_i)\text{'s} \\ \text{intersect one another}\end{matrix}\right\} \tag{19.2c}$$

These properties are illustrated graphically (using the Euclidean metric for d) in Fig. 19.1. Note that properties (19.2) relate to tangency and intersection of closed balls *containing* the clusters, rather than to the clusters themselves. G thus depends on the compactness (via the $\{r_i\}$) and separation (via the $\{c_{ij}\}$) of c closed balls containing the hard clusters U identified in X. To reiterate, each of these closed balls has (at least) one data point on its surface. It is clear that U identifies increasingly more distinctive substructure in X as G approaches unity. This observation forms the basis for Gunderson's cluster validity strategy, which applies to pairs $(U, \mathbf{v}) \in M_c \times \mathbb{R}^{cp}$: let Ω_{cd} denote candidate pairs $(U, \mathbf{v})$ for fixed c and fixed metric d. Take as the most valid *hard* clustering and prototypical representatives of X the pair $(U^*, \mathbf{v}^*)$ with c^* clusters that solves

$$\max_{2 \leq c \leq n} \left\{ \max_{\Omega_{cd}} \{G(U; \mathbf{v}; c; X, d)\} \right\} \tag{19.3}$$

This validity strategy is operationally similar to previous ones in that one tries to press as close to a fixed bound as possible. It differs importantly from previous schemes in its rationale, since maximizing G corresponds directly to optimizing a well-defined geometrical property of the clusters involved. It should be further emphasized that (19.3) affords a means for comparing clusters obtained by *all* hard clustering algorithms: given U, e.g., calculate $\mathbf{v}$ with (9.8a), choose d, and compute G. This also points up the fact that G cannot be used directly for *fuzzy* cluster validity.

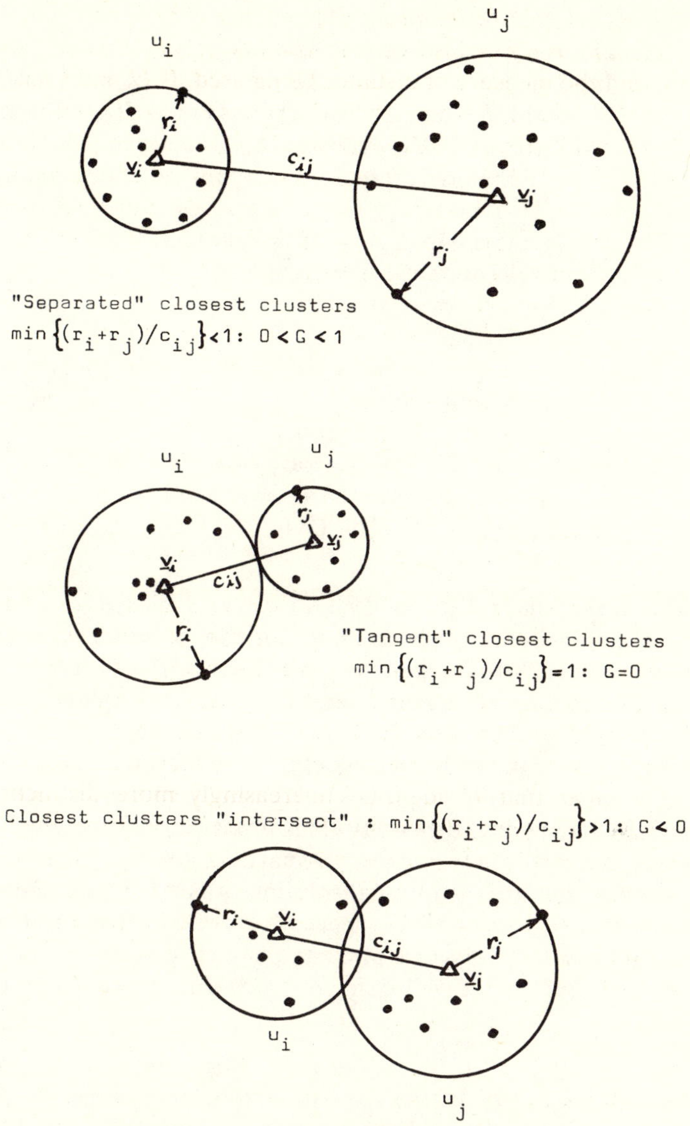

Figure 19.1. Geometry of the separation coefficient.

Given a fuzzy $U \in M_{fc}$, there are many ways one can convert it to a hard c-partition of X in order to implement (19.3). The ad hoc method suggested by Gunderson is (A19.1); interested readers will enjoy posing variations that may be more effective for other applications.

(A19.1) *Algorithm 19.1* (*Gunderson*[52])

(A19.1a) Let $c = 2$. Find $U \in M_{f2}$ using (A11.1).

(A19.1b) For $1 \leq i \leq c$, find the core of u_i at threshold α via (D11.2); let u_i' be the characteristic function of $C(u_i: \alpha) \subset X$; array the $\{u_i'\}$ as rows of a $c \times n$ matrix, and then delete its zero columns (in effect, reduce X to X', $|X'| = n'$, by ignoring all data points which had no membership as great as threshold α): let $U' \in M_c$ be the resultant hard c-partition of X'.

(A19.1c) For $1 \leq i \leq c$, calculate $\mathbf{v}_i'$, the "effective" center of u_i', with (9.8a).

(A19.1d) For $1 \leq i \leq c$ and $1 \leq k \leq n$, calculate membership u_{ik}'' of all n $\mathbf{x}_k$'s using $\{\mathbf{v}_i'\}$ and (11.3a). Call this fuzzy c-partition of X the matrix U''.

(A19.1e) Convert U'' to U''', a hard c-partition of X''', $|X'''| = n'''$; the ith row of U''' is the characteristic function of $C(u_i''; \beta)$. Calculate $G(U'''; \mathbf{v}'; c; X'''; d)$.

Several Observations about (*A19.1*). First, U at (A19.1a) could come from *any* fuzzy partitioning scheme; and updating $\mathbf{v}_i$'s and u_{ik}'s need not employ (9.8a) or (11.3a). The pair (U''', $\mathbf{v}'$) is *not* related to any clustering criterion (in particular, to J_m). And finally, X''' need not be all of the original data—threshold β pares points from X that do not achieve at least this membership in one of the fuzzy clusters in U''—so that strategy (19.3) is not formally applicable for outputs from (A19.1). In (A19.1) then, maximizing G may produce from initial fuzzy clusters in X a hard c^*-clustering of some *subset* of X. If more than a few of the original data are discarded, it is not at all obvious what relation the optimal solution of (19.3) has to the original problem. For this reason, (A19.1) is perhaps more accurately called a data sharpening or "pruning" method, in contradistinction to the general applicability of (19.3), which is a well-defined (hard) cluster validity strategy.

(E19.1) *Example 19.1* (*Gunderson*[52]). Spacecraft require navigational systems based on extraterrestrial reference frames. A possible method for establishing spatial reference is to scan the starfield in the immediate vicinity of the vehicle, and identify its position by matching displayed starfield clusters with cataloged data. Gunderson investigated the feasibility of (A19.1) as a potentially useful method of processing star-tracker data using the data set shown in Fig. 19.2, which depicts $X = 51$ bright stars near Polaris (*UMi* α, the North Star) projected onto (the plane of) the Celestial Equator.

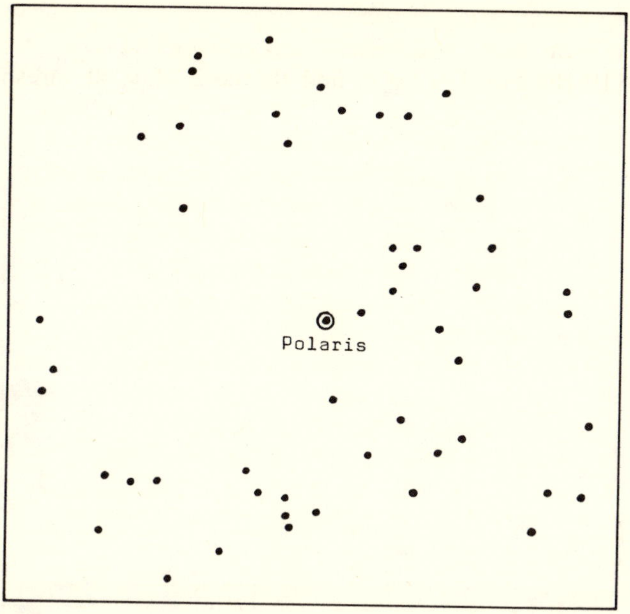

Figure 19.2. Fifty-one unlabeled bright stars near Polaris: (E19.1).

To imagine the difficulty these data cause for clustering algorithms, try to guess the number and linkage of starfield clusters in Fig. 19.2 before reading on.

These stars have three real spatial dimensions, and form the nine apparent (visual) clusters (the ones established by early astronomers) shown in Fig. 19.3 only when a fourth variable—intensity—is added. In other words, stars which form the apparent clusters in Fig. 19.3 could be quite well separated spatially. This raises again the ephemeral nature of "clusters"; certainly one cannot expect algorithms which depend primarily on distances to produce the clusters of Fig. 19.3 if intensity is disregarded! Ling demonstrated in (74), however, that graph-theoretic methods could be successfully applied to a 60-point superset of X using only (x, y)-coordinates on the equatorial plane. Conversely, fuzzy c-means did quite poorly with a 48-point subset of X; this failure was attributed to several characteristics of the data, viz., unequal populations, elongated chains, and the lack of intensity data discussed above.[10]

Gunderson suggested that an interpretative variation of the output of (A11.1) might alleviate the difficulties reported in (10). Following the scheme outlined in (A19.1), he processed X(x–y coordinates only) with (A11.1) at $m = 2$, $\varepsilon_L = 0.01$, $\|\cdot\|$ = Euclidean, and $c = 2, 3$, and 4. For each

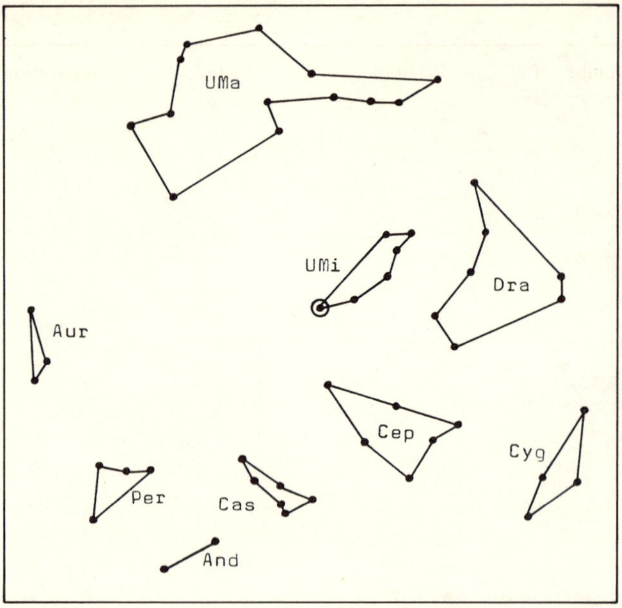

Figure 19.3. Nine visually apparent starfield clusters: (E19.1).

c, (A19.1) was applied to the optimal (U, $\mathbf{v}$) emanating from fuzzy c-means [$\mathbf{v}$ was disregarded: one could use this $\mathbf{v}$ instead of $\mathbf{v}'$ in (A19.1c)!]. Values for the partition coefficient and entropy listed in Table 19.1 were calculated on the U'' matrices in (A19.1d).

In (10), c was allowed to range from 2 to 10, in hopes that F would indicate $c^* = 9$; instead, F decreased from the maximum at $c = 2$ to a "local" minimum at $c = 5$; and then progressed to $c = 10$ at a nearly constant low value. The same type of (mis)behavior for H over this range prompted Gunderson to adopt a different strategy—viz., sequential separation—wherein primary or raw substructure is extracted from X at a low value of c, followed by secondary clustering of each primary cluster to identify finer substructures. This also attempts to circumvent the monotonic tendencies of F and H, which both seem to favor low values for c^*. The results of primary separation are listed in Table 19.1.

Both F and H indicate $c^* = 2$; observe however, that $F(U; 2) = 0.75$ and $H(U; 2) = 0.39$ are nearly in the *middle* of their ranges ([0.50, 1.00] and [0.00, 0.69], respectively). From this, Gunderson inferred that the fuzzy 2-partition producing these values lacked enough character to force H and F close to their (hard) boundaries. On the other hand, G maximizes at $c^* = 3$ with a value which is "closer" to its bound (relative to the range of G). This

Table 19.1. Primary Validity Indicants for 51 Stars: (E19.1)

Number of clusters c	Partition coefficient F	Partition entropy H	Separation coefficient G
2	0.75 ←	0.39 ←	−0.90
3	0.70	0.54	−0.29 ←
4	0.65	0.68	−0.53

led Gunderson to adopt the hard clusters shown in Fig. 19.4 by dotted lines as the primary substructure of X. We denote these three primary star clusters as X_1, X_2, and X_3: $X''' = X_1 \cup X_2 \cup X_3$. Note that X''' from (A19.1e) contains 48 stars: Cepheus γ, Draco χ, and Draco δ are pruned from X by threshold $\beta = 0.50$. Threshold α in (A19.1b) was $\alpha = 0.85$, the metric d used for G was the Euclidean metric. Having this primary decomposition of X, Gunderson reprocessed X_1, X_2, and X_3 separately, using exactly the same procedure as described above. Table 19.2 reports validity indicators for each of these. The results of maximizing G (again different from using F or H except for X_3) are also illustrated graphically in Fig. 19.4: secondary clusters are indicated by edge links.

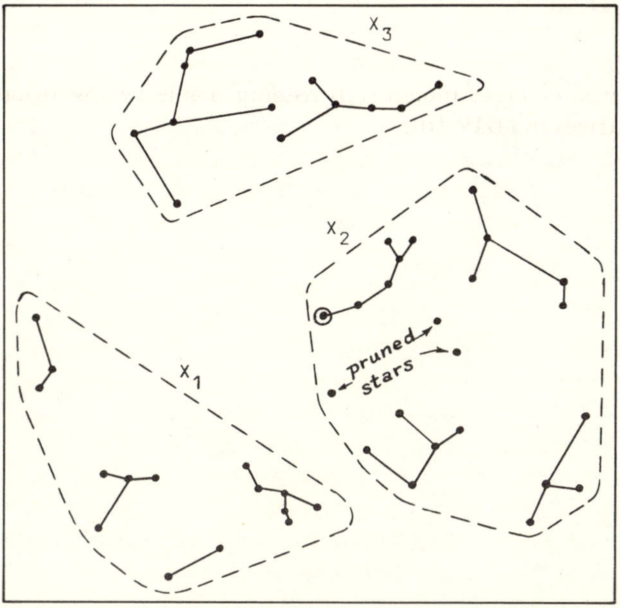

Figure 19.4. Three primary and ten secondary starfield clusters obtained with (A19.1): (E19.1).

Table 19.2. Secondary Validity Indicants for (E19.1)

Data set X	Number of clusters c	Partition coefficient F	Partition entropy H	Separation coefficient G
X_1	2	0.83	0.27	−0.10
	3	0.83	0.33	−0.08
	4	0.79	0.43	0.14 ←
X_2	2	0.77	0.37	−0.44
	3	0.66	0.60	−0.40
	4	0.64	0.70	−0.28 ←
X_3	2	0.80	0.32	−0.14 ←
	3	0.71	0.52	−0.22
	4	[a]	[a]	[a]

[a] Not recorded: 3 of 4 runs resulted in a cluster with no memberships greater than $\alpha = 0.85$.

Gunderson's secondary separation results in $c^* = 10$ hard clusters. Of these, six are apparent clusters, and Cepheus and Draco are correct except for the "pruned" stars mentioned above. The division of Ursa Major into two clusters accounts for the only significant deviation between apparent and algorithmic star clusters. Counting the 6-star subtree of Ursa Major and three pruned stars as errors yields a clustering error via (13.7) of $E(U) = 0.176$, or 17.6%. There are several ways to interpret this. One can correctly argue that none of these validity indicators provides a reliable mechanism for rejection of the cluster hypothesis. On the other hand, the principal algorithmic clustering of Ursa Major (X_3 in Fig. 19.4) was consistently correct. Gunderson infers from this that apparent astronomical clusters are not necessarily the best ones to use in a star-tracker cluster identification catalog.

Remarks

Three things stand out in Gunderson's work. First, G measures a well-defined geometric property of the clusters it recommends, because the data set X is explicitly linked to its values. Secondly, the idea of paring or pruning fuzzy clusters to find distinctive "cores" is intriguing; many authors have discussed hard clustering methods for data sharpening—see Duda and Hart[34] for references in this direction. Reference (11) contains an example of sequential separation of the iris data (E15.1) using $F(U; c)$, with pruning by the maximum membership rule of (D9.2). Finally, Gunderson's observation concerning the efficacy of apparent star clusters as a basis for a star-tracker catalog is important. If the computer consistently identifies *algorithmic* clusters which would provide an accurate data base, why should they be regarded as "incorrect"? From an operational standpoint, it might be fruitful to consider apparent clusters as one useful interpretation (to astronomers) of the (visually perceived) starfield, and the algorithmic clusters as a more useful interpretation for the computerized navigation discussed in (E19.1).

Several interesting questions arise concerning the sensitivity of G to changes in d, the metric used in (19.1); and to $\{\mathbf{v}_i\}$, the centers used as a basis for calculating G. One also wonders how changes in the thresholds α and β of (A19.1) affect the utility of this pruning process.

Exercises

H19.1. Give proofs for each of equations (19.2).

H19.2. Apply (A19.1) to the four 2-partitions of Ruspini's butterfly data which are listed in Figs. 9.1, 9.2, 11.2, and 11.3. Use $\alpha = \beta = 0.90$, and the Euclidean metric for d. Compare these four hard 2-partitions to the ones exhibited in Fig. 11.4 derived by the α-core method of (E11.5).

S20. Separation Indices and Fuzzy c-Means

A pair of indices assessing hard cluster validity which are similar to the separation coefficient above were proposed by Dunn in (36). These measures also depend on X and the topological structure induced upon $\mathbb{R}^p$ by an arbitrary metric d. However, no prototypes are involved; the geometric properties of optimal clusters are instead tied to cluster shapes and interfacial distances. In the following definitions, dia(A) denotes the usual diameter of a subset $A \subset \mathbb{R}^p$; dis(A, B) is the distance between two subsets of $\mathbb{R}^p$; and d is any metric induced by an inner product on $\mathbb{R}^p$.

(D20.1) *Definition 20.1* (*Separation Indices*). Let $U \in M_c$ be a hard c-partition of X. The CS (compact, separated) and CWS (compact, well-separated) indices of U are, respectively,

$$D_1(U; c; X, d) = \min_{i+1 \leq j \leq c-1} \left\{ \min_{1 \leq i \leq c} \left\{ \frac{\text{dis}(u_i, u_j)}{\max\limits_{1 \leq k \leq c} \{\text{dia}(u_k)\}} \right\} \right\} \tag{20.1a}$$

$$D_2(U; c; X, d) = \min_{i+1 \leq j \leq c} \left\{ \min_{1 \leq i \leq c-1} \left\{ \frac{\text{dis}(u_i, \text{conv}(u_j))}{\max\limits_{1 \leq k \leq c} \{\text{dia}(u_k)\}} \right\} \right\} \tag{20.1b}$$

Replacing u_j in (20.1a) by its convex hull in (20.1b) renders D_2 a somewhat more stringent index than D_1. These two indices are not independent: in fact, it is shown in (36) that

$$D_1 \geqslant D_2 \geqslant D_1 - 1 \tag{20.2}$$

for every $U \in M_c$. Dunn suggested the following terminology in connection with D_1 and D_2.

(D20.2) *Definition 20.2* (*CS and CWS clusters*). Let $U \in M_c$ be a hard c-partition of X. We say that U contains c compact separated (CS) clusters relative to d if $D_1 > 1$; the clusters are compact and well separated (CWS) if $D_2 > 1$:

$$U \text{ is a CS partition of } X \Leftrightarrow D_1(U; c; X, d) > 1 \tag{20.3a}$$

$$U \text{ is a CWS partition of } X \Leftrightarrow D_2(U; c; X, d) > 1 \tag{20.3b}$$

In view of (20.2), any CWS partition is a CS partition. Furthermore, the following theorem shows that there is at most one CS c-partition of X.

(T20.1) *Theorem 20.1* (*Dunn*[36]). For every c, $2 \leqslant c < n$, if there is a $U \in M_c$ such that $D_1(U; c; X, d) > 1$, U is unique.

Analysis. This result shows that CS partitions are distinguished by uniqueness whenever they exist. Since there is at most one such hard clustering of X at each c, a validity strategy based on maximizing D_1 over M_c is well defined via this unique limit. The proof itself depends on a simple observation: since X is fixed, any pair of distinct c-partitions of X must intersect.

Proof. Let $U = \{u_1, u_2, \ldots, u_c\}$ and $V = \{v_1, v_2, \ldots, v_c\}$ be two arbitrary distinct hard c-partitions of X. Since X is fixed, at least one hard subset in U, say u_i, intersects at least two hard clusters in V, say v_j and v_k. Then

$$\text{dia}(u_i) \geqslant \text{dis}(v_j, v_k)$$

so

$$\text{(A)} \quad \max_{1 \leqslant t \leqslant c} \{\text{dia}(u_t)\} \geqslant \min_{q+1 \leqslant s \leqslant c} \{\min_{1 \leqslant q \leqslant c-1} \{\text{dis}(v_q, v_s)\}\}$$

In an entirely analogous fashion, we may reverse the roles of U and V to obtain

$$\text{(B)} \quad \max_{1 \leqslant t \leqslant c} \{\text{dia}(v_t)\} \geqslant \min_{q+1 \leqslant s \leqslant c} \{\min_{1 \leqslant q \leqslant c-1} \{\text{dis}(u_q, u_s)\}\}$$

Combining (A) and (B) yields

$$\frac{1}{D_1(U; c; X, d)} \geq D_1(V; c; X, d)$$

so

$$D_1(U; c; X, d) > 1 \Rightarrow D_1(V; c; X, d) < 1$$

Since $V \in M_c$ is arbitrary, there is at most one CS partition in M_c at each c.

(C20.1) *Corollary 20.1.* For fixed c, $2 \leq c < n$, define

$$\bar{D}_1(c) = \max_{U \in M_c} \{D_1(U; c; X, d)\} \tag{20.4}$$

Then

$$X \text{ contains } c \text{ CS clusters} \Leftrightarrow \bar{D}_1(c) > 1 \tag{20.5}$$

Equation (20.5) lends a certain authority to D_1 that all of our previous validity indicators lack, because it makes a direct statement about the existence and uniqueness of data substructure which has well-defined geometrical properties that depend only on X and d. The main drawback with direct implementation of (20.4) is computational: calculation of D_1 for even a few U's in M_c becomes expensive as c and n increase; search for $\bar{D}_1(c)$ is out of the question. Thus, use of this index requires an algorithm that generates partitions which may solve (20.4). Dunn discovered some interesting relationships between D_1, D_2 and the c-means algorithms of S9 and S11. In particular, it was shown in (36) that when $D_2(U; c, X, d) > 1$, so that U is the unique CS partition of X, U must be a solution of (9.9)—i.e., (the only) fixed point of hard c-means algorithm (A9.2). We shall see below that D_1 is related to local extrema of J_m, the fuzzy c-means functional, and (A11.1). The distinction between D_1 and D_2 is most significant when D_1 is near 1. Since D_1 is large if and only if D_2 is [via (20.2)], the distinction between them diminishes as D_2 grows. Note that $D_2 > 1 \Rightarrow D_1 > 1$, so there is at most one CWS partition of X, and if we put

$$\bar{D}_2(c) = \max_{U \in M_c} \{D_2(U; c; X, d)\} \tag{20.6}$$

then

$$X \text{ contains } c \text{ CWS clusters} \Leftrightarrow \bar{D}_2(c) > 1 \tag{20.7}$$

Since $\bar{D}_1(c) > 2 \Rightarrow \bar{D}_2(c) > 1$, this means that the distinction between CS and CWS clusters is useful only when $\bar{D}_2(c) \leq 1$ and $1 < \bar{D}_1(c) \leq 2$, i.e., when X contains unique CS clusters but not CWS clusters.

Dunn conjectured that if $\bar{D}_1(c)$ became sufficiently large, the membership functions of an optimal fuzzy c-partition of X generated by FCM (A11.1) should closely approximate the characteristic functions of the $U \in M_c$ solving (20.4). The proof of this conjecture begins with the following theorem.

(T20.2) *Theorem 20.2* (*Dunn*[36]). Let $U \in M_c$ be a hard c-partition of X, $|X| = n$. Let $D_1 = D_1(U; c; X, d)$, $2 \leq c < n$; let $m \in (1, \infty)$ be fixed, and suppose $U^* \in M_{fc}$ is part of an optimal pair for J_m. Then

$$\sum_{i=1}^{c} \left\{ \min_{1 \leq j \leq c} \left\{ \sum_{\mathbf{x}_k \notin u_j} (u_{ik}^*)^m \right\} \right\} \leq \frac{2n}{(D_1)^2} \tag{20.8}$$

Analysis. The proof of (20.8) depends upon a factorization of $J_m(U^*, \mathbf{v}^*)$ which eliminates $\mathbf{v}^*$ using (11.3b), and the fact that the optimal cluster centers for J_1 using $U \in M_c$ are centroids as in (9.8a). The result relates fuzzy c-partitions of X generated by (A11.1) to Dunn's CS index D_1 (cf. H11.3).

Proof. Let $(U^*, \mathbf{v}^*)$ denote an optimal pair for J_m, so that equations (11.3) prevail for U^* and $\mathbf{v}^*$. Since this pair is a local solution of (11.2) and $U \in M_c \subset M_{fc}$,

$$\begin{aligned} J_m(U^*, \mathbf{v}^*) &\leq J_m(U, v) \\ &= \sum_{k=1}^{n} \sum_{i=1}^{c} (u_{ik})^m \|\mathbf{x}_k - \mathbf{v}_i\|^2 \\ &= \sum_{i=1}^{c} \sum_{\mathbf{x}_k \in u_i} \|\mathbf{x}_k - \mathbf{v}_i\|^2 \end{aligned}$$

Since $U \in M_c$, the $\mathbf{v}_i$'s here are centroids of the hard u_i's: in particular, $\mathbf{v}_i \in \text{conv}(u_i)\ \forall i$, so $\|\mathbf{x}_k - \mathbf{v}_i\|^2 \leq [\text{dia}\,(u_i)]^2\ \forall \mathbf{x}_k \in u_i$. Letting n_i denote the number of elements in u_i, we thus have

$$\begin{aligned} J_m(U^*, \mathbf{v}^*) &\leq \sum_{i=1}^{c} n_i [\text{dia}\,(u_i)]^2 \\ &\leq n\left(\max_{1 \leq t \leq c} \{[\text{dia}\,(u_t)]^2\}\right) \\ \text{(A)} \qquad &\leq n\left(\max_{1 \leq t \leq c} \{\text{dia}\,(u_t)\}\right)^2 \end{aligned}$$

Now, we factor $J_m(U^*, \mathbf{v}^*)$ as follows[36]:

$$J_m(U^*, \mathbf{v}^*) = \frac{1}{2}\left\{\sum_{i=1}^{c}\left[\sum_{k=1}^{n}(u_{ik}^*)^m\right]^{-1}\sum_{s=1}^{n}\sum_{t=1}^{n}(u_{is}^* u_{it}^*)^m \|\mathbf{x}_s - \mathbf{x}_t\|^2\right\}$$

$$= \frac{1}{2}\left\{\sum_{i=1}^{c}\left[\sum_{k=1}^{n}(u_{ik}^*)^m\right]^{-1} \times \sum_{p=1}^{c}\sum_{j=1}^{c}\sum_{\mathbf{x}_s \in u_j}\sum_{\mathbf{x}_t \in u_p}(u_{is}^* u_{it}^*)^m \|\mathbf{x}_s - \mathbf{x}_t\|^2\right\}$$

$$\geqslant \frac{1}{2}\left\{\sum_{i=1}^{c}\left[\sum_{k=1}^{n}(u_{ik}^*)^m\right]^{-1} \times \sum_{p=1}^{c}\sum_{\substack{j=1 \\ j \neq p}}^{c}\sum_{\mathbf{x}_s \in u_j}\sum_{\mathbf{x}_t \in u_p}(u_{is}^* u_{it}^*)^m [\text{dis}\,(u_j, u_p)]^2\right\}$$

(B)
$$\geqslant \frac{1}{2}\left[\sum_{i=1}^{c}\min_{1 \leqslant p \leqslant c}\sum_{\mathbf{x}_k \notin u_p}(u_{ik}^*)^m\right] \times \min_{j+1 \leqslant p \leqslant c}\left\{\min_{1 \leqslant j \leqslant c-1}\{[\text{dis}\,(u_p, u_j)]^2\}\right\}$$

Inequality (20.8) now follows from (A), (B), and the definition of D_1, equation (20.1a).

(C20.2) *Corollary 20.2.* Suppose $\hat{U} \in M_c$ solves (20.4), so that $\bar{D}_1(c) = D_1(\hat{U}; c; X, d)$. Then for $1 \leqslant i \leqslant c$,

$$0 \leqslant \min_{1 \leqslant j \leqslant c}\left\{\sum_{\mathbf{x}_k \notin \hat{u}_j}(u_{ik}^*)^m\right\} \leqslant \left\{\frac{2n}{[\bar{D}_1(c)]^2}\right\} \tag{20.9}$$

Now let $\sigma(i)$ identify the hard cluster in $\hat{U}$ satisfying

$$\sum_{\mathbf{x}_k \notin \hat{u}_{\sigma(i)}}(u_{ik}^*)^m = \min_{1 \leqslant j \leqslant c}\left\{\sum_{\mathbf{x}_k \notin \hat{u}_j}(u_{ik}^*)^m\right\} \tag{20.10}$$

Then from (20.9) we have the bound

$$\mathbf{x}_k \notin \hat{u}_{\sigma(i)} \Rightarrow 0 \leqslant u_{ik}^* \leqslant \left\{\frac{2n}{[\bar{D}_1(c)]^2}\right\}^{1/m} \tag{20.11}$$

Evidently, u_i^* becomes arbitrarily small on $X - \hat{u}_{\sigma(i)}$ as $\bar{D}_1$ grows without bound. Conversely, (20.11) suggests that $u_{ik}^* \to 1$ as $\bar{D}_1$ increases for $\mathbf{x}_k \in \hat{u}_{\sigma(i)}$ if σ is a bijection of $\{1, 2, \ldots, c\}$ into itself. Using a lemma displayed in (36), we come at last to the main result.

(T20.3) *Theorem 20.3* (*Dunn*[36]). Let $0 < \delta < 1$; let $\delta:\{1, 2, \ldots, c\} \to \{1, 2, \ldots, c\}$. If, for $1 \leq i \leq c$,

$$\mathbf{x}_k \notin \hat{u}_{\sigma(i)} \Rightarrow 0 \leq u_{ik}^* \leq \delta/c \tag{20.12}$$

where $\bar{D}_1(c) = D_1(\hat{U}; c; X, d)$ solves (20.4), $\hat{U} \in M_c$, and $U^* \in M_{fc}$ satisfies (11.3a) for $m > 1$, and σ satisfies (20.10), then σ is one-to-one, and for $1 \leq i \leq c$,

$$\mathbf{x}_k \in \hat{u}_{\sigma(i)} \Rightarrow 1 - \frac{(c-1)\delta}{c} \leq u_{ik}^* \leq 1 \tag{20.13}$$

Analysis. (T20.3) shows that the membership functions $\{u_i^*\}$ generated by fuzzy c-means will approximate the characteristic functions (of some permutation) of $\hat{U} \in M_c$, the maximally separated hard c-partition of X in the sense of D_1, to an arbitrary degree of accuracy as $\bar{D}_1$ grows without bound. In particular, FCM generates arbitrarily good approximants to the unique CS clusters in X as $\bar{D}_1$ grows positively above 1, and likewise for the unique CWS clusters as $\bar{D}_1$ grows above 2. The proof hinges on σ being one-to-one and on the inequality (20.11).

Proof. If $\{u_1, u_2, \ldots, u_c\}$ is *any* family of c subsets of X, and σ is any map on $\{1, 2, \ldots, c\} \to \{1, 2, \ldots, c\}$, then[36]

$$\bigcap_{i=1}^{c} (X - u_{\sigma(i)}) = X - \bigcup_{i=1}^{c} (u_{\sigma(i)})$$

Suppose σ is not one-to-one: then for the hard clusters $\hat{u}_{\sigma(i)}$, $\bigcup_{i=1}^{c} \hat{u}_{\sigma(i)} \neq X$, so $\bigcap_{i=1}^{c} (X - \hat{u}_{\sigma(i)}) \neq \varnothing$. Let $\mathbf{x}_k \in \bigcap_{i=1}^{c}(X - \hat{u}_{\sigma(i)})$: from hypothesis (20.12) then, for this $\mathbf{x}_k$, $\sum_i u_{ik}^* < \sum_i(\delta/c) = \delta < 1$, which contradicts (5.6b), so σ is one-to-one, and for $1 \leq i \leq c$, we have

$$\bigcap_{\substack{j=1 \\ j\neq 1}}^{c} (X - \hat{u}_{\sigma(j)}) = X - \bigcup_{\substack{j=1 \\ j\neq 1}}^{c} (\hat{u}_{\sigma(j)}) = \hat{u}_{\sigma(i)}$$

Equations (5.6b) and (20.12) then imply that for $1 \leq i \leq c$; $1 \leq k \leq n$,

$$\mathbf{x}_k \in \hat{u}_{\sigma(i)} \Rightarrow 1 \geq u_{ik}^* = 1 - \sum_{\substack{j=1 \\ j\neq 1}}^{c} u_{jk}^*$$

$$\geq 1 - \frac{(c-1)\delta}{c}$$

The implication of (T20.3) is this: if $\bar{D}_1 > 1$, X contains unique CS clusters; if $\bar{D}_1 > 2$, X contains unique CWS clusters. As $\bar{D}_1 \to \infty$, fixed points of J_m (for every m) become arbitrarily good approximations of this unique hard structure. Thus, if CS or CWS clusters are an appealing type of substructure to identify, algorithm (A11.1) may provide an effective means for finding *approximate* CS or CWS clusters. To *implement* this strategy, one finds $U^* \in M_{fc}$ via (A11.1), converts U^* to $U_m \in M_c$ via (D9.2)—or any other reasonable expedient—and calculates D_1 on U. If Ω_{cd} denotes candidates $U \in M_c$ for a fixed metric d, then the $U^* \in M_{c^*}$ that solves

$$\max_{2 \leq c \leq n} \{\max_{\Omega_{cd}} \{D_1(U; c; X, d)\}\} = \max_{2 \leq c \leq n} \{\bar{D}_1(c)\} \tag{20.14}$$

is assumed to be the most valid clustering of X. If this number is greater than 1, the unique CS clusters have been found; if greater than 2, they are unique CWS clusters. Because D_1 involves only distances and diameters (no convex hulls), $\bar{D}_1$ is calculable for the few U's in each Ω_{cd}, so (20.14) is a tractable strategy. Observe that D_1 is valued in $(0, \infty)$. However, the "benchmark" problem of previous strategies is not inherent with (20.14), because D_1 identifies *unique* hard clusters in X via the known CS and CWS thresholds, 1 and 2, respectively. Whether (A11.1) has generated a good approximation to one of these unique substructures or not is validated simply by calculating D_1 for the hard c-partition suggested by an optimal fuzzy one. Further, there is no reason to restrict Ω_{cd} to candidates generated by fuzzy c-means: D_1 is well defined for hard c-partitions generated by any method whatsoever; (T20.3) does suggest, however, that (A11.1) is a reasonable way to look for CWS clusters.

(E20.1) *Example 20.1.* Figure 20.1 depicts $n = 21$ two-dimensional data points arranged in $c = 3$ visually apparent clusters. Table 20.1 lists the coordinates of each $\mathbf{x}_k \in X$. The FCM algorithm (A11.1) using the Euclidean norm, $m = 2.00$, $\varepsilon_L = 0.005$, and the max norm on cluster center components $\{v_{ij}\}$ was applied to X for $c = 2, 3, 4, 5$, and 6, using a single initialization for $U^{(0)}$. The resultant MM clusters obtained by converting $U^* \in M_{fc}$ to $U_m \in M_c$ via (D9.2) are listed in columns 3–7 of Table 20.1, labeled as clusters numbered 1 to c for each c. As is evident in Table 20.1, maximum memberships hold up well only for $c = 3$ (the lowest being 0.94

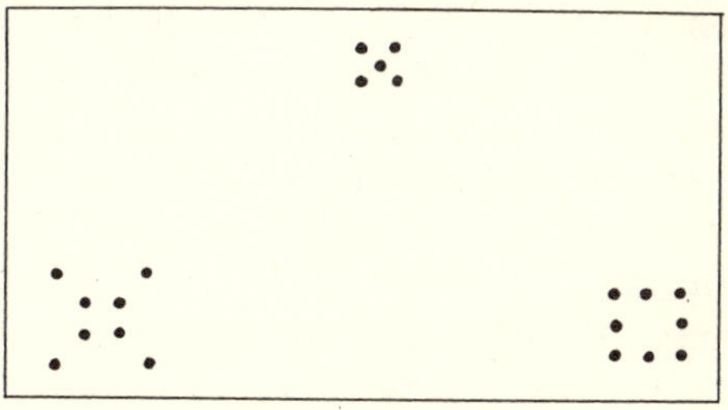

Figure 20.1. Data for (E20.1).

Table 20.1. MM Clusters for (E20.1)

Data X		MM clusters from (D9.2) applied to output of (A11.1)[a]				
k	$\mathbf{x}_k$	$c = 2$	$c = 3$	$c = 4$	$c = 5$	$c = 6$
1	(0, 0)	1	1	1	1	1
2	(0, 3)	1	1	1 [0.51]	1 [0.70]	1 [0.56]
3	(1, 1)	1	1	1	1	1
4	(1, 2)	1	1 [0.94]	1	1	1
5	(2, 1)	1	1	2	2	2
6	(2, 2)	1 [0.53]	1	2 (0.51]	2 [0.78]	3 [0.77]
7	(3, 0)	1	1	2	2	2 [0.52]
8	(3, 3)	1	1	2	2	3
9	(10, 9)	1	2	3	3	4
10	(10, 10)	1	2	3	3	4
11	(10.5, 9.5)	2	2 [0.99]	3 [0.98]	3 [0.98]	4 [0.98]
12	(11, 9)	2	2	3	3	4
13	(11, 10)	2	2	3	3	4
14	(18, 0)	2	3	4	4	5
15	(18, 1)	2	3	4	4	5
16	(18, 2)	2 [0.51]	3	4	4 [0.61]	5 [0.63]
17	(19, 0)	2	3 [0.98]	4 [0.97]	5	6
18	(19, 2)	2	3	4	4	5
19	(20, 0)	2	3	4	5 [0.61]	6 [0.63]
20	(20, 1)	2	3	4	5	6
21	(20, 2)	2	3	4	5	6

[a] Bracketed numbers are minimum maximum fuzzy memberships.

at $\mathbf{x}_8$). Hard configurations suggested by fixed points of FCM at every other c are derived from U^*'s having "very fuzzy" columns. Now, does X contain CS or CWS clusters? Fixing d for (D20.1) as the Euclidean metric, both $\bar{D}_1$ and $\bar{D}_2$ were calculated for the five hard c-partitions of X reported in Table 20.1. These values appear in Table 20.2, along with values for F, H, and P on the fuzzy U^* from which each U_m was derived.

From Table 20.2 we see that the hard 3-partition of X displayed as column 4 of Table 20.1 *is* the unique CS and CWS hard clustering of X (these follow from $\bar{D}_1 > 1$ and $\bar{D}_2 > 1$, respectively). In other words, the three visually apparent clusters of Figure 20.1 are unique CS and CWS clusters. We *cannot* infer from Table 20.2 that unique CS or CWS clusters do not exist at other values of c: we know only that the hard partitions in Table 20.1 suggested by FCM (A11.1) for $c \neq 3$ are not of these types. Observe that F and H both identify $c^* = 3$ quite strongly, but the proportion exponent P indicates $c^* = 6$! This further corroborates Windham's

Table 20.2. Validity Indicators for (E20.1)

No. of clusters	Partition coefficient F	Partition entropy H	Proportion exponent P	CS index $\bar{D}_1$	CWS index $\bar{D}_2$
c		using $U^* \in M_{fc}$		using $U_m \leftarrow U^*$	
2	0.85	0.23	62.3	0.05	0.03
3	→ 0.97	→ 0.09	179.4	→ 2.17	→ 2.17
4	0.85	0.26	225.0	0.33	0.33
5	0.75	0.43	207.0	0.33	0.21
6	0.72	0.50	→ 278.9	0.33	0.21

supposition that F and H *seem* more reliable than P if a large majority of membership values in U^* are relatively hard.

Remarks

Dunn's indices rely on a minimum amount of mathematical apparatus for their substance, and possess a number of other attractive features. They identify *unique* cluster substructure having well-defined properties that depend only on the data X and a measure of distance on $\mathbb{R}^p$. The dependence of D_1 and D_2 on d essentially relates them to the *shapes* of CS and CWS clusters. It would be of great value to know whether (and to what extent) solutions of (20.14) are sensitive to changes in d. Another important aspect of Dunn's work is that (T20.3) lends additional credibility to clustering with fuzzy c-means (A11.1). Other algorithms may generate approximations to CWS structure; at present, this is not known. Another unanswered theoretical question of practical value is whether a fixed data set X and metric d admit unique CWS clusters for more than one value of c.

Since D_1 and D_2 depend only on X and d (for fixed U and c), they are somewhat more general than Gunderson's coefficient, which has cluster centers as an additional variable. In view of the geometric similarities underlying definitions 19.1c and 20.1, G, D_1, and D_2 are probably related to each other, perhaps in the way of equivalent norms, by bounding one another above and below; this has yet to be established.

As a validity indicator, maximizing D_1 or D_2 has the same analytical status as all of the previous methods in the sense that relative partition qualities are ordered against an upper bound: the important difference, of course, is that when $\bar{D}_1$ crosses 1 or 2, a definitive statement about the corresponding structure can be made. Dunn's indices do not offer advice concerning the ambiguous null hypothesis—does X contain clusters?

Instead, D_1 and D_2 afford means for answering less ambitious, but mathematically well-defined questions—does X contain CS or CWS clusters? This is perhaps the most significant feature of the separation indices: cluster "validity" is a well-defined term.

Exercises

H20.1. Verify the inequalities in (20.2).[36]

H20.2. Determine which, if any, of the four hard 2-partitions of Fig. 11.4 are CS clusters. Use the Euclidean metric.

H20.3. Repeat H20.2 for CWS clusters.

H20.4. X any set, $\{Y_1, \ldots, Y_c\}$ any c subsets of X. If $\sigma: \{1, 2, \ldots, c\} \to \{1, 2, \ldots, c\}$, show that[36]

$$\bigcap_{i=1}^{c} (X - Y_{\sigma(i)}) = X - \bigcup_{i=1}^{c} Y_{\sigma(i)}$$

H20.5. $U^* \in M_{fc} \subset V_{cn}$ and δ as in (T20.3). Show that $F(U^*, c)$ is bounded below as follows:

$$\{1 - [\delta(c-1)/c]\}^2 \leq F(U^*; c).$$

H20.6. $X = \{0, 1, 2\} \cup \{100, 101, 102\} \cup \{200, 201, 202\} \cup \{300, 301, 302\}$ is partitioned into four hard clusters.

(i) Calculate $D_2(4)$ for this U with the Euclidean metric. (Answer: 49.)
(ii) Is $D_2(4) = \bar{D}_2(4)$? (Answer: Yes.)
(iii) Calculate Gunderson's separation coefficient for this U, using (9.8a) and the Euclidean metric. (Answer: 49/50.)
(iv) Let s be the interface distance between the four clusters of X: in (i), $s = 98$. Imagine reducing s continuously: e.g., when $X = \{0, 1, 2\} \cup \{20, 21, 22\} \cup \{40, 41, 42\} \cup \{60, 61, 62\}$, $s = 18$. What is the smallest s for which X contains CWS clusters? What is the smallest s for which $G > 0$? (Answer: $D_2 = s/2 > 1 \Leftrightarrow s > 2$; $G > 0\ \forall s > 0$.)
(v) Show that $D_1 = D_2\ \forall s$ in this example. Why is this not true in general?
(vi) Express D_2 as a function of G. Why can this not be done in general?

H20.7. Let $U \in M_c$ label X, $|X| = n$. Suppose there are two clusters u_i and u_j of X so that $\text{conv}(u_1) \cap \text{conv}(u_2) \neq \varnothing$. Can X have CS clusters? CWS clusters?

5

Modified Objective Function Algorithms

In this chapter we discuss three modifications of previous algorithms which have been proposed in an attempt to compensate for the difficulties caused by variations in cluster *shape.* The basic dilemma is that "clusters" defined by criterion functions usually take mathematical substance via metrical distances in data space. Each metric induces its own unseen but quite pervasive topological structure on $\mathbb{R}^p$ due to the geometric shape of the open balls it defines. This often forces the criterion function employing d to unwittingly favor clusters in X having this basic shape—even when none are present! In S21, we discuss a novel approach due to Backer which "inverts" several previous strategies. S22 considers an interesting modification of the FCM functional J_m due to Gustafson and Kessel,[54] which uses a different norm for each cluster! S23 and S24 discuss generalization of the fuzzy c-means algorithms (A11.1) in a different way—the prototypes $\{\mathbf{v}_i\}$ for $J_m(U, \mathbf{v})$ become r-dimensional linear varieties in $\mathbb{R}^p$, $0 \leqslant r \leqslant p-1$.

S21. Affinity Decomposition: An Induced Fuzzy Partitioning Approach

This section describes one of a number of innovative approaches to fuzzy partitioning due to Backer,[4] who has developed several families of clustering algorithms which combine some of our previous ideas in a rather novel way. The general premise of Backer's approach is this: if one accepts the basic heuristic of S15–S17, viz., that "good" fuzzy clusters are in fact not very fuzzy (as indicated by a validity functional such as F, H, or P), why not use a validity functional itself as the objective function? In other words, the two-stage process developed in Chapters 3 and 4, wherein objective functionals, say $J(U)$, are used to generate optimality candidates $U \in M_{fc}$, followed by optimization of a validity functional, say $V(U; c)$, to choose the best of the optimal U's identified by J; is replaced by a one-stage strategy, wherein $V(U; c)$ is both the objective function controlling the algorithm and the validity functional evaluating its output [recall, however, that

(E14.1) illustrates the possibility that objective functions are not necessarily good validity indicators per se]. Moreover, Backer's algorithms simultaneously generate a coupled *pair* of optimal hard and (induced) fuzzy c-partitions of X, so that conversions via expedients such as (D9.2) are unnecessary. Finally, a number of interesting options at various stages are available, so that many kinds of data sets may be amenable to one of Backer's algorithms.

Given $X = \{\mathbf{x}_1, \mathbf{x}_2, \ldots, \mathbf{x}_n\} \subset \mathbb{R}^p$, the point of departure for Backer's approach is an initial *hard* c-partition $U \in M_c$ of X. Letting u_i denote the ith hard subset of U, we put $n_i = |u_i|$ and call $p_i = n_i/n$ the relative size of u_i for $1 \leq i \leq c$. Using this hard c-partition, a $c \times n$ matrix of *affinity decompositions* is generated: we let $A(U) = [a_{ik}(U)]$ denote this matrix. Element $a_{ik}(U)$ is the affinity data point $\mathbf{x}_k$ has for hard subset u_i of U—not necessarily the membership of $\mathbf{x}_k$ in u_i. Moreover, each $\mathbf{x}_k$ is assigned a point to set affinity $A_k(U)$ between itself and X. Backer requires that

$$a_{ik}(U) \geq 0 \qquad \forall i, k \tag{21.1a}$$

and

$$A_k(U) = \sum_{i=1}^{c} p_i a_{ik}(U) \qquad 1 \leq k \leq n \tag{21.1b}$$

Three systems of affinity assignment are delineated in (4): these are based on the concepts of distance, neighborhood, and probability. Once $U \in M_c$ and $A(U)$ are chosen, a fuzzy c-partition of X is created via the following definition.[4]

(D21.1) *Definition 21.1* (*Induced Fuzzy c-Partition*). Let $U \in M_c$ be a hard c-partition of X, $|X| = n$; $|u_i| = n_i$, $p_i = n_i/n$ $\forall i$. The fuzzy c-partition U^* in M_{fc} induced by U and an affinity matrix $[a_{ik}(U)]$ satisfying (21.1) has elements defined $\forall i, k$, by

$$u_{ik}^* = \frac{p_i a_{ik}(U)}{A_k(U)} = \frac{p_i a_{ik}(U)}{\sum_{j=1}^{c} p_j a_{jk}(U)} \tag{21.2}$$

This equation has the functional form of Bayes' rule. Note that a slightly stronger requirement than (21.1a) is necessary to ensure that (21.2) is well defined—namely, that $\forall k$, $\exists i$ so that $a_{ik}(U) > 0$; failing this, $A_k(U)$ could be zero for some k. The membership functions $\{u_i^*\}$ defined by (21.2) take their particular form from the affinity matrix $A(U)$. For our purposes it suffices to exemplify the general idea with a specific case. Let $d_{kj} = \|\mathbf{x}_k - \mathbf{x}_j\|$ be the Euclidean distance between $\mathbf{x}_k$ and $\mathbf{x}_j$, and define the affinity for $\mathbf{x}_k$ relative to u_i as

$$a_{ik}(U) = 1 - \left(\frac{\beta}{n_i} \sum_{t=1}^{n} u_{it} d_{tk}\right) \tag{21.3}$$

where β is some multiple of the maximum distance among the $\{d_{tk}\}$. For this choice of $A(U)$, the memberships induced by (21.2) are

$$u_{ik}^* = \frac{n_i - \beta(\sum_{t=1}^n u_{it}d_{tk})}{n - \beta(\sum_{t=1}^n d_{tk})} \tag{21.4}$$

Having obtained an induced fuzzy c-partition $U^* \in M_{fc}$, the next step is to assess the amount of uncertainty it represents. Backer considers several functionals on $M_{fc} \to \mathbb{R}$ based on one of three ideas: amount of (pairwise) fuzzy cluster overlap; average (pairwise) inter-fuzzy-cluster distance; and average fuzzy set structure. The measure defined below is the one used in (E21.1); see (4) for others.

(D21.2) *Definition 21.2 (Average Partition Separability).* Let $U \in M_{fc}$ be a fuzzy c-partition of X, $|X| = n$. The average (pairwise) separability (of the fuzzy clusters in) U is, for $2 \leq c < n$,

$$B(U;c) = 1 - \frac{c}{(c-1)}[1 - F(U;c)] \tag{21.5a}$$

where $F(U;c)$ is the partition coefficient of (D15.2).

The terminology in (21.5) arises from the functional form of B, which can be written as

$$B(U;c) = \frac{1}{(c-1)} \sum_{j=i+1}^{c} \sum_{i=1}^{c-1} \left[\frac{1}{n} \sum_{k=1}^{n} (u_{ik} - u_{jk})^2\right] \tag{21.5b}$$

This emphasizes that B assesses the fuzziness of U by summing the variances between rows of U. From another point of view, B is simply a normalization of F, which was related to the total content of pairwise fuzzy intersections in U in S15. Theorem 15.1 yields the following theorem.

(T21.1) *Theorem 21.1.* Let $U \in M_{fc}$ be a fuzzy c-partition of n data points. For $2 \leq c < n$,

$$0 \leq B(U;c) \leq 1 \tag{21.6a}$$

$$B(U;c) = 1 \Leftrightarrow U \in M_{co} \text{ is hard} \tag{21.6b}$$

$$B(U;c) = 0 \Leftrightarrow U = [1/c] \tag{21.6c}$$

Analysis and Proof. Results (21.6) follow immediately from the corresponding parts of equations (15.6). B simply transforms the range of F from $[1/c, 1]$ to $[0, 1]$. •

B can be viewed as a measure of cluster validity; in view of (21.4), its behavior in this respect will be quite analogous to that of F. The novelty of

Backer's approach is to regard B [and other functions discussed in (4)] as an *objective function* for Picard iteration through affinity decompositions of hard c-partitions of X. The next step in the computational strategy is to reclassify each $\mathbf{x}_k \in X$. To accomplish this, the induced $U^* \in M_{fc}$ is used to define a $c \times n$ reclassification matrix $R = [r_{ij}]$, the columns of which are calculated as follows: for each $\mathbf{x}_k \in u_i$ (recall that hard $U \in M_c$ *induces* fuzzy $U^* \in M_{fc}$), let

$$ {}_k u_i^- = u_i - \{\mathbf{x}_k\} \tag{21.7a} $$

$$ {}_k u_j^+ = u_j \cup \{\mathbf{x}_k\}, \qquad 1 \leq j \leq c; j \neq i \tag{21.7b} $$

The idea here is that u_i and u_j are old hard clusters; ${}_k u_i^-$ and ${}_k u_j^+$ are (parts of) new hard clusters formed by transferring $\mathbf{x}_k$ from u_i to some $u_j \neq u_i$. Transfer decisions or membership reallocation is based on the columns of R: for each k, $1 \leq k \leq n$, define

$$ r_{ik} = \frac{1}{n_i}\left[(u_{ik}^*)^2 - \frac{1}{n_i - 1} \sum_{\mathbf{x}_t \in {}_k u_i^-} (u_{it}^*)^2\right] \tag{21.8a} $$

$$ r_{jk} = \frac{1}{n_j}\left[(u_{jk}^*)^2 - \frac{1}{n_j - 1} \sum_{\mathbf{x}_t \in {}_k u_j^+} (u_{jk}^*)^2\right], \qquad 1 \leq j \leq c; j \neq i \tag{21.8b} $$

The necessity of using these particular reclassification functions is established in (4), where it is shown that the following decision rule iteratively optimizes B.

(T21.2) *Theorem 21.2* (*Backer*[4]). Let $U \in M_c$ be a hard c-partition of X, $|X| = n$. Let $U^* \in M_{fc}$ be induced via (21.4); and let $R \in V_{cn}$ be constructed from (U, U^*) via (21.8). Let r_k denote the maximum entry in the kth column of R:

$$ r_k = \bigvee_{j=1}^{c} r_{jk}, \qquad 1 \leq k \leq n \tag{21.9} $$

If $r_k = r_{ik}$ with $\mathbf{x}_k \in u_i$ for $1 \leq k \leq n$, then U^* is a constrained stationary point of B. Otherwise, there is at least one k so that $\mathbf{x}_k \in u_i$ and $r_{ik} < r_k$.

Analysis and Proof. This result is a special case of Ruspini's theorem (T9.1) for the particular choice $J(U) = B(U; c)$: interested readers are referred to (4) for the details. The result itself yields the reclassification rule required for iterative optimization of B: if $r_{ik} < r_k$, transfer $\mathbf{x}_k$ to u_j, where $r_{jk} = r_k$. •

Now we are ready to state Backer's algorithm. We emphasize again that this is but one of a number of similar algorithms analyzed in (4) which are realized by exercising various options within the basic loop.

(A21.1) *Algorithm 21.1* (*Backer*[4])

(A21.1a) Fix $c, 2 \leq c < n$, and a dissimilarity measure d. Let $d_{tk} = d(\mathbf{x}_t, \mathbf{x}_k)$ be the dissimilarity between $\mathbf{x}_t$ and $\mathbf{x}_k$. Calculate $\beta = \max_{t,k}\{d_{tk}\}$. Initialize $U^{(0)} \in M_c$. Then at step $l, l = 0, 1, \ldots$:

(A21.1b) Use the hard c-partition $U^{(l)}$ to induce the fuzzy c-partition $(U^*)^{(l)}$ via (21.4). Calculate $B((U^*)^{(l)}; c)$.

(A21.1c) Construct the transfer matrix $R^{(l)}$ from $U^{(l)}$ and $(U^*)^{(l)}$ using (21.8). For $k = 1, \ldots, n$, update $U^{(l)}$ as follows:

(i) for i so that $u_{ik}^{(l)} = 1$, if $r_k = r_{ik}$, then set $u_{jk}^{(l+1)} = u_{jk}^{(l)}$; $1 \leq j \leq c$

(ii) for i so that $u_{ik}^{(l)} = 1$, if $r_k = r_{ik} > r_{tk}$, then set

$$u_{jk}^{(l+1)} = \begin{cases} 1, & j = t \\ 0, & 1 \leq j \leq c; j \neq t \end{cases}$$

(A21.1d) If $U^{(l+1)} = U^{(l)}$, Stop. Otherwise, set $l = l + 1$ and return to (A21.1b).

The basis for updating in (A21.1c) is (T21.2), which implies that the *next* induction $[(U^*)^{(l+1)}$ from $U^{(l+1)}]$ will improve (i.e., increase) B towards its absolute maximum. The stopping criterion in (A21.1d) implies that Backer's algorithm always terminates at a fixed $\hat{U} \in M_c$ after finitely many steps. Backer has shown that this is indeed the case. In fact, algorithm (A21.1) does *not* search over all of M_{fc}, because there are exactly as many fuzzy $(U^*)^{(l)}$'s to consider as hard $U^{(l)}$'s to generate them from—namely, $|M_c|$, the number given at (6.1). Thus, for a given data set (or matrix of dissimilarities $\{d_{kj}\}$ derived therefrom), there is a pair $(\hat{U}, \hat{U}^*) \in (M_c \times M_{fc})$ so that $\hat{U}$ induces $\hat{U}^*$ via affinity decomposition, and $B(\hat{U}^*; c) \geq B(U^*; c) \; \forall U^*$ induced by $U \in M_c$. Backer proves in (4) that (A21.1) generates a sequence $\{(U^{(l)}, (U^*)^{(l)}\}$ which has the ascent property relative to B [i.e., $B((U^*)^{(l+1)}; c) \geq B((U^*)^{(l)}; c) \; \forall l$]: and since every such sequence is finite, (A21.1) must ultimately terminate at $\hat{U}^*$. In other words, since $(U^*)^{(l)}$ is uniquely defined via a specified type of affinity decomposition, (A21.1) is guaranteed to terminate at a global solution of the constrained optimization problem

$$\max_{U^* \in M_{fc}^* \subset M_{fc}} \{B(U^*, c)\} \tag{21.10}$$

where M_{fc}^* is the image of M_c under the one-to-one mapping $U \to U^*$ given by a particular membership assignment rule derived from a specified affinity decomposition. Therefore, the only effect of altering $U^{(0)}$, the initialization

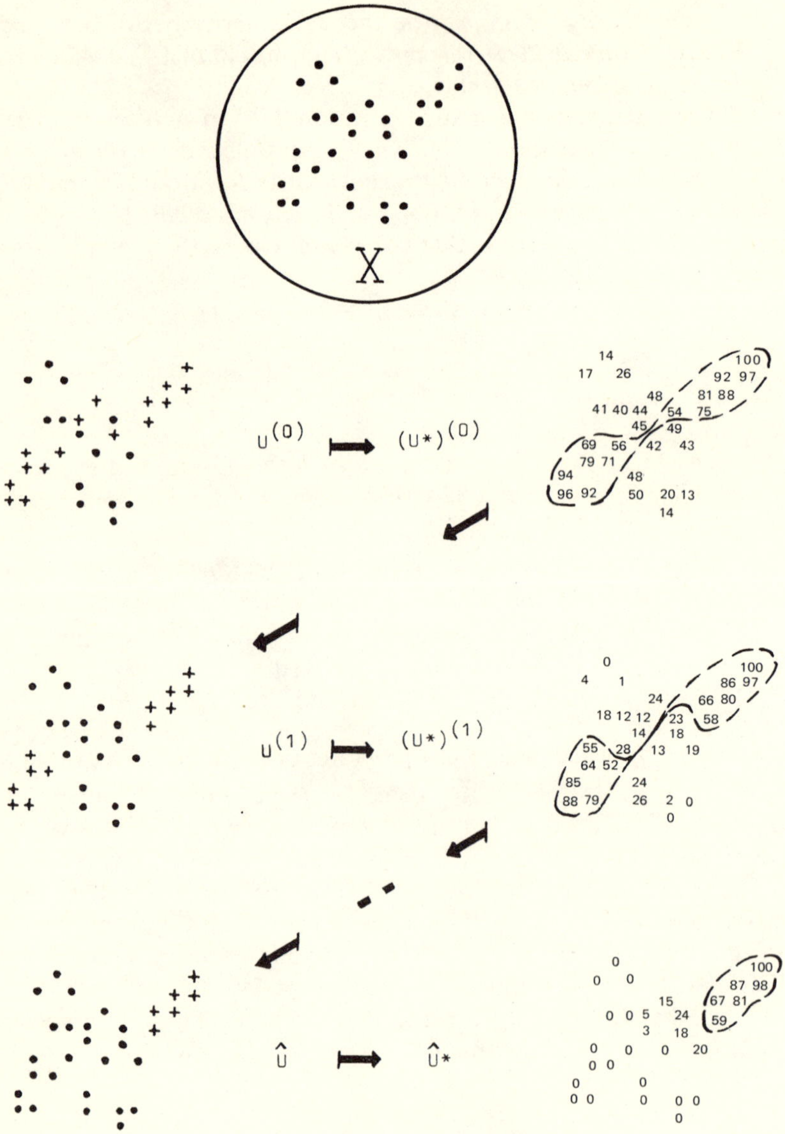

Figure 21.1. The operation of Backer's algorithm: (E21.1).

in (A21.1a), is to alter the number of steps it takes to reach $\hat{U}^*$. This is illustrated in the following theorem.

(E21.1) *Example 21.1* (*Backer*[4]). The inset of Fig. 21.1 shows $n = 30$ two-dimensional data vectors. There is no particular substructure evident in X: Backer's intent was simply to illustrate the operation of (A21.1). Choosing $c = 2$, d the Euclidean norm metric, and $\beta = \max_{k,j}\{d_{kj}\}/5$, (A21.1) was initialized by the $U^{(0)}$ shown graphically in Fig. 21.1, where $u_1^{(0)}$ has $n_1^{(0)} = 16$ (the +'s); and $u_2^{(0)}$ has $n_2^{(0)} = 14$ (the •'s). The fuzzy c-partition $(U^*)^{(0)}$ induced via (21.4) has membership functions as shown (without leading decimals), with $B((U^*)^{(0)}; 2) = 0.318$. Reallocation of hard memberships via (A21.1c) leads to the $U^{(1)}$ shown: four $\mathbf{x}_k$'s have been transferred from (+) to (•). $(U^*)^{(1)}$ induced by $U^{(1)}$ is also depicted, with $B((U^*)^{(1)}; 2) = 0.504$. The last two illustrations illustrate $(\hat{U}, \hat{U}^*)$, which were achieved at $l = 2$ (i.e., $U^{(2)} = U^{(3)} = U^{(l)}\ \forall l \geq 3$); only six points remain in u_1, and the solution of (21.10) is $B(\hat{U}^*; 2) = 0.802$, the global maximum of B over M_{fc}^* for this X, d, β, and $A(U)$.

Because of the nature of (A21.1), we know that (A21.1) will converge to the same $(\hat{U}, \hat{U}^*)$ regardless of $U^{(0)}$. Backer reported convergence to $(\hat{U}, \hat{U}^*)$ in one iteration using an initial guess which (roughly) divided X by a diagonal line from left to right. Since $|M_2| = 2^{29} - 1$, it is not impossible (but perhaps improbable!) that for some undistinguished $U^{(0)}$, (A21.1) could iterate for quite a while before arriving at the optimal solution.

It is clear from Fig. 21.1 that $U^{(l+1)}$ is *not* the MM matrix $U_m^{(l)}$ derived from $(U^*)^{(l)}$ and (D9.2); ostensibly at least, the maximum membership matrix has no direct connection with iterates of (A21.1). On the other hand, note that $\hat{U}_m^{(l)}$ derived from $(\hat{U}^*)^{(l)}$ *is* $\hat{U}^{(l+1)}$; one is tempted to conjecture that (A21.1) terminates precisely when applying (D9.2) to $(U^*)^{(l)}$ results in the $U^{(l+1)}$ calculated at (A21.1c). •

The optimal c-partition of X shown in Fig. 21.1 is substantially different from those one could obtain using a FCM algorithm (A11.1), because at any m with $c = 2$, the cluster centers $\mathbf{v}_1$ and $\mathbf{v}_2$ (which provide the location of zones of maximum membership for u_1^* and u_2^*) would tend towards the geometric center of X. This is illustrated in the following example.

Example 21.2. The data set X was processed using FCM algorithm (A11.1) with the Euclidean norm, $c = 2$, $\varepsilon_L = 0.005$, and the max norm applied to components of the centers $\{v_{ij}\}$ as a cutoff criterion. The weighting exponent was $m = 2.00$, and only one initialization for $U^{(0)}$ was used. Fuzzy 2-means terminated at the partition $\hat{U}$ shown in Fig. 21.2 after 15 iterations, with $B(\hat{U}; 2) = 0.446$ [or $F(\hat{U}; c) = 0.723$]. The results are quite predictable! The Euclidean norm favors circular zones of equimembership which

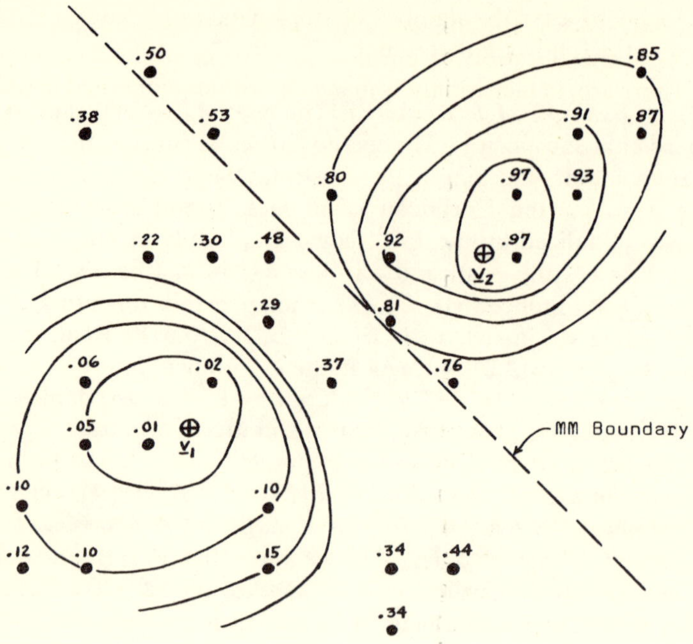

Figure 21.2. A FCM partitioning of Backer's data: (E21.2).

maximize at the cluster centers $\mathbf{v}_i = (2.60, 3.27)$ and $\mathbf{v}_2 = (7.12, 6.49)$. The essential shapes favored are clearly evident by looking at plots of several level curves of each membership function, as depicted in Fig. 21.2. In contradistinction to Backer's algorithm, the MM hard 2-partition derived from this solution divides X along the diagonal boundary indicated in Fig. 21.2, which has $n_1 = 18$, or $n_2 = 12$ points, respectively, as opposed to Backer's solution, wherein $n_1 = 24$, or $n_2 = 6$. Note further that the highest memberships in the terminal partition of (E21.1) occur at the boundaries of conv(X), whereas fuzzy 2-means memberships peak near $\mathbf{v}_1$ and $\mathbf{v}_2$, well into the interior of conv(X). Thus, (A11.1) using the Euclidean metric favors—as it always will—membership surfaces in $\mathbb{R}^{p+1}$ that would look like (the graph of) a mixture of c p-variate Gaussian probability density functions. (A21.1) can, on the other hand, generate membership surfaces that peak near the "edges" of the data: in this respect, the performance of Backer's algorithm in (E21.1) is similar to Ruspini's (A9.1), (E9.1), where the edges of the butterfly receive the highest memberships. •

We close this section with some general observations concerning (A21.1). First, it differs from previous ones in that the optimal fuzzy c-partition is *always* associated with a unique hard U via (A21.1c). Another important difference is the convergence aspect discussed above. Not dealt with so far is the iteration of problem (21.10) over the parameters of (A21.1). There are, in fact, infinitely many algorithms embodied in (A21.1), because one may take $\beta = m(\max_{k,j}\{d_{kj}\})$; $0 < m < \infty$. The threshold β makes its presence felt at (21.4) in the calculation of the induced u_{ik}^*'s, and, in this respect, has the same algorithmic status as does m in (A11.1). Moreover, each choice for $A(U)$ and d leads to a different M_{fc}^*; and finally, one may iterate c itself. Letting Ω_c again denote all of the fuzzy optimality candidates one cares to obtain by variation of these parameters in (A21.1), we may presume that if $(\hat{U}^*; c^*)$ solves

$$\max_{2 \leq c < n} \{ \max_{U^* \in \Omega_c} \{B(U^*; c)\}\} \tag{21.11}$$

then $\hat{U}^*$ is the most attractive (in the sense of B) fuzzy c^*-partition of X. Here, of course, there will be a corresponding *hard* c^*-partition $\hat{U}$ via (A21.1c)—in all probability, the MM membership matrix obtained from $\hat{U}^*$ via (D9.2). Note that (21.11) does not concern itself with the null hypothesis (that X does not *have* clustered substructure: $c^* = 1$ or $c^* = n$). In this respect, (21.11) is quite analogous to (and thus suffers the disadvantages of) the validity schemes of Sections 15, 16, and 17. In fact, solutions of (21.11) and (15.13) are almost certainly identical if Ω_c is the same in both cases: the real difference, of course, is that B is used in (A21.1) to generate "its own" Ω_c for (21.11), while F, H, and P "borrow" Ω_c's from other algorithms [possibly including (A21.1)!].

The actual property of optimal $\hat{U}^*$'s generated by (A21.1) is obscured by the equations defining its steps. Equations (21.8), which determine the transfer matrix, depend on B; other equations would result if one used an objective other than B. Specifically, one wonders how (A21.1) would perform if modified to minimize $H(U; c)$ or maximize $P(U; c)$. It seems clear that Backer's algorithms, like those of Ruspini,[(89)] may enjoy greater success than those of previous sections when clusters are not essentially hyperellipsoidal in shape; determination of geometric properties identified by his criteria would greatly enhance their attractiveness in this regard.

Remarks

(A21.1) and variations thereof are discussed and exemplified at length in Backer's thesis, (4). Several real data sets are discussed, including the Iris data (E15.4), a set of speech data, a set of digital image data, and a randomly generated mixture of two bivariate normal distributions. It seems safe to assert that Backer's algorithms generate substantially different interpretations of data substructure than those of previous sections. Moreover,

Backer[5] has recently developed another class of fuzzy clustering algorithms based on fuzzy relations. Finally, a recent paper by Backer and Jain[6] attempts to compare the utility of various *hard* clustering algorithms using $B(U; c)$ and the induced fuzzy partition approach. This interesting idea is in some sense antithetical to Sections 19 and 20; coefficients G and D—measures of hard cluster validity—were nonetheless used to assess the quality of fuzzy c-partitions of X; here B—a measure of fuzziness—is used to rank the attractiveness of hard clusters in X.

Exercises

Let $X = \{x_1, \ldots, x_n\}$, let $c_A: X \to \{0, 1\}$ be the characteristic function of any hard subset $A \subset X$, and let $u: X \to [0, 1]$ be a fuzzy subset of X. A function $i: [0, 1] \to \mathbb{R}$ is a *measure of fuzziness* [see (4)] if

(i) $i(0) = i(1) = 0$

(ii) $i(\lambda) = i(1 - \lambda)\ \forall \lambda \in [0, 1/2]$

(iii) $0 \leqslant x \leqslant y \leqslant 1/2 \Rightarrow i(x) \leqslant i(y)$

Given any measure of fuzziness, the amount of fuzziness in u relative to i is

(iv) $I(u; i) = \sum_{k=1}^{n} i(u(x_k))/n$

H21.1. Prove that $I(u; i) = 0 \Leftrightarrow u = c_A$ for some $A \subset X$.

H21.2. Prove that $I(u; i) \leqslant I(u_{0.5}; i)$, where $u_{0.5}(x_k) = 0.5\ \forall k$; with equality $\Leftrightarrow u = u_{0.5}$.

H21.3. u^* is a sharpening of u in case: $u^* \geqslant u\ \forall k$ so that $u(x_k) > (1/2)$; $u^* \leqslant u\ \forall k$ so that $u(x_k) < (1/2)$; and $u^* = u$ whenever $u(x_k) = (1/2)$. Prove that $I(u^*; i) \leqslant I(u; i)$ whenever u^* sharpens u.

H21.4. Prove that $I(u; i) = I(\tilde{u}; i)$.

H21.5. Let u, w be fuzzy subsets of X. Show that

$$I(u \wedge w; i) + I(u \vee w; i) = I(u; i) + I(w; i)$$

H21.6. Let

$$i_h(x) = \begin{Bmatrix} -x \log x - (1 - x)\log(1 - x), & 0 < x < 1 \\ 0, & x = 0, 1 \end{Bmatrix}$$

(i) Show that i_h is a measure of fuzziness.

(ii) Show that $I(u; i_h) = h_f(u, \tilde{u})$, the normalized entropy of DeLuca and Termini[32] at (16.3).

(iii) Show that $\sum_{j=1}^{c} I(u_j; i_h) = H(U; c)$, the partition entropy of $U \in M_{fc} \subset V_{cn}$ defined at (16.5).

H21.7. Let $A_{0.5} = \{x_k \in X | u(x_k) \geq 0.5\}$, with $c_{0.5}$ its characteristic function, and let $i_d(x, u) = i_d(x)$. Show that

$$i_d(x) = |u(x) - c_{0.5}(x)|$$

is a measure of fuzziness.

Using H21.7, one can interpret $I(u, i_d)$ as the amount of uncertainty that arises upon converting fuzzy set u to its "nearest" hard set $A_{0.5}$ via maximum memberships. This leads to a measure of the "average information" lost when $U \in M_{fc}$ is converted via (D9.2) to $U_m \in M_c$, namely, $\sum_{j=1}^{c} I(u_j, i_d)/c$.

H21.8. For each row of $U \in M_{fc} \subset V_{cn}$, show that

$$I(u_j; i_d) = \frac{1}{n} \sum_{k=1}^{n} [u_{jk} \wedge (1 - u_{jk})]$$

H21.9 For $U \in M_{fc} \subset V_{cn}$, show that

(i) $$I\left(\bigwedge_{j=1}^{c} u_j; i_d\right) = \frac{1}{n} \sum_{k=1}^{n} \left(\bigwedge_{j=1}^{c} u_{jk}\right)$$

(ii) $$I\left(\bigvee_{j=1}^{c} u_j; i_d\right) = \frac{1}{n} \sum_{k=1}^{n} \left\{\bigvee_{j=1}^{c} u_j \wedge \left[\bigwedge_{s=1}^{c} (1 - u_{sk})\right]\right\}$$

(iii) $$I\left(\bigwedge_{j=1}^{c} u_j; i_d\right) \leq I\left(\bigvee_{j=1}^{c} u_j; i_d\right)$$

(iv) $$\bigvee_{j=1}^{c} I(u_j; i_d) \leq I\left(\bigvee_{j=1}^{c} u_j; i_d\right)$$

(v) $$\bigwedge_{j=1}^{c} I(u_j; i_d) \geq I\left(\bigwedge_{j=1}^{c} u_j; i_d\right)$$

H21.10. Verify that the two forms of $B(U; c)$ shown as (21.5) are equivalent.

H.21.11. Calculate $B(U; 2)$ and $B(V; 2)$ with U, V in H15.5. Compare with H15.5, H16.2, H17.1, and H18.1.

S22. Shape Descriptions with Fuzzy Covariance Matrices

Gustafson and Kessel[54] proposed an interesting modification of the fuzzy c-means algorithms of S11 which attempts to recognize the fact that different clusters *in the same data set* X may have differing geometric shapes (cf. Fig. 8.1f). Since the norm of (11.1d) controls the basic shape of all c clusters identified with $J_m(U, \mathbf{v})$ via the topological structure of open sets in the norm metric it induces [as in (E21.2) above], perhaps *local* variation of the norm in (11.1d) would allow a modified objective function to identify clusters of various shapes which are *locally* compatible with different topological structures in the same data set. Mathematical realization of this idea in (54) is accomplished by considering the class of inner product norms induced on $\mathbb{R}^p$ by symmetric, positive-definite matrices in V_{pp}. Let us denote by $\mathbf{A}$ a c-tuple of such matrices, $\mathbf{A} = (A_1, A_2, \ldots, A_c)$, and, as in (11.5), let the weighted inner product induced on $\mathbb{R}^p$ by A_i be $\langle \mathbf{x}, \mathbf{x} \rangle_{A_i} = \|\mathbf{x}\|_{A_i}^2 = \mathbf{x}^T A_i \mathbf{x}$: the distance between $\mathbf{x}, \mathbf{y} \in \mathbb{R}^p$ in this weighted norm is $\|\mathbf{x} - \mathbf{y}\|_{A_i}$. With these ideas fixed, we extend the argument list of J_m in (11.1a) to include the vector of matrices $\mathbf{A}$ as follows: define, for $X \subset \mathbb{R}^p$, $|X| = n$:

$$\hat{J}_m(U, \mathbf{v}, \mathbf{A}) = \sum_{k=1}^{n} \sum_{i=1}^{c} (u_{ik})^m \|\mathbf{x}_k - \mathbf{v}_i\|_{A_i}^2 \tag{22.1}$$

where U, $\mathbf{v}$, m satisfy (11.1b), (11.1c), and (11.1e), respectively. If $A_i = A$ is constant $\forall i$, then $\hat{J}_m(U, \mathbf{v}, \mathbf{A}) = J_m(U, \mathbf{v}, A)$, the functional of (11.6) for the particular norm induced by A on $\mathbb{R}^p$ via (11.5). In this sense, $\hat{J}_m$ generalizes J_m: on the other hand, J_m at (11.1a) is well defined, and (A11.1) applies, for *any* inner product norm—not just those induced by some A, so J_m is more general than $\hat{J}_m$ in this way.

The clustering criterion employed by $\hat{J}_m$ is exactly the same as that of J_m. The basic difference between $\hat{J}_m$ and J_m is that *all* the distances $\{d_{ik}\}$ in (11.1a) are measured by a prespecified norm; whereas (possibly) c different norm metrics—one for each $u_i \in U$—are being *sought* for the functional $\hat{J}_m$. The optimization problem suggested by (22.1) is formalized by letting PD denote the set of symmetric, positive definite matrices in V_{pp}, and PD^c its c-fold Cartesian product. Then solutions of

$$\underset{M_{fc} \times \mathbb{R}^{cp} \times \mathrm{PD}^c}{\text{minimize}} \{\hat{J}_m(U, \mathbf{v}, \mathbf{A})\} \tag{22.2}$$

are least-squared-error stationary points of $\hat{J}_m$. An infinite family of clustering algorithms—one for each $m \in (1, \infty)$—is obtained via necessary conditions for solutions of (22.2). A little thought about $\hat{J}_m$ should make it clear that (T11.1) applies to (22.2) in its entirety, with the obvious exception that $d_{ik} = \|\mathbf{x}_k - \mathbf{v}_i\|$ must be replaced by $d_{ik} = \|\mathbf{x}_k - \mathbf{v}_i\|_{A_i}$ wherever it appears. In other words, for fixed $\mathbf{A}$ in (22.2), conditions necessary for minimization of $\hat{J}_m(U, \mathbf{v}, (\cdot))$ coincide with those necessary for $J_m(U, \mathbf{v})$. All that remains to be done is fix $(U, \mathbf{v}) \in M_{fc} \times \mathbb{R}^{cp}$, and derive the necessary form each $A_i \in \mathrm{PD}$ must have at local minima of $\hat{J}_m$.

To render minimization of $\hat{J}_m$ with respect to $\mathbf{A}$ tractable, each A_j is constrained by requiring the determinant of A_j, $\det(A_j)$, to be fixed. Specification of $\det(A_j) = \rho_j > 0$ for each $j = 1$ to c amounts to constraining the volume of cluster u_j along the jth axis. Allowing A_j to vary while keeping its determinant fixed thus corresponds to seeking an optimal cluster *shape* fitting the $\mathbf{x}_k$'s to a fixed volume for each u_j. The extension of (A11.1) proposed by Gustafson and Kessel is based on the results of the following theorem.

(T22.1) *Theorem 22.1* (*Gustafson and Kessel*[54]). Let $\eta : \text{PD}^c \to \mathbb{R}$, $\eta(\mathbf{A}) = \hat{J}_m(U, \mathbf{v}, \mathbf{A})$, where $(U, \mathbf{v}) \in M_{fc} \times \mathbb{R}^{cp}$ are fixed and satisfy equations (11.3) under the hypotheses of (T11.1). If $m > 1$ and for each j, $\det(A_j) = \rho_j$ is fixed, then $\mathbf{A}^*$ is a local minimum of η only if

$$A_j^* = [\rho_j \det(S_{fj})]^{(1/p)}(S_{fj}^{-1}), \qquad 1 \leq j \leq c \tag{22.3}$$

where

$$S_{fj} = \sum_{k=1}^{n} (u_{jk})^m (\mathbf{x}_k - \mathbf{v}_j)(\mathbf{x}_k - \mathbf{v}_j)^T$$

is the fuzzy scatter matrix of u_j appearing in (D11.3).

Analysis and Proof. The proof of (22.3) is a straightforward application of the Lagrange multiplier method. Note that (22.3) does *not* provide, with equations (11.3), a necessary condition for local solutions of (22.2), because $\det(A_j) = \rho_j = \text{const}$ reduces the domain of η to a subset of PD^c. It does, however, within the volume constraints imposed by these requirements, allow one to search for a set of c norm-inducing matrices that attempt to correlate differing cluster shapes to criterion $\hat{J}_m$ via Picard iteration. To prove (22.3), we form the Lagrangian of η, say

$$F(\boldsymbol{\beta}, \mathbf{A}) = \sum_{k=1}^{n} \sum_{i=1}^{c} (u_{ik})^m (\mathbf{x}_k - \mathbf{v}_i)^T A_i (\mathbf{x}_k - \mathbf{v}_i) - \sum_{i=1}^{c} \beta_i[\det(A_i - \rho_i)]$$

and put its gradient in all variables equal to the appropriate zero vector. At $(\boldsymbol{\beta}^*, \mathbf{A}^*)$, it is necessary that

$$\text{(A)} \quad \nabla_{\boldsymbol{\beta}} F(\boldsymbol{\beta}^*, \mathbf{A}^*) = \{-[\det(A_1^*) - \rho_1], \ldots, -[\det(A_c^*) - \rho_c]\} = \boldsymbol{\theta} \in \mathbb{R}^c$$

For A_j^*, we use the facts that $\nabla_{A_j}(\mathbf{x}^T A_j \mathbf{x}) = \mathbf{x}\mathbf{x}^T$; and $\nabla_{A_j}[\det(A_j)] = \det(A_j) \cdot A^{-1}$, to find

$$\begin{aligned}\text{(B)} \quad \nabla_{A_j} F(\boldsymbol{\beta}^*, \mathbf{A}^*) &= \sum_{k=1}^{n} (u_{ik})^m (\mathbf{x}_k - \mathbf{v}_j)(\mathbf{x}_k - \mathbf{v}_j)^T - \beta_j^* \det(A_j^*)(A_j^*)^{-1} \\ &= S_{fj} - \beta_j^*[\det(A_j^*)](A_j^*)^{-1} = \boldsymbol{\theta} \in V_{pp}\end{aligned}$$

(B) and the fact that $\det A_j^* = \rho_j \; \forall j$ yield

$$\text{(C)} \qquad S_{fj} = \beta_j^* \rho_j (A_j^*)^{-1}$$

so

$$\beta_j^* I = (\rho_j)^{-1}(S_{fj} A_j^*)$$

where I is the $p \times p$ identity matrix. Taking determinants of this last equation yields

$$(\beta_j^*)^p = (\rho_j)^{-p}\rho_j \det S_{fj}$$

thus

$$\beta_j^* = (\rho_j)^{-1}(\rho_j \det S_{fj})^{(1/p)}$$

Substitution of this into (C) now yields

$$S_{fj} = (\rho_j)^{-1}(\rho_j)(\rho_j \det S_{fj})^{(1/p)}(A_j^*)^{-1}$$

whence

$$A_j^* = (\rho_j \det S_{fj})^{(1/p)}(S_{fj}^{-1})$$

Note that $\det(A_j^*) = \rho_j$, and that $A_j^* \in \text{PD}$, as required. •

Using (22.3), fuzzy c-means algorithm (A11.1) is augmented as follows.

(A22.1) *Algorithm 22.1* (*Gustafson and Kessel*[54])

(A22.1a) Fix c, $2 \leq c < n$, fix $m \in (1, \infty)$, and fix the c volume constraints $\rho_j \in (0, \infty)$, $1 \leq j \leq c$. Initialize $U^{(0)} \in M_{fc}$. Then at step l, $l = 0, 1, 2, \ldots$:

(A22.1b) Calculate the c fuzzy cluster centers $\{\mathbf{v}_i^{(l)}\}$ with (11.3b) and $U^{(l)}$.

(A22.1c) Calculate the c fuzzy scatter matrices $\{S_{fi}^{(l)}\}$ with (11.8b) and $(U^{(l)}, \mathbf{v}^{(l)})$. Calculate their determinants, and their inverses.

(A22.1d) Calculate the norm-inducing matrices $\{A_j^{(l)}\}$ with (22.3).

(A22.1e) Update $U^{(l)}$ to $U^{(l+1)}$ using (11.3a), $\{\mathbf{v}^{(l)}\}$; distance $d_{ik}^{(l)}$ in (11.3a1) is $d_{ik}^{(l)} = \|\mathbf{x}_k - \mathbf{v}_i^{(l)}\|_{A_i}$; $1 \leq i \leq c$, $1 \leq k \leq n$.

(A22.1f) Compare $U^{(l)}$ to $U^{(l+1)}$ in a convenient matrix norm: if $\|U^{(l+1)} - U^{(l)}\| \leq \varepsilon_L$, stop. Otherwise, return to (A22.1b) with $l = l + 1$.

As noted above, if $A_j = A\ \forall j$, $\hat{J}_m(U, \mathbf{v}, \mathbf{A}) = J_m(U, \mathbf{v}, A)$, and (A22.1) reduces to (A11.1). The appearance of the fuzzy scatter matrices $\{S_{fi}\}$ in this algorithm suggested the following terminology to Gustafson and Kessel.

(D22.1) *Definition 22.1* (*Fuzzy Covariance Matrix*). Assume $m \in (1, \infty)$; $X = \{\mathbf{x}_1, \mathbf{x}_2, \ldots, \mathbf{x}_n\} \subset \mathbb{R}^p$; and $(U, \mathbf{v}) \in M_{fc} \times \mathbb{R}^{cp}$. The fuzzy covariance matrix of cluster u_i is

$$C_{fi} = \frac{\sum_{k=1}^n (u_{ik})^m (\mathbf{x}_k - \mathbf{v}_i)(\mathbf{x}_k - \mathbf{v}_i)^T}{\sum_{k=1}^n (u_{ik})^m} = \frac{S_{fi}}{\sum_{k=1}^n (u_{ik})^m} \tag{22.4}$$

Just as S_{fi} reduces to the hard scatter matrix of u_i when $U \in M_c$, C_{fi} reduces to the (sample) covariance matrix

$$C_i = \sum_{\mathbf{x}_k \in u_i} (\mathbf{x}_k - \mathbf{v}_i)(\mathbf{x}_k - \mathbf{v}_i)^T \Big/ n_i$$

of the n_i points in hard cluster u_i having centroid $\mathbf{v}_i$. Accordingly, Gustafson and Kessel refer to (A22.1) as "fuzzy covariance" clustering algorithms.

Now suppose $(U, \mathbf{v}, \mathbf{A})$ is optimal for $\hat{J}_m$. Let

$$\lambda_{jk} = \frac{(u_{jk})^m [\rho_j \det(S_{fj})]^{(1/p)}}{\sum_{t=1}^{n} (u_{jt})^m}, \qquad 1 \le j \le c; \qquad 1 \le t \le n$$

At this optimal point, the value of $\hat{J}_m$ in terms of the $\{\lambda_{jk}\}$ is

$$\hat{J}_m(U, \mathbf{v}, \mathbf{A}) = \sum_{i=1}^{c} \left[\sum_{k=1}^{n} \lambda_{ik} (\mathbf{x}_k - \mathbf{v}_i)^T C_{fi}^{-1} (\mathbf{x}_k - \mathbf{v}_i) \right] \tag{22.5}$$

If $U \in M_c$, the number $(\mathbf{x}_k - \mathbf{v}_i)^T C_i^{-1} (\mathbf{x}_k - \mathbf{v}_i)$ is the squared Mahalonobis distance between $\mathbf{x}_k \in u_i$ and its (sub) sample mean $\mathbf{v}_i$, C_i^{-1} being the inverse of the sample covariance matrix of the points in u_i. Thus, it seems appropriate to call $(\mathbf{x}_k - \mathbf{v}_i)^T C_{fi}^{-1} (\mathbf{x}_k - \mathbf{v}_i)$ the squared *fuzzy* Mahalonobis distance between $\mathbf{x}_k$ and fuzzy cluster center $\mathbf{v}_i$. Form (22.5) then exhibits the property of optimal points for $\hat{J}_m$: loosely speaking, memberships are distributed to minimize the overall "fuzzy scattering volume" of c fuzzy clusters having fixed individual "volumes" (ρ_i = const), by varying the *shapes* of the hyperellipsoids in $\mathbb{R}^p$ defined by the level sets of the weighted fuzzy Mahalonobis distance functions. Stated somewhat differently, (22.3) shows that $\hat{J}_m$ tries to identify each A_j^* by scaling the inverse of the fuzzy covariance matrix C_{fj}^{-1} with a multiple of the scalar $\det(S_{fj})$, which is in some sense a measure of (the square of) a generalized scatter "volume" due to assignment of partial membership of all n $\mathbf{x}_k$'s in fuzzy cluster u_j. The following example illustrates some differences between algorithms (A9.2), (A11.1), and (A22.1).

(E22.1) (*Gustafson and Kessel*[54]). The artificial data set X shown in the inset of Fig. 22.1 consists of $n = 20$ points in $\mathbb{R}^2$. These data form two visually apparent *linear* clusters in the shape of a cross; the (x, y)-coordinates of each point are listed in column 1 of Table 22.1. X was generated by plotting at right angles two samples of size 10 from a bivariate uniform distribution over $[-10, 10] \times [-1, 1]$. X has the geometric characteristics responsible for the failure of J_1 illustrated in Fig. (8.2b), and of J_2 in (E19.1), viz., elongated, chainlike structure which seems quite incompatible with the topological structure of $\mathbb{R}^2$ induced by the Euclidean norm.

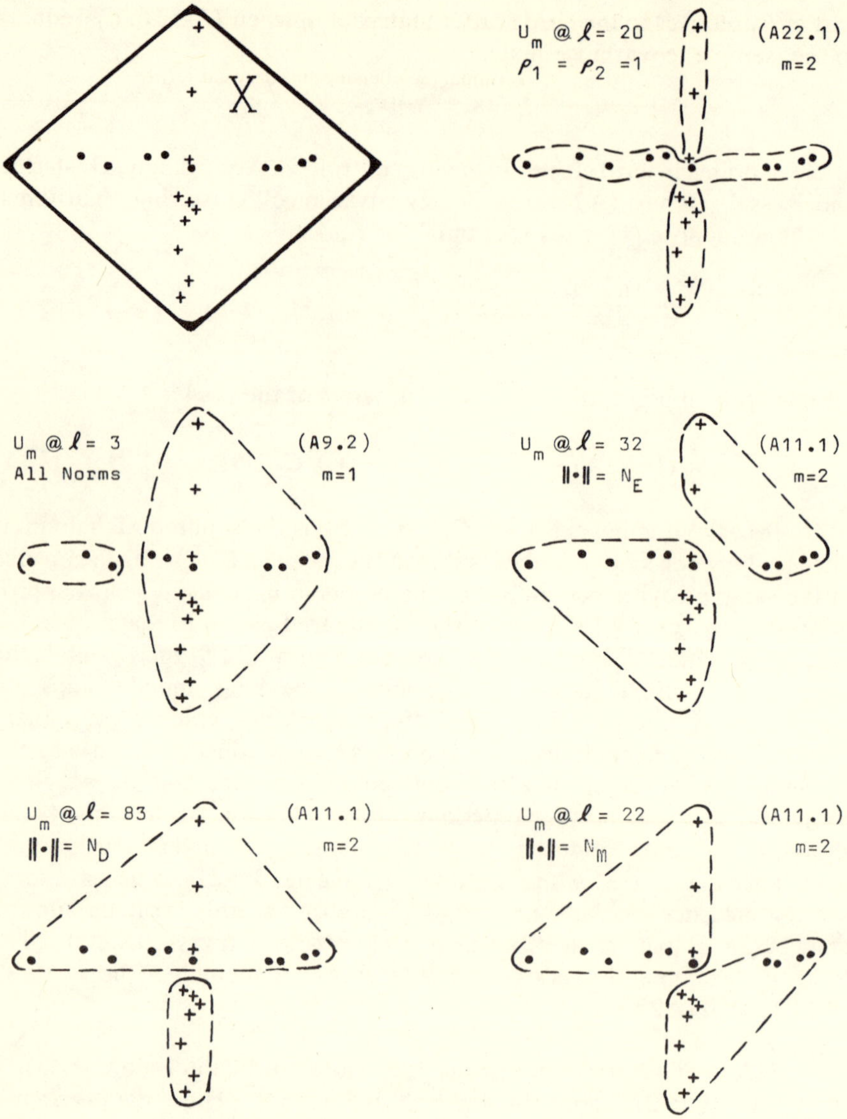

Figure 22.1. Gustafson's cross: (E22.1) and (E22.2).

Fixing $c = m = 2.00$, and $\rho_1 = \rho_2 = 1$, Gustafson and Kessel processed X with (A22.1). The fuzzy 2-partition of X satisfying $\max_{i,k}\{|u_{ik}^{(l+1)} - u_{ik}^{(l)}|\} \leq \varepsilon_L = 0.001$ exhibited as column 2 of Table 22.1 was attained in $l = 20$ iterations; the maximum membership matrix U_m derived from $U^{(20)}$ via (D9.2) is depicted graphically in Fig. 22.1. Evidently $\hat{J}_2$ adopts two norms

Table 22.1. Terminal Fuzzy Clusters for (E22.1) and (E22.2)

	Terminal memberships in fuzzy cluster u_1				
Data $\mathbf{x}_k$	(A22.1) $m = 2$	(A9.2) 3 Norms $m = 1$	(A11.1) N_E $m = 2$	(A11.1) N_D $m = 2$	(A11.1) N_M $m = 2$
(−9.75, −0.15)	0.03	0.00	0.25	0.45	0.30
(−6.44, 0.34)	0.03	0.00	0.18	0.38	0.20
(−4.69, −0.30)	0.03	0.00	0.11	0.39	0.15
(−2.04, 0.37)	0.02	1.00	0.08	0.16	0.02
(−1.24, 0.45)	0.32	1.00	0.10	0.08	0.05
(0.33, −0.08)	0.02	1.00	0.22	0.17	0.45
(5.04, −0.21)	0.03	1.00	0.98	0.48	0.76
(5.86, −0.25)	0.03	1.00	0.97	0.49	0.74
(7.54, 0.16)	0.03	1.00	0.91	0.49	0.69
(7.67, 0.24)	0.03	1.00	0.91	0.49	0.69
(−0.30, −8.07)	0.97	1.00	0.30	0.77	0.72
(0.13, −7.13)	0.97	1.00	0.29	0.80	0.75
(−0.37, −5.18)	0.97	1.00	0.19	0.89	0.80
(0.03, −3.33)	0.97	1.00	0.13	0.99	0.89
(0.35, −2.63)	0.98	1.00	0.13	0.99	0.93
(0.23, −2.68)	0.98	1.00	0.12	0.99	0.92
(−0.05, −2.00)	0.98	1.00	0.07	0.89	0.84
(0.41, 0.37)	0.74	1.00	0.29	0.08	0.36
(0.69, 4.75)	0.97	1.00	0.57	0.18	0.26
(0.74, 8.87)	0.97	1.00	0.57	0.29	0.34
Iterations	20	3	32	83	22

induced by fuzzy covariance matrices $C_{f1}^{(20)}$ and $C_{f2}^{(20)}$ which successfully label all 20 data points correctly (although the memberships of points 6 and 18 at the center of the cross are admittedly fortuitous). The same terminal partition was achieved in even less iterations ($l = 14$) using a slightly different initialization of (A22.1). In this instance then, there is little doubt that (A22.1), guided by the modified 2-means functional $\hat{J}_2$, performs quite adequately: adaptation to account for variances in cluster shape seems to proceed as theory predicts. As a matter of interest, we list the terminal fuzzy covariance and actual hard (sample) covariance matrices of each class:

$$C_1 = \begin{bmatrix} 33.92 & -0.03 \\ -0.03 & 0.07 \end{bmatrix}, \qquad C_{f1}^{(20)} = \begin{bmatrix} 36.33 & 0.88 \\ 0.88 & 0.05 \end{bmatrix}$$

$$C_2 = \begin{bmatrix} 0.13 & 1.50 \\ 1.50 & 24.44 \end{bmatrix}, \qquad C_{f2}^{(20)} = \begin{bmatrix} 0.06 & 0.72 \\ 0.72 & 25.65 \end{bmatrix}$$

The diagonal entries of C_1 and C_2 accurately reflect the predominantly linear structure of classes 1 and 2 in X, class 1 having large variance along the x-axis, while class 2 has large variance along the y-axis. The determinants of C_1 and C_2 are 2.48 and 0.84, respectively, whereas those of C_{f1} and C_{f2} are fixed at 1 (up to roundoff errors). The entries of the fuzzy covariance matrices leave little doubt that (A22.1) detects the *trends* that the hard variance–covariance structure exhibits. Some adjustments in m for $\hat{J}_m$ would probably align these estimates with the C_i's even more closely (cf. E23.1). •

A comparison of the results of (E22.1) with algorithms (A9.2) and (A11.1) using the Euclidean norm was given in (54). As a more extensive backdrop, the data of Table 22.1 were processed using (A9.2) and (A11.1) with each of the three norms labeled N_E (Euclidean), N_D (diagonal), and N_M (Mahalonobis) in equations (15.14), to see if changing the global topological (norm-metrical) structure underlying J_1 or J_2 could at least partially compensate for the difficulties caused to J_m by the linearity of cluster shapes in Gustafson's cross. The variance–covariance structure of the data suggests that the diagonal norm, N_D at (15.14b), should be more effective than N_E or N_M. This was in fact true only for the fuzzy c-means algorithms.

(E22.2) *Example 22.2.* Fixing $c = 2$, $\varepsilon_L = 0.02$, and using the maximum norm on $(U^{(l+1)} - U^{(l)})$ as the termination criterion, six runs were made on the data set X of (E22.1), one run of algorithms (A9.2) and (A11.1) using each of the three norms N_E, N_D, and N_M described by equations (15.14). Only one initialization of $U^{(0)}$ was used; the weighting exponent for (A11.1) was $m = 2$. The memberships of each data point in cluster u_1 at termination are reported as columns 3 through 6 in Table 22.1. Note first that hard 2-means algorithm (A9.2) converged to the *same* 2-partition of X in $l = 3$ iterations *for all three norms*! Evidently, changes in topological structure due to variation of the norm for J_1 are insufficient to override algorithm (A9.2)'s preference for this hard 2-partition (exhibited graphically on Fig. 22.1). Other starting guesses do lead to different terminal 2-partitions [cf. (54)].

Terminal *fuzzy* 2-partitions of X listed in Table 22.1 are quite clearly a function of the norm used in (11.6). Maximum membership partitions derived from each of columns 4, 5, and 6 via (D9.2) are illustrated graphically in Fig. 22.1. Using the more generous of the two possible ways to allocate clustering errors for each of these results in error counts of 8 for N_E; 3 for N_D; and 7 for N_M.

The overall conclusion suggested by Table 22.1 is that the Euclidean and Mahalonobis norms do not induce topological structures that are very compatible with the localized structure apparent in X. Column 5 of Table

22.1 suggests that the diagonal norm—when used with J_2—is much more amenable to the structure of the data than N_E or N_M. In view of the variance–covariance structure of X, this is not too surprising. What *is* surprising, however, is that J_1 with N_D found exactly the same hard 2-partition of X as did J_1 with N_E or N_M. Since J_2 with N_D reduced the clustering error to (3/20) or 15%, it seems plausible to attribute this substantial improvement to the fuzziness allowed by J_2. •

Remarks

Considering the results of (E22.2), Gustafson and Kessel's algorithm (A22.1) performed very well in (E22.1). The combined effect of these two examples is to suggest that *no* change in global topology induced by changes in the fixed norm for $J_m (m \geq 1)$ will be as effective as $\hat{J}_m$, which seems able to accommodate local shape variations by using a combination of different norms in the same criterion function. Algorithm (A22.1) raises some interesting issues. Convergence properties have yet to be studied; it seems reasonable to expect that an analysis similar to that of S12 is possible. The choice of each ρ_i for equation (22.3) seems crucial: there may be a way to optimize this choice based on some measure of clustering performance. More generally, there may be constraints other than this which lead to new algorithms based on minimizing $\hat{J}_m$. Another facet of (A22.1) is its potential utility as an estimator of (sub-) sample covariance matrices. This was not emphasized in (E22.1), but has been briefly discussed in (54). A more general discussion of the use of fuzzy algorithms for estimation of *statistical* parameters which includes this idea will be given in Section 25.

Exercises

H22.1. Let $A \in V_{pp}$ be symmetric, $\mathbf{x}^T = (x_1, \ldots, x_p) \in \mathbb{R}^p$. Show that
(i) $\nabla_A(\mathbf{x}^T A\mathbf{x}) = \mathbf{x}\mathbf{x}^T$
and
(ii) $\nabla_A[\det(A)] = \det(A)(A^{-1})$, whenever A is invertible.

H22.2. Let $\mathbf{x}^T = (x_1, x_2) \in \mathbb{R}^2$. Define the functions $N_i : \mathbb{R}^2 \to \mathbb{R}$ as

$$N_1(\mathbf{x}) = |x_1| + |x_2|$$

$$N_2(\mathbf{x}) = (x_1^2 + x_2^2)^{1/2}$$

$$N_\infty(\mathbf{x}) = |x_1| \vee |x_2|$$

(i) Show that each N_i satisfies the norm properties in H15.1.
(ii) Describe $\bar{B}(\mathbf{\theta}, 1)$, the closed ball about (0, 0) of radius 1 using each of the three norms. (Answer: circle, diamond, square.)
(iii) Describe a probability distribution which might generate clusters of each of these shapes as samples.

H22.3. Let PD(k) be the subset of V_{pp} consisting of all real positive-definite $p \times p$ matrices with fixed determinant k, and let PD be the positive-definite matrices with arbitrary determinant.
(i) $f:\text{PD} \to \mathbb{R}$, $f(A) = \det A$. Is f continuous on PD?
(ii) $g:\text{PD} \to V_{pp}$, $g(A) = A^{-1}$. Is $A^{-1} \in \text{PD}$ whenever A is? (Answer: yes.)
(iii) g as in (ii): is g continuous on PD?
(iv) $h:\text{PD} \to \text{PD}(k)$, $h(A) = (k \det A)^{(1/p)}A^{-1}$. Is h continuous on PD?

H22.4. Let $(U, \mathbf{v}) \in M_{fc} \times \mathbb{R}^{cp}$, and let $F_i(U, \mathbf{v}) = S_{fi}$, where S_{fi} is the fuzzy scatter matrix of (D11.3), with $X = \{\mathbf{x}_k\} \subset \mathbb{R}^p$ a fixed data set with n elements. Is F_i continuous?

H22.5. What is the rank of the matrix $\mathbf{x}\mathbf{x}^T$, $\mathbf{x} \in \mathbb{R}^p$, $\mathbf{x} \neq \mathbf{0}$? (Answer: 1.)

H22.6. Is the fuzzy scatter matrix S_{fi} in (D11.3) necessarily positive definite?

H22.7. Argue that the convergence proof for (A11.1) given in S12 *cannot* be extended directly to (A22.1).

S23. The Fuzzy *c*-Varieties Clustering Algorithms

Gustafson and Kessel attempt to improve the ability of J_m to detect different cluster shapes in a fixed data set by locally varying the metric topology around a fixed kind of prototype—namely, the cluster centers $\{\mathbf{v}_i\}$, which are prototypical data points in $\mathbb{R}^p$. Another attempt to enhance the ability of J_m to detect nonhyperellipsoidally shaped substructure takes an approach which is in some sense opposite to that embodied by Gustafson and Kessel's J_m. In the fuzzy c-varieties (FCV) functionals defined below, Bezdek *et al.*[16] retain the global metric structure induced by a fixed matrix A as in (11.5), but allow the c prototypes $\{\mathbf{v}_i\}$ to be r-dimensional linear varieties, $0 \leq r \leq p-1$, rather than just points in $\mathbb{R}^p$. As might be suspected, this type of objective functional is most amenable to data sets which consist essentially of c clusters, all of which are drawn from linear varieties of the same dimension. S24 contains a modification of these functionals which attempts to reconcile the ideas underlying S22 and S23 via convex combinations. To begin, we define the notion of linear variety.

(D23.1) *Definition 23.1 (Linear Variety).* The *linear variety* of dimension r, $0 \leq r \leq p$ through the point $\mathbf{v} \in \mathbb{R}^p$, spanned by the linearly independent vectors $\{\mathbf{s}_1, \mathbf{s}_2, \ldots, \mathbf{s}_r\} \subset \mathbb{R}^p$, is the set

$$V_r(\mathbf{v}; \mathbf{s}_1, \mathbf{s}_2, \ldots, \mathbf{s}_r) = \left\{ \mathbf{y} \in \mathbb{R}^p \,\middle|\, \mathbf{y} = \mathbf{v} + \sum_{j=1}^{r} t_j \mathbf{s}_j;\ t_j \in \mathbb{R} \right\} \tag{23.1}$$

If in (23.1) $\mathbf{v}$ is the zero vector, then V_r is just the linear hull or span of the $\{\mathbf{s}_i\}$, an r-dimensional linear subspace through the origin *parallel* to the set in (23.1). A common way to denote this is by writing (23.1) as

$$V_r(\mathbf{v}; \{\mathbf{s}_i\}) = \{\mathbf{v}\} + \operatorname{span}(\{\mathbf{s}_i\}) \tag{23.2}$$

wherein V_r is called the translate of span($\{\mathbf{s}_i\}$) away from $\boldsymbol{\theta}$ by $\mathbf{v}$. Figure 6.3 depicts these ideas in a different context for the vector space V_{cn}. Linear varieties of all dimensions are thought of as "flat" sets in $\mathbb{R}^p$: r is the number of directions in which this "flatness" extends. Certain linear varieties have special names and notations:

$$V_0(\mathbf{v}; \varnothing) = \mathbf{v} \qquad \sim \text{"points"} \tag{23.3a}$$

$$V_1(\mathbf{v}; \mathbf{s}) = L(\mathbf{v}; \mathbf{s}) \qquad \sim \text{"lines"} \tag{23.3b}$$

$$V_2(\mathbf{v}; \mathbf{s}_1, \mathbf{s}_2) = P(\mathbf{v}; \mathbf{s}_1, \mathbf{s}_2) \sim \text{"planes"} \tag{23.3c}$$

$$V_{p-1}(\mathbf{v}; \{\mathbf{s}_i\}) = \mathrm{HP}(\mathbf{v}; \{\mathbf{s}_i\}) \sim \text{"hyperplanes"} \tag{23.3d}$$

Thus, we call V_0 a point; V_1 a line through $\mathbf{v}$ parallel to $\mathbf{s}$; V_2 a plane through $\mathbf{v}$ parallel to the plane through $\boldsymbol{\theta}$ spanned by $\{\mathbf{s}_1, \mathbf{s}_2\}$; and V_{p-1} a hyperplane through $\mathbf{v}$ parallel to the $(p-1)$-dimensional vector subspace through $\boldsymbol{\theta}$ spanned by the $\{\mathbf{s}_i\}$.

A fuzzy clustering criterion which recognizes varietal shapes can be based on distances from data points to prototypical linear varieties. Specifically, the orthogonal (OG) distance (in the A norm on $\mathbb{R}^p$) from $\mathbf{x}$ to V_r, when $\{\mathbf{s}_i\}$ are an orthonormal basis for their span (i.e., $\langle \mathbf{s}_i, \mathbf{s}_j\rangle_A = \delta_{ij}$), is

$$D_A(\mathbf{x}, V_r) = \left[\|\mathbf{x}-\mathbf{v}\|_A^2 - \sum_{j=1}^{r} (\langle \mathbf{x}-\mathbf{v}, \mathbf{s}_j\rangle_A)^2\right]^{1/2} \tag{23.4}$$

This equation follows by projecting $(\mathbf{x}-\mathbf{v})$ onto span $(\{\mathbf{s}_i\})$, and then calculating the length of $(\mathbf{x}-\mathbf{v})$ minus its best least-squares approximation. Distance (23.4) is easiest to visualize with $A = I_p$ and, say, $r = 1$, in which case it is just the shortest distance from a point $\mathbf{x}$ to the line $L(\mathbf{v}; \mathbf{s})$ in the ordinary Euclidean sense. In what follows, we deal with sets of c linear varieties of dimension r; this would necessitate a formalism such as

$$\mathbf{V}_r = (V_{r1}, V_{r2}, \ldots, V_{rc}) \tag{23.5}$$

where $\forall i$, $V_{ri} = V_{ri}(\mathbf{v}_i; \mathbf{s}_{i1}, \mathbf{s}_{i2}, \ldots, \mathbf{s}_{ir})$ as in (23.1). We ease the notational burden by suppressing dimension r, and write $D_A(\mathbf{x}_k, V_{ri})$ more briefly as

$$D_A(\mathbf{x}_k, V_{ri}) = \left[\|\mathbf{x}_k-\mathbf{v}_i\|_A^2 - \sum_{j=1}^{r} (\langle \mathbf{x}_k-\mathbf{v}_i, \mathbf{s}_{ij}\rangle_A)^2\right]^{1/2} \doteq D_{ik} \tag{23.6}$$

Note that (23.6) reduces to the distance $d_{ik} = \|\mathbf{x}_k - \mathbf{v}_i\|_A$ when $r = 0$.

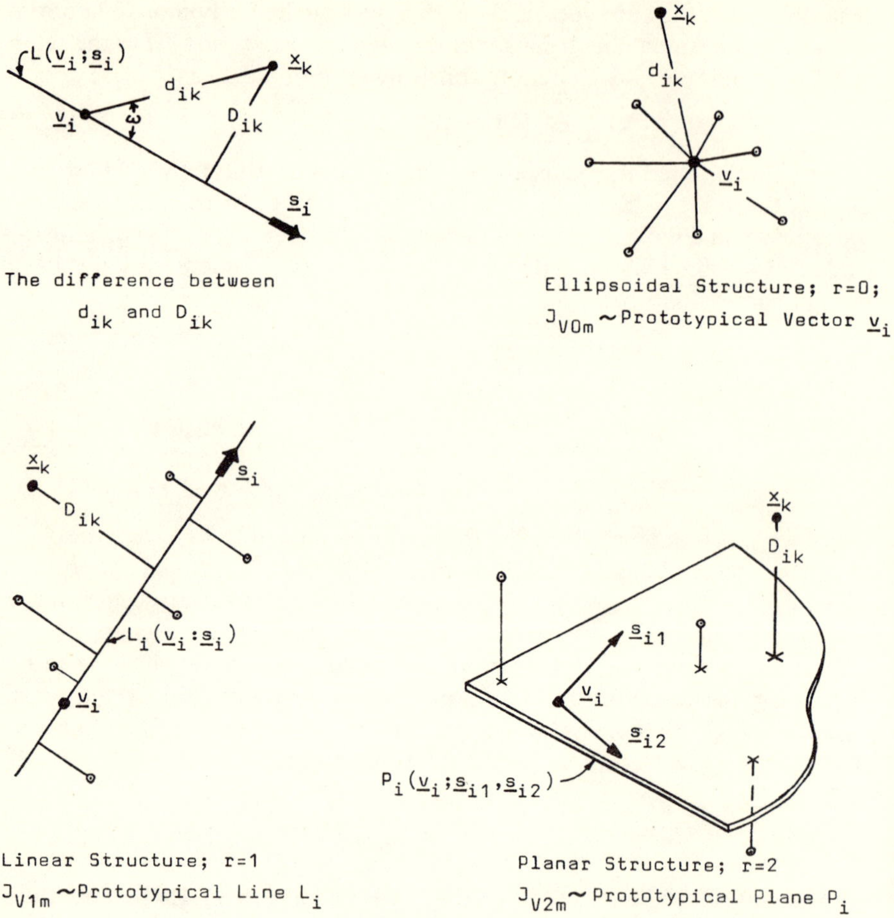

Figure 23.1. A geometric interpretation for criterion $J_{V_{rm}}$.

Otherwise, D_{ik} is the OG A-distance from $\mathbf{x}_k$ to the linear variety V_{ri}. These distances are illustrated graphically for $r = 0, 1, 2$ in Fig. 23.1, with $A = I_p$.

The weighted objective function to be optimized in this section is the natural extension of the functional J_m of S11 which measures the total weighted sum of squared OG errors from each $\mathbf{x}_k \in X = \{\mathbf{x}_1, \mathbf{x}_2, \ldots, \mathbf{x}_n\} \subset \mathbb{R}^p$ to each of c r-dimensional linear varieties. Specifically, let

$$J_{Vrm}(U, \mathbf{V}) = \sum_{k=1}^{n} \sum_{i=1}^{c} (u_{ik})^m (D_{ik})^2, \qquad 1 \leq m < \infty \tag{23.7}$$

where $U \in M_{fc}$ is a fuzzy c-partition of X, $\mathbf{V} = \mathbf{V}_r$ is defined at (23.5), and

D_{ik} at (23.6). Note that for $r = 0$, J_{V0m} reduces to J_m in (11.6); for $r = 1$, we call $J_{V1m} = J_{Lm}$; and in general, we put, for $0 \le r < p$;

$$\overbrace{\underbrace{J_{V0m} = J_{Mm}}_{\substack{\text{fuzzy}\\ c\text{-means}\\ \text{(FCM)}}}}^{r=0} \Subset \overbrace{\underbrace{J_{V1m} = J_{Lm}}_{\substack{\text{fuzzy}\\ c\text{-lines}\\ \text{(FCL)}}}}^{r=1} \Subset \overbrace{\underbrace{J_{V2m} = J_{Pm}}_{\substack{\text{fuzzy}\\ c\text{-planes}\\ \text{(FCP)}}}}^{r=2} \Subset \cdots \Subset \overbrace{\underbrace{J_{V(p-1)m}}_{\substack{\text{fuzzy}\\ \text{hyperplanes}\\ \text{(FHP)}}}}^{r=p-1} \tag{23.8}$$

where $\Subset$ in (23.8) means "is extended by." J_{Vrm} is called hereafter a fuzzy c-varieties (FCV) functional. Taking $m = 1$ in (23.8) results in a unique ($\forall r$) hard c-varieties functional: in particular, J_{V01} is the classical within-group sum-of-squared-errors criterion, J_w at (9.7a), when $U \in M_c$ is hard.

The problem of optimizing J_{Vrm} over the Cartesian product $M_{fc} \times R^{cp} \times (R^{cp})^r$ will be attacked in stages. In general, we assume throughout that m, c, and A are fixed, and begin with the following proposition.

(P23.1) *Proposition 23.1.* Let $\hat{\mathbf{V}} \in R^{cp} \times (R^{cp})^r$ be fixed. If $1 < m < \infty$ and $\hat{D}_{ik} > 0\ \forall i, k$, then $\hat{U} \in M_{fc}$ is a strict local minimum of $\phi(U) = J_{Vrm}(U, \hat{\mathbf{V}})$ if and only if

$$\hat{u}_{ik} = \left\{ \sum_{j=1}^{c} \left(\frac{\hat{D}_{ik}}{\hat{D}_{jk}} \right) \right\}^{-2/(m-1)} \qquad \forall i, k \tag{23.9}$$

Proof. Identical to (P12.1). See (16) for details •

By analogy to (P12.2), we next fix $\hat{U} \in M_{fc}$ and minimize J_{Vrm} over the c-varieties $\{V_i\}$.

(T23.1) *Theorem 23.1.* Let $\hat{U} \in M_{fc}$ be fixed. If $1 < m < \infty$, then $\hat{\mathbf{V}} \in R^{cp} \times (R^{cp})^r$ is a local minimum of $\psi(\mathbf{V}) = J_{Vrm}(\hat{U}, \mathbf{V})$ if and only if each prototype $\hat{V}_i = \hat{V}_{ri}$ satisfies

$$\hat{V}_i = V_i(\hat{\mathbf{v}}_1; \hat{\mathbf{s}}_{i1}, \ldots, \hat{\mathbf{s}}_{ir})$$

as in (23.1), where

$$\hat{\mathbf{v}}_i = \frac{\sum_{k=1}^{n} (\hat{u}_{ik})^m \mathbf{x}_k}{\sum_{k=1}^{n} (\hat{u}_{ik})^m} \qquad \forall i \tag{23.10a}$$

and

$$\mathbf{s}_{ij} = A^{-1/2} \mathbf{v}_{ij} \tag{23.10b}$$

where $\mathbf{s}_{ij}$ is the jth unit eigenvector of the generalized (via A) fuzzy scatter matrix

$$\hat{S}_{fiA} = A^{1/2}\left[\sum_{k=1}^{n} (\hat{u}_{ik})^m (\mathbf{x}_k - \hat{\mathbf{v}}_i)^T (\mathbf{x}_k - \hat{\mathbf{v}}_i)\right] A^{1/2} = A^{1/2}\hat{S}_{fi}A^{1/2} \tag{23.10c}$$

corresponding to its jth largest eigenvalue. $\hat{S}_{fi}$ is the fuzzy scatter matrix defined at (11.8b): note that $\hat{S}_{fiA} = \hat{S}_{fi}$ when $A = I_p$, and further reduces to $\hat{S}_i$, the scatter matrix of hard cluster $\hat{u}_i \in U \in M_c$ if $m = 1$, $A = I_p$.

Analysis. This theorem is analogous to (P12.2). Since minimization of ψ is an unconstrained problem, one's first inclination is to set the gradient of ψ equal to zero at $\hat{\mathbf{V}}$: however, this leads to a hopelessly tangled system of nonlinear equations. Instead, a geometric argument is given, based on the following *guess*: each "center of mass" $\hat{\mathbf{v}}_i$ of $\hat{V}_{ri}$ should continue to satisfy (11.3b), since $J_{Vrm} = J_m$ if $r = 0$. Acting on this suspicion leads to the results desired. Just as in (A22.1), equations (23.10b) and (23.10c) simply "add on" an additional step in the basic fuzzy c-means loop of (A11.1). Further elaboration concerning the reappearance of the fuzzy scatter matrices $\{S_{fi}\}$ will be given below.

Proof. Although the proof for any $r > 1$ is not much more difficult, we give the proof for $r = 1$, because the geometric idea can be clearly illustrated in this special case. Details of the general case may be found in (16). Accordingly, fix $r = 1$. Our objective is thus to minimize $\psi(\mathbf{V}) = \psi(\mathbf{v}, \mathbf{s})$ over $\mathbb{R}^{cp} \times \mathbb{R}^{cp}$, where $(\mathbf{v}_i, \mathbf{s}_i)$ defines the line $L(\mathbf{v}_i, \mathbf{s}_i)$, $1 \leqslant i \leqslant c$. Now it suffices to minimize

$$\psi_i(\mathbf{v}, \mathbf{s}) = \sum_{k=1}^{n} (\hat{u}_{ik})^m (D_{ik})^2, \qquad 1 \leqslant i \leqslant c \tag{23.11}$$

so let $\hat{L}_i = \hat{L}_i(\hat{\mathbf{v}}_i; \hat{\mathbf{s}}_i)$ be the straight line in $\mathbb{R}^p$ that minimizes ψ_i. Denote by $\hat{H}_i$ the $(p-1)$-dimensional hyperplane through the origin of $\mathbb{R}^p$ which is orthogonal (in the weighted inner product $\langle \cdot, \cdot \rangle_A$) to $\hat{L}_i$. Further, let $\tilde{\mathbf{x}}_k$ be the orthogonal projection of $\mathbf{x}_k$ into $\hat{H}_i$, and let $\tilde{D}_{ik}$ be the distance in $\hat{H}_i$ between $\mathbf{x}_k$ and $\hat{\mathbf{p}}_i = \hat{H}_i \cap \hat{L}_i$. Figure 23.2a illustrates the geometric situation.

With $(\hat{\mathbf{v}}_i, \hat{\mathbf{s}}_i)$ chosen, ψ_i at (23.11) is now a function of $\hat{\mathbf{p}}_i \in \mathbb{R}^p$, say

$$\begin{aligned}\gamma_i(\hat{\mathbf{p}}_i) = \psi_i(\hat{\mathbf{v}}_i, \hat{\mathbf{s}}_i) &= \sum_{k=1}^{n} (\hat{u}_{ik})^m (\hat{D}_{ik})^2 \\ &= \sum_{k=1}^{n} (\hat{u}_{ik})^m (\tilde{D}_{ik})^2 = \sum_{k=1}^{n} (\hat{u}_{ik})^m \|\tilde{\mathbf{x}}_k - \hat{\mathbf{p}}_i\|_A^2\end{aligned}$$

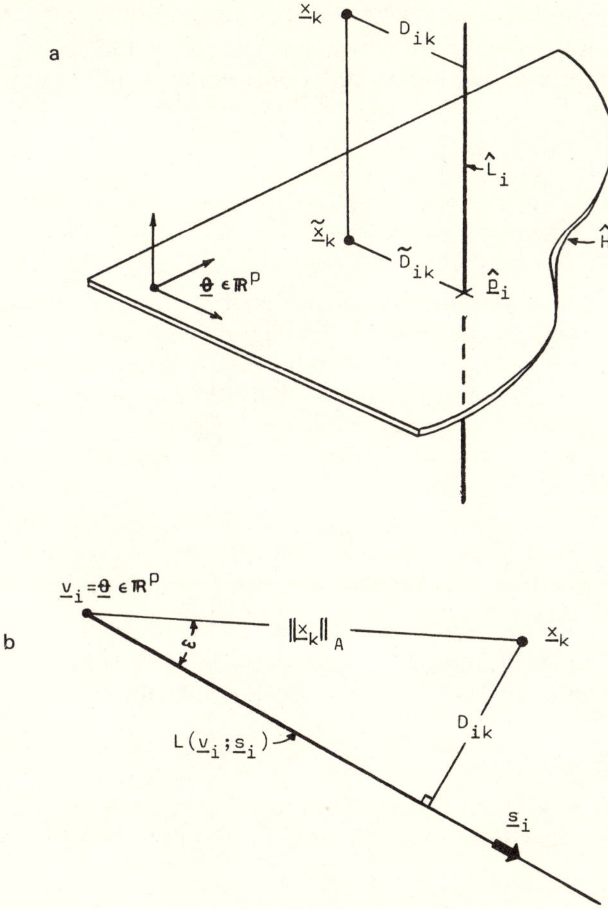

Figure 23.2. The geometry of Theorem 23.1.

We minimize γ_i over R^p by zeroing its gradient:

$$\nabla\gamma_i(\hat{\mathbf{p}}_i) = \boldsymbol{\nabla} \sum_{k=1}^{n} (\hat{u}_{ik})^m(\|\tilde{\mathbf{x}}_k - \hat{\mathbf{p}}_i\|_A^2) = -2A \sum_{k=1}^{n} (\hat{u}_{ik})^m(\tilde{\mathbf{x}}_k - \hat{\mathbf{p}}_i) = \boldsymbol{\theta}$$

Since A is positive definite, A^{-1} exists, so we can apply $-\frac{1}{2}A^{-1}$ to this last equation; thus $\hat{\mathbf{p}}_i$ may minimize γ_i only if

$$\sum_{k=1}^{n} (u_{ik})^m(\tilde{\mathbf{x}}_k - \hat{\mathbf{p}}_i) = \boldsymbol{\theta} \tag{23.12}$$

Rearranging (23.12) yields for $\hat{\mathbf{p}}_i$

$$\hat{\mathbf{p}}_i = \sum_{k=1}^{n} (\hat{u}_{ik})^m\tilde{\mathbf{x}}_k \Big/ \sum_{k=1}^{n} (\hat{u}_{ik})^m \tag{23.13}$$

Note that the denominator in (23.13) is positive because of (23.9). Consider the coordinate system for $\mathbb{R}^p$ in which axis j is parallel to line $\hat{L}_i$. In this system the coordinates of $\mathbf{x}_k$ and $\tilde{\mathbf{x}}_k$ differ only in the jth place, so for $t \neq j$,

$$\hat{p}_{it} = \sum_{k=1}^{n} (\hat{u}_{ik})^m \tilde{x}_{kt} \Big/ \sum_{k=1}^{n} (\hat{u}_{ik})^m$$

$$= \sum_{k=1}^{n} (\hat{u}_{ik})^m x_{kt} \Big/ \sum_{k=1}^{n} (\hat{u}_{ik})^m$$

For $t = j$, $\hat{L}_i$ contains all points of the jth coordinate of $\mathbf{x}_k$ and $\tilde{\mathbf{x}}_k$, so in particular, it contains the point whose jth coordinate is

$$\sum_{k=1}^{n} (\hat{u}_{ik})^m x_{kj} \Big/ \sum_{k=1}^{n} (\hat{u}_{ik})^m$$

Thus (23.13) is equivalent to

$$\hat{\mathbf{p}}_i = \sum_{k=1}^{n} (\hat{u}_{ik})^m \mathbf{x}_k \Big/ \sum_{k=1}^{n} (\hat{u}_{ik})^m \tag{23.14}$$

But (23.14) and (23.10a) are identical, and (23.14) is independent of any coordinate system. Therefore, $\hat{\mathbf{p}}_i$ is optimal for $\gamma_i \Leftrightarrow (\hat{\mathbf{v}}_i, \hat{\mathbf{s}}_i)$ is optimal for $\psi_i \Leftrightarrow \hat{L}_i(\hat{\mathbf{v}}_i; \hat{\mathbf{s}}_i)$ passes through $\hat{\mathbf{v}}_i$ at (23.10a). So much for $\hat{\mathbf{v}}_i$.

Without loss, assume $\hat{\mathbf{v}}_i = \mathbf{\theta}$ and $\|\mathbf{s}_i\|_A = 1$. Let ω be the angle in $\mathbb{R}^p$ between $\mathbf{x}_k$ and $\mathbf{s}_i$, so that

$$\cos^2(\omega) = \left(\frac{\langle \mathbf{x}_k, \mathbf{s}_i \rangle_A}{\|\mathbf{x}_k\|_A}\right)^2 = 1 - \sin^2(\omega)$$

$$\sin^2(\omega) = 1 - \left(\frac{\langle \mathbf{x}_k, \mathbf{s}_i \rangle_A}{\|\mathbf{x}_k\|_A}\right)^2 \tag{23.15}$$

$$\|\mathbf{x}_k\|_A^2 \sin^2(\omega) = \|\mathbf{x}_k\|_A^2 - (\langle \mathbf{x}_k, \mathbf{s}_i \rangle_A)^2 = (D_{ik})^2$$

This is just equation (23.4) when $\mathbf{x} = \mathbf{x}_k$, $\mathbf{v} = \mathbf{\theta}$, and $r = 1$. The geometric content of (23.15) is illustrated in Fig. 23.2b. Equation (23.15) suppresses the dependency of D_{ik} on $\hat{\mathbf{v}}_i$; we let

$$\tau_i(\mathbf{s}_i) = \sum_{k=1}^{n} (\hat{u}_{ik})^m [\|\mathbf{x}_k\|_A^2 - (\langle \mathbf{x}_k, \mathbf{s}_i \rangle_A)^2]$$

and minimize τ_i over the unit ball (in the A norm),

$$\delta B(\mathbf{\theta}, 1) = \{\mathbf{y} \in \mathbb{R}^p \mid \|\mathbf{y}\|_A = 1\}$$

Now

$$\tau_i(\mathbf{s}_i) = \sum_{k=1}^{n} (\hat{u}_{ik})^m \|\mathbf{x}_k\|_A^2 - \sum_{k=1}^{n} (\hat{u}_{ik})^m (\langle \mathbf{x}_k, \mathbf{s}_i \rangle_A)^2$$

Because $\hat{U}$ and X are fixed, the first term is constant relative to $\mathbf{s}_i$, so that τ_i is minimum on $\delta B(\boldsymbol{\theta}, 1) \Leftrightarrow$ the last term maximizes, i.e., in case

$$\xi_i(\mathbf{s}_i) = \sum_{k=1}^{n} (\hat{u}_{ik})^m (\langle \mathbf{x}_k, \mathbf{s}_i \rangle_A)^2$$

is maximum over $\delta B(\boldsymbol{\theta}, 1)$. Now

$$\xi_i(\mathbf{s}_i) = \sum_{k=1}^{n} (\hat{u}_{ik})^m (\mathbf{x}_k^T A \mathbf{s}_i)(\mathbf{x}_k^T A \mathbf{s}_i)$$

$$= \sum_{k=1}^{n} (\hat{u}_{ik})^m \mathbf{s}_i^T (A \mathbf{x}_k \mathbf{x}_k^T A) \mathbf{s}_i$$

$$= \mathbf{s}_i^T \left\{ A^{1/2} \left(\sum_{k=1}^{n} (\hat{u}_{ik})^m A^{1/2} \mathbf{x}_k \mathbf{x}_k^T A^{1/2} \right) A^{1/2} \right\} \mathbf{s}_i$$

where we have used the fact that every positive-definite matrix A has a unique "square root" $A^{1/2}$, $A = (A^{1/2})(A^{1/2})$, which is also positive definite. Consequently, $\mathbf{s}_i$ minimizes $\tau_i(\mathbf{s}_i)$ if and only if $\mathbf{s}_i$ solves the equivalent problem

$$\max_{\|\mathbf{s}_i\|_A^2 = 1} \left\{ (A^{1/2}\mathbf{s}_i)^T \left[A^{1/2} \left(\sum_{k=1}^{n} (\hat{u}_{ik})^m \mathbf{x}_k x_k^T \right) A^{1/2} \right] (A^{1/2}\mathbf{s}_i) \right\} \quad (23.16)$$

The solution of (23.16) is well known: letting $\hat{\mathbf{y}}_i$ be a (unit) eigenvector of

$$\tilde{S}_{fiA} \doteq A^{1/2} \left[\sum_{k=1}^{n} (\hat{u}_{ik})^m \mathbf{x}_k \mathbf{x}_k^T \right] A^{1/2}$$

corresponding to its largest eigenvalue, then

$$\hat{\mathbf{s}}_i = A^{-1/2} \hat{\mathbf{y}}_i \quad (23.17)$$

which is (23.10b) when $r = 1$ for the normalized ($\hat{\mathbf{v}}_i = \boldsymbol{\theta}$) fuzzy scatter matrix $\tilde{S}_{fiA}$.

Translation of $\boldsymbol{\theta}$ back to $\hat{\mathbf{v}}_i$ in the above argument completes the proof, as $\tilde{S}_{fiA}$ becomes $\hat{S}_{fiA}$. •

Because singularities ($\hat{D}_{ik} = 0$) *can* occur, it is necessary to see how (23.9) is modified in this eventuality. Theorem 23.2 corroborates one's suspicion that tie-breaking is arbitrary up to constraint (5.6b) in this instance.

(T23.2) *Theorem 23.2.* Let $I = \{1, 2, \ldots, c\}$. For each k in (T23.1) such that one or more $\hat{D}_{ik}$'s $= 0$, let $I_k = \{i \in I \mid \hat{D}_{ik} = 0\}$, and $\tilde{I}_k = I - I_k$. Then $\hat{U}$ may be optimal for J_{Vrm}, $1 < m < \infty$, $0 \leq r < p$, only if

$$\left\{ \begin{array}{ll} \hat{u}_{ik} = 0 & \forall i \in \tilde{I}_k \\ \sum_{i \in I_k} \hat{u}_{ik} = 1 & \end{array} \right\} \tag{23.18}$$

Analysis and Proof. Taking $\|\mathbf{x}_k - \hat{\mathbf{v}}_i\| = D_{ik}$ in (T11.1) gives the result. Note that column k in (23.18) is actually *hard* (all zeros and a 1) if and only if the singularity is unique (I_k is a singleton). Otherwise, nonzero fuzzy weights can be assigned arbitrarily in the singular rows of column k. •

Note that (23.9) remains necessary in columns of $\hat{U}$ without singularities. Equations (23.10) are still necessary in the singular case. Theorems 23.1 and 23.2 constitute a basis for Picard iteration through necessary *joint* conditions for local minima of J_{Vrm} via iterative optimization. The fuzzy c-varieties algorithms are, for each r, $0 \leq r < p$, the infinite family of clustering strategies described in the following algorithm.

(A23.1) *Algorithm 23.1* [*Fuzzy c-Varieties* (*FCV*), *Bezdek et al.*[16]]

(A23.1a) Fix c, $2 \leq c < n$; r, $0 \leq r < p$; m, $1 < m < \infty$. Choose any positive-definite matrix A in V_{pp}. Initialize $U^{(0)} \in M_{fc}$. Then at step l, $l = 0, 1, 2, \ldots$:

(A23.1b) Calculate the c fuzzy cluster centers $\{\mathbf{v}_i^{(l)}\}$ using $U^{(l)}$ and (23.10a).

(A23.1c) Calculate the c generalized fuzzy scatter matrices $\{S_{fiA}^{(l)}\}$ using $U^{(l)}$, $\{\mathbf{v}_i^{(l)}\}$, A, and (23.10c).

(A23.1d) Extract from each $S_{fiA}^{(l)}$ its r principal unit eigenvectors $\{\mathbf{y}_{ij}^{(l)} \mid 1 \leq j \leq r\}$, and construct from them the vectors $\{\mathbf{s}_{ij}^{(l)} \mid 1 \leq r \leq r\}$ using (23.10b). At this stage $\{\mathbf{v}_i^{(l)}; \mathbf{s}_{ij}^{(l)}; 1 \leq i \leq c; 1 \leq j \leq r\}$ determine the lth estimate of the c linear varieties $V_{ri}^{(l)}$.

(A23.1e) Update $U^{(l)}$ to $U^{(l+1)}$ using (23.9) and the distances $\{D_{ik}^{(l)}\}$ at (23.6).

(A23.1f) Compare $U^{(l)}$ to $U^{(l+1)}$ in a convenient matrix norm. If $\|U^{(l+1)} - U^{(l)}\| \leq \varepsilon_L$, stop. Otherwise, put $l = l + 1$ and return to (A23.1b).

Algorithm 23.1 reduces to (A11.1) if $r = 0$. Otherwise, it generates simultaneously an estimate of $U \in M_{fc}$ and c linear varieties $\{V_{ri}\}$ of dimension r which best fit data set X in the sense of the weighted least-squared error clustering criterion J_{Vrm}. For $m = 1$, hard c-varieties algorithms ensue, using necessary conditions delineated in (17). If one chooses $A = I, c = m = r = 1, U \in M_c = M_1$, (A23.1) is nothing more than principal components analysis[103] to find the unique line in $\mathbb{R}^p$ upon which X can be projected to account for a maximal amount of its sample variance. The appearance of S_{fi} when $A = I$ suggests that (A23.1) is somehow related to factor analysis; this connection, however, remains to be established. As a first example, we reconsider the data of (E22.1) and (E22.2) in the following.

(E23.1) *Example 23.1.* The data set X illustrated in Fig. 22.1 (Gustafson's cross) was processed with the fuzzy c-lines (FCL) algorithm (A23.1) using $c = 2$, $m = 2$, $r = 1$, $A = I \Rightarrow \|\cdot\|_A =$ Euclidean, $\varepsilon_L = 0.02$, and the maximum norm on $(U^{(l+1)} - U^{(l)})$ for (A23.1f). Recall that (A22.1), using fuzzy covariance matrices to accommodate different shapes of individual clusters, generated a fuzzy 2-partition of X in (E22.1) whose maximum memberships correctly labeled all 20 points in X. Further recall (E22.2) that neither hard nor fuzzy 2-means could be adjusted by changes in norm to yield results as satisfactory as those obtained by Gustafson and Kessel. In this example (A23.1) attempts to identify the linear substructure in X by prescribing a search for prototypical *lines* in $\mathbb{R}^2$ about which the data cluster closely in the sense of J_{V12}. Figure 23.3 illustrates the results. Fuzzy 2-lines converged to this solution in three iterations. Table 23.1 lists the terminal membership functions $\{u_{ik}^{(3)}\}$, centers of mass $\{\mathbf{v}_i^{(3)}\}$, and direction vectors $\{\mathbf{s}_i^{(3)}\}$, that correspond to $(U^{(3)}, \mathbf{V}^{(3)})$ which locally minimizes J_{V12}.

Note first that this partition is in fact *nearly hard* even at $m = 2$! These values should be compared to column 4 of Table 22.1, which gives the terminal partition of these data achieved by fuzzy 2-means (A11.1) using exactly the same data and algorithmic parameters *except* r: there $r = 0$; here

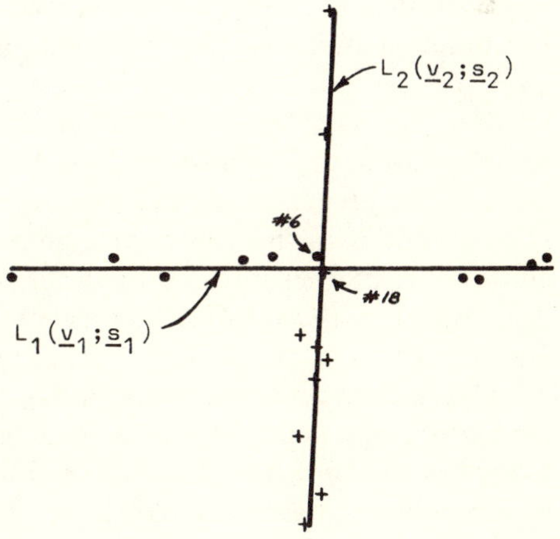

Figure 23.3. Fuzzy 2-lines applied to Gustafson's cross: (E23.1). Equation of prototype L_1: $y_1 = -0.0005x + 0.0001$; equation of prototype L_2: $y_2 = 16.2203x + 4.7521$.

Table 23.1. Fuzzy 2-Lines Output for Gustafson's Cross: (E23.1)

Data point	$u_{1k}^{(3)}$ Class 1 membership	$u_{2k}^{(3)}$ Class 2 membership
1	1.00	0.00
2	1.00	0.00
3	0.99	0.01
4	0.98	0.02
5	0.94	0.06
6	0.07	0.93
7	1.00	0.00
8	1.00	0.00
9	1.00	0.00
10	1.00	0.00
11	0.00	1.00
12	0.00	1.00
13	0.00	1.00
14	0.00	1.00
15	0.00	1.00
16	0.00	1.00
17	0.01	0.99
18	0.08	0.92
18	0.00	1.00
20	0.00	1.00
$\mathbf{v}_i^{(3)}$	(0.24, 0.07)	(0.19, −1.60)
$\mathbf{s}_i^{(3)}$	(−1.00, 0.00)	(0.06, 1.00)

$r = 1$. With the exception of point $\mathbf{x}_6$, (A23.1) produces exactly the same maximum membership partition of X as (A22.1). Vectors $\mathbf{x}_6$ and $\mathbf{x}_{18}$ (cf. Fig. 23.3) are at the centers of the cross: precisely the two points which (A22.1) labeled correctly by chance. Thus, it seems fair to assert that the FCL solution displayed here produces—in *memberships*—the same results as (A22.1). There is, however, a significant difference between these two solutions: it is in the prototypes generated by each. In (E22.1), (A22.1) clustered X correctly even though the prototypes for $\hat{J}_2$ in (22.1) are vectors $\{\mathbf{v}_i\}$—that is, *points* in $\mathbb{R}^2$. It is a measure of credit for Gustafson and Kessel's algorithm that $\hat{J}_2$ successfully forced adoption of local norms A_1^* and A_2^* which generate accurate memberships, even though the prototypes are of the "wrong dimension"! On the other hand, the prototypical lines $\{L_i(\hat{\mathbf{v}}_i; \hat{\mathbf{s}}_i)\}$ identified by the fuzzy 2-lines algorithm not only help J_{V12} find accurate membership, but provide an explicit *representation* of the prototypical linear

Table 23.2. Covariance Estimates via (A22.1) and (A23.1): (E23.1)

Matrix	Class 1 $\{\mathbf{x}_1, \ldots, \mathbf{x}_{10}\}$		Class 2 $\{\mathbf{x}_{11}, \ldots, \mathbf{x}_{20}\}$	
Sample covariance matrix C_i	33.92	−0.03	0.13	1.50
	−0.03	0.07	1.50	24.44
Fuzzy covariance matrix C_{fi} (A22.1)	36.33	0.88	0.06	0.72
	0.88	0.05	0.72	25.65
Fuzzy covariance matrix C_{fi} (A23.1)	38.24	−0.02	0.12	1.41
	−0.02	0.08	1.41	22.95

substructure in X. The equations of these linear approximants are given in Fig. 23.3: the lines were actually drawn *as shown* by a graphics plotter using a clipper routine to truncate the infinite extent of each L_i based on the values $\{u_{ik}^{(3)}\}$. Visually at least, this pair of lines does seem to "best fit" the two linear clusters in X.

Finally, we compare the fuzzy covariance matrices $\{C_{fi}\}$ generated by algorithms (A22.1) and (A23.1) with the (sub)sample covariance matrices. Table 23.2 contains these matrices. The numbers speak for themselves! Both algorithms provide a rather accurate means for estimation of actual subclass covariance structure. On the face of it, Table 23.2 suggests little advantage for either method. (A23.1) is, however, a much more stable and computationally efficient process, since the extraction of r eigenvalues and eigenvectors from each S_{fiA} is much more efficient (and numerically reliable) than taking determinants and inverses. On the other hand, J_{Vrm} is restricted to c linear varieties of equal *dimension* (and therefore similar shapes), whereas $\hat{J}_m$ ostensibly will adapt to different subshapes in the same data! In any case (E23.1) provides convincing evidence to support the conjecture that these algorithms may be highly successful for certain statistical estimation problems. This point is emphasized at length in S26. •

The analysis of convergence for the FCV algorithms (A23.1) with $r \geq 1$ is completely analogous to the theorems of S12 which deal with the case $r = 0$ in all respects except one. This exception concerns the extension of (T12.3) to the FCV operator T_{rm}: since (A23.1d) requires the extraction of eigenvectors during each iteration of (A23.1), the formalization of T_{rm} as a composition of four functions includes one factor that is a point to *set* mapping. Consequently, (T12.1) cannot be applied to (A23.1) directly, because (12.5b) is insufficient for the results (12.6a) and (12.6b). Strict convergence of J_{Vrm} on the iterates of T_{rm} proceeds exactly as in S12: if S^* is the set of strict local minima of J_{Vrm}, then J_{Vrm} is a descent functional for (T_{rm}, S^*). Further, the iterates of T_{rm} are confined to a compact set, namely, $M_{fc} \times [\text{conv}(X)]^c \times [\delta B(\mathbf{0}, 1)]^c$, $\delta B(\mathbf{0}, 1)$ being the (surface of) the unit ball in $\mathbb{R}^p$. However, the extension of (T12.3) to the present case requires verification that T_{rm} is *closed* (in the sense of Zangwill[(125)]). It is shown in (17) that T_{rm} is closed if and only if the extraction of principal components of each $S_{fiA}^{(l)}$ is closed. Following the plan of (T23.1), we present the proof for $r = 1$; (17) contains details in the general case.

Let Ω be any set, and let $P(\Omega)$ denote its power set. A point-to-set map $g: \Omega \to P(\Omega)$, has, for $x \in \Omega$, values $g[x] \in P(\Omega)$ which are *subsets* of Ω. If d and d' denote metrics on Ω, $P(\Omega)$ respectively, then convergence of sequences in metric spaces (Ω, d) and $(P(\Omega), d')$ is well defined. Under these circumstances, we say that g is *closed* at $x_0 \in \Omega$ in case

$$\left.\begin{cases} \text{(i)} & \{x_k\} \to x_0 \\ \text{(ii)} & y_k \in g[x_k]\ \forall k \\ \text{(iii)} & \{y_k\} \to y_0 \end{cases}\right\} \Rightarrow y_0 \in g[x_0] \tag{23.19}$$

If g is a point-to-point map, (23.19) follows whenever g is continuous; T_{0m} in S12 was closed because each of its factors was a continuous map. The closedness of T_{rm} hinges upon verifying that the functions $\{g_i\}$ which operate on S_{fiA} are closed. This is the content of the following theorem.

(T23.3) *Theorem 23.3.* Let $E_i^{(l)}$ denote the eigenspace of $S_{fiA}^{(l)}$ associated with its largest eigenvalue. Let g_i be the function mapping $S_{fiA}^{(l)}$ onto this eigenspace:

$$g_i = \{S_{fiA}^{(l)} \to E_i^{(l)}\} \tag{23.20}$$

Then g_i is closed on the set Ω of all bounded, symmetric matrices in V_{pp}.

Analysis and Proof. This result ensures that each iteration of (A23.1) is "well behaved" in the sense (23.19). The proof depends on one crucial fact: viz., that the eigenvalues of any matrix are continuous functions of its entries. Let $\{B_k\}$ be a sequence in Ω which converges to $B \in \Omega$; let λ_k, λ be (one of) the eigenvalues of B_k, B, respectively, of maximum value; and let $E[B_k]$, $E[B]$ denote eigenspaces of B_k, B for λ_k, λ. Finally, let $\mathbf{s}_k \in E[B_k]$ be a unit eigenvector of B_k $\forall k$, and suppose $\{\mathbf{s}_k\} \to \mathbf{s}$. We assume $\{B_k\} \to B$ in the sup norm on Ω, which is consistent with the vector norm on $\mathbb{R}^p$ in which $\{\mathbf{s}_k\} \to \mathbf{s}$. (23.19) requires us to show that $\mathbf{s} \in E[B]$, i.e., that $B\mathbf{s} = \lambda\mathbf{s}$.

$$0 \leqslant \|B\mathbf{s} - \lambda\mathbf{s}\| \leqslant \left\{\begin{matrix} \|B\mathbf{s} - B_k\mathbf{s}\| \\ + \\ \|B_k\mathbf{s} - B_k\mathbf{s}_k\| \\ + \\ \|B_k\mathbf{s}_k - \lambda_k\mathbf{s}_k\| \\ + \\ \|\lambda_k\mathbf{s}_k - \lambda_k\mathbf{s}\| \\ + \\ \|\lambda_k\mathbf{s} - \lambda\mathbf{s}\| \end{matrix}\right\} \leqslant \left\{\begin{matrix} \|B - B_k\|\cdot\|\mathbf{s}\| & \text{(a)} \\ + & \\ \|B_k\|\cdot\|\mathbf{s} - \mathbf{s}_k\| & \text{(b)} \\ + & \\ \|B_k\mathbf{s}_k - \lambda_k\mathbf{s}_k\| & \text{(c)} \\ + & \\ |\lambda_k|\cdot\|\mathbf{s}_k - \mathbf{s}\| & \text{(d)} \\ + & \\ |\lambda_k - \lambda|\cdot\|\mathbf{s}\| & \text{(e)} \end{matrix}\right.$$

Now as $k \to \infty$, we have for the terms (a)–(e)

(a) $\{\|B - B_k\|\cdot\|\mathbf{s}\|\} \to 0\cdot 1 = 0$.
(b) $\{\|B_k\|\cdot\|\mathbf{s} - \mathbf{s}_k\|\} \to \|B\|\cdot 0 = 0$ because $\|B\| < \infty$ in sup norm.
(c) $\|B_k\mathbf{s}_k - \lambda_k\mathbf{s}_k\| = 0$ because $\mathbf{s}_k \in E[B_k]\forall k$.
(d) $\{|\lambda_k|\cdot\|\mathbf{s}_k - \mathbf{s}\|\} \to |\lambda|\cdot 0 = 0$.
(e) $\{|\lambda_k - \lambda|\cdot\|\mathbf{s}\|\} \to 0\cdot 1 = 0$, because the eigenvalues of any matrix are continuous functions of the elements of the matrix: in particular, $\{B_k\} \to B \Rightarrow \{\lambda_k\} \to \lambda$ in $\mathbb{R}$.

Thus, as $k \to \infty$, $0 \leqslant \|B\mathbf{s} - \lambda\mathbf{s}\| \leqslant 0$, so $\lambda \in E[B]$, and g_i is closed at B for any $B \in \Omega$ such that $\|B\|_{\sup} < \infty$. In particular, each $S_{fiA}^{(l)}$ in (23.10c) is so bounded, because X and $U^{(l)}$ always are. •

Combining this result with the fact that all additional terms in the decomposition of T_{rm} are continuous point-to-point maps leads to the following result for (A23.1).

(T23.4) *Theorem 23.4.*[17] Let $2 \leqslant c < n$, $m > 1$, $0 \leqslant r < p$. If $D_{ik}^{(l)} > 0\ \forall i, k$, and l, and $X = \{\mathbf{x}_1, \mathbf{x}_2, \ldots, \mathbf{x}_n\}$ is bounded in $\mathbb{R}^p$, then for any initial $(U^{(0)}, \mathbf{V}^{(0)}) \in M_{fc} \times [\text{conv}(X)]^c \times [\delta B(\boldsymbol{\theta}, 1)]^{cr}$, the iterates of (A23.1) either converge to a strict local minimum of J_{Vrm}, or, at worst, the limit of every convergent subsequence generated by (A23.1) is a strict local minimum of J_{Vrm}.

Analysis and Proof. This is a direct generalization of (T12.5); its proof can be found in detail in (17). We emphasize that the disclaimers concerning (T12.5) apply here. Specifically, readers are again reminded that local minima of J_{Vrm} need *not* be "good" clusterings of X. Moreover, the *number* of local minima of J_{Vrm} (and hence number of potentially misleading algorithmically suggested "optimal" clusterings) seems to increase dramatically with r—a point we return to in (S24). Nonetheless, (T23.4) provides some theoretical reassurance concerning the loop in (A23.1). •

Although the fuzzy c-line algorithms ($r = 1$) have occupied the spotlight so far, this by no means precludes interest in algorithms (A23.1) for $r > 1$. At $r = 2$, the prototypical linear varieties $\{V_{ri}\}$ are planes in $\mathbb{R}^p$. The data in (E23.2) are essentially a three-dimensional analog of Gustafson's cross.

(E23.2) *Example 23.2.* Data set X for (E23.2) consists of $n = 20$ vectors $\{\mathbf{x}_k\} \subset \mathbb{R}^3$. These vectors are depicted graphically in Fig. 23.4 in all three dimensions. The lower portion of Fig. 23.4 portrays their projections

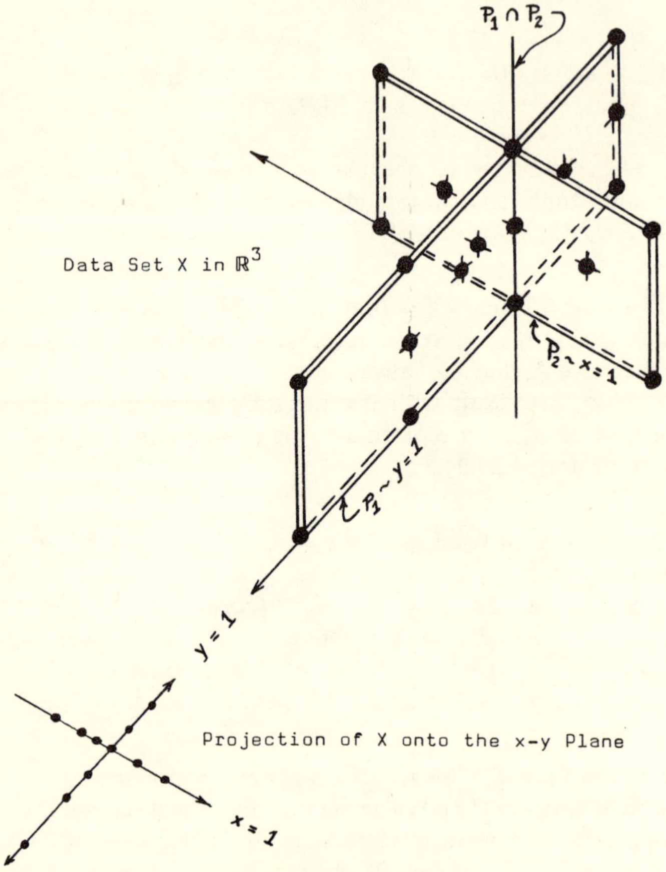

Figure 23.4. Two planar clusters in $\mathbb{R}^3$: (E23.2).

Table 23.3 Planar Data and 2-Partitions: (E23.2)

				Terminal memberships in cluster $u_1^{(l)}$		
Data vector $\mathbf{x}_k$				(A11.1) Means $r = 0$ $l = 8$	(A23.1) Lines $r = 1$ $l = 12$	(A23.1) Planes $r = 2$ $l = 2$
k	x_k	y_k	z_k			
1	1	0	0	0.66	0.79	1.00
2	1	0	1	0.65	0.87	1.00
3	1	0.5	0.5	0.82	0.70	1.00
4	1	2	0	0.63	0.99	1.00
5	1	2	1	0.62	0.23	1.00
6	1	1.5	0.5	0.79	0.77	1.00
7	1	0.75	0.25	0.88	0.89	1.00
8	1	1	1	0.79	0.98	0.50
9	1	1	0	0.82	0.92	0.50
10	1	1	0.5	0.93	0.40	0.50
11	0	1	0	0.85	0.02	0.00
12	0	1	1	0.84	0.02	0.00
13	0	1	0.5	0.88	0.09	0.00
14	0.5	1	0.5	0.98	0.20	0.00
15	2	1	0	0.12	0.13	0.00
16	2	1	1	0.11	0.20	0.00
17	2	1	0.5	0.00	0.00	0.00
18	1.5	1	0.5	0.35	0.05	0.00
19	3	1	0	0.17	0.07	0.00
20	3	1	1	0.17	0.04	0.00

onto the (x, y)-plane; this shows that X is quite similar to the data of (E22.1), except that each point gains substance in a third direction. The data are listed numerically in Table 23.3: analytically, vectors 1 to 10 lie in the plane $P_2 \sim x = 1$; vectors 11 to 20 lie in the plane $P_1 \sim y = 1$; and vectors 8, 9, 10 also lie in the line of intersection $(x = y = 1)$ of P_1 and P_2. These data, then, consist of two *planar* clusters in $\mathbb{R}^3$, so the prototypes generated by FCM $(r = 0)$ and FCL $(r = 1)$ would seem to have the wrong number of dimensions, or "degrees of freedom" for representation of the apparent substructure in Fig. 23.4. To test the supposition that fuzzy 2-planes is more amenable to X than 2-means or 2-lines, (A23.1) was applied to X with $c = m = 2$; $A = I$, the 3×3 identity matrix; $\varepsilon_L = 0.01$ and

$$\|U^{(l+1)} - U^{(l)}\| = \sum_{k=1}^{20} \sum_{i=1}^{2} |u_{ik}^{(l+1)} - u_{ik}^{(l)}| \leq \varepsilon_L$$

as the termination criterion. The initial guess for $U^{(0)} \in M_{fc}$ assigned membership $u_{1k}^{(0)} = 0.70$ to $\mathbf{x}_1$ through $\mathbf{x}_{10}$; memberships $u_{2k}^{(0)} = 0.30$ to $\mathbf{x}_{11}$ through $\mathbf{x}_{20}$. One run each of fuzzy 2-means ($r = 0$); 2-lines ($r = 1$); and 2-planes ($r = 2$) using these parameters was made. The results are reported numerically in Table 23.3. Fuzzy 2-means converged in $l = 8$ iterations to a partition whose maximum memberships group together the first 14 points: the last six points are the "forward" portion of plane $y = 1$ in Fig. 23.3. Points 11–14 are mislabeled: further, points 8, 9, and 10 do not indicate equal preference for the two (visual) planar clusters, as suggested by their geometric placement on the line of intersection of P_1 and P_2.

The fuzzy 2-lines algorithm produced in $l = 12$ iterations the memberships listed in column 6 of Table 23.3. The maximum membership partitioning of X derived from these numbers has but two mislabeled points, $\mathbf{x}_5$ and $\mathbf{x}_{10}$. Since $\mathbf{x}_{10}$ is in both planes, however, only $\mathbf{x}_5$ is really misleading. Consequently, it can be asserted that FCL does a much better job than FCM for these data.

Finally, the fuzzy 2-planes algorithm converged in $l = 2$ iterations to the partition exhibited in column 7 of Table 23.3. This partition clearly indicates exactly the apparent visual planar substructure in X. Since $m = 2$, the occurrence of 17 hard memberships in this U suggests that the partition and prototypical planes found are a quite "stable" local minimum of J_{V22}. This seems to corroborate the theoretical presumption that when data do have c clusters of linear varietal form for some r, $0 \leq r < p$, J_{Vrm} will show a much stronger affinity for tight clusters about prototypical varieties of the correct dimension than will J_{Vqm}, $q \neq r$. Note finally that points 8, 9, and 10 receive precisely the memberships they seem to deserve: 0.50 in *each* of the two planar clusters. For the record, the two prototypical planes characterizing the planar structure in X were

$$P_1(\mathbf{v}_1; \mathbf{s}_{11}, \mathbf{s}_{12}) \sim \left\{ \begin{aligned} \mathbf{v}_1 &= (1, 0.97, 0.47) \\ \mathbf{s}_{11} &= (0, -0.01, 0.99) \\ \mathbf{s}_{12} &= (0, -0.99, -0.01) \end{aligned} \right\}$$

$$P_2(\mathbf{v}_2; \mathbf{s}_{21}, \mathbf{s}_{22}) \sim \left\{ \begin{aligned} \mathbf{v}_2 &= (1.37, 1, 1.06) \\ \mathbf{s}_{21} &= (-0.84, 0, 0.54) \\ \mathbf{s}_{22} &= (0.54, 0, 0.84) \end{aligned} \right\}$$

Remarks

Algorithms (A22.1) and (A23.1) represent two attempts to mitigate the difficulties geometrically differing shapes cause for objective functions that measure (dis)similarities with norm metrics on $\mathbb{R}^p$. (A23.1) is computationally superior to (A22.1), and provides prototypical characterizations (the c-varieties) which can be used to design nearest prototype classifiers (cf. Chapter 6): on the other hand, (A22.1) seems more flexible when a fixed data set has locally different shapes—e.g., one spherical cluster, two linear ones, and a planar structure all in the same data. Both of these algorithms appear potentially useful for a variety of applications, in particular, for character recognition and parametric estimation in mixture analysis. The estimation of covariance matrices is especially intriguing, as is the possible connection between factor analysis and fuzzy c-varieties. The question of cluster validity with respect to c needs to be studied, and for (A23.1), with respect to r as well. Thus, given data in $\mathbb{R}^p$ with $p \geq 3$, how do we *suspect* that clusters are r-dimensional? It is worth mentioning that other numerical examples of fuzzy c-lines are given in (16) for $c > 2$, including one 400-point data set involving the count rate of atomic oxygen as a function of altitude in the upper atmosphere. Finally, an empirical observation: the *number* of local trap states for $\hat{J}_m$ seems large: the number for J_{Vrm} seems to grow significantly with r. Consequently, the problems of initialization and "stability" for (A22.1) and (A23.1) are significant ones that warrant further study. Moreover, singularities ($D_{ik} = 0$) are much more prevalent for (A23.1) than for (A11.1); this necessitates greater attention to numerical programming details than fuzzy c-means requires.

Exercises

H23.1. Verify formula (23.4).

H23.2. Describe a physical process which generates linear clusters; planar clusters.

H23.3. Let $\{\mathbf{x}_1, \mathbf{x}_2, \ldots, \mathbf{x}_r\} \subset \mathbb{R}^p$; $\mathbf{x}_k \in \mathbb{R}^p\ \forall k$. Let A be any symmetric matrix in V_{pp}, with $p \geq r$; and let $\{(\lambda_j, \mathbf{y}_j)\}$ be the eigenvalue–unit-eigenvector pairs from A for its r largest eigenvalues. Prove that $Q(\mathbf{y}_1, \mathbf{y}_2, \ldots, \mathbf{y}_r) = \sum_{j=1}^r \lambda_j$ is the maximum of $Q(\mathbf{x}_1, \mathbf{x}_2, \ldots, \mathbf{x}_r) = \sum_{j=1}^r \mathbf{x}_j^T A \mathbf{x}_j$, over all vectors $\{\mathbf{x}_j\} \subset \mathbb{R}^p$ such that $\langle \mathbf{x}_j, \mathbf{x}_k \rangle = \delta_{jk} = 1$, $j = k$; 0 otherwise.[16]

H23.4. Let $A: \mathbb{R} \to P(\mathbb{R})$ be the point-to-set map

$$A(y) = \left\{\begin{matrix} [0, y], & 0 < y \leq 1 \\ 0, & y = 0 \end{matrix}\right\}$$

Take $S^* = \{0\}$, and show that

(i) $g(y) = y$ is a descent function for (A, S^*);

(ii) $\{y_{n+1} = A(y_n)\} \to (1/2)$ for $y_0 = 1$;

(iii) A is not closed on $\mathbb{R} - \{0\}$.

S24. Convex Combinations: Fuzzy c-Elliptotypes

The examples of S23 suggest that the fuzzy c-varieties algorithms are reliable for the detection and characterization of substructure in X when all c clusters are essentially r-dimensional and linear varietal in shape. There may be, however, a disadvantage of fuzzy c-varieties (of any dimension > 1) which is intrinsic to the variety itself, namely, "size." For example, lines (varieties of dimension 1) have infinite length, and it may be the case that collinear clusters which are in fact widely separated in $\mathbb{R}^p$ will not be identified as such by a FCL algorithm. This situation is depicted graphically in Fig. 24.1, where four visually apparent linear clusters can be fitted quite accurately by $c = 2$ prototypical straight lines.

This difficulty suggests that the utility of J_{V1m} might be considerably improved by "leavening" its propensity for linear structure, by forcing each cluster to contain a center of mass ($\mathbf{v}_i$) in or near its convex hull. Although the natural supposition stemming from these remarks would be to form a convex combination of J_{V0m} and J_{V1m}, it is a remarkable fact that *arbitrary* convex combinations of all the J_{Vrm}'s are minimized over choices of dimensionally different varieties simply by using the linear varieties which are necessary for minimization of the individual term of highest dimension. To formulate the general result, let

$$\{r_i\} \text{ be } t \text{ distinct integers in } \{0, 1, 2, \ldots, p-1\} \tag{24.1a}$$

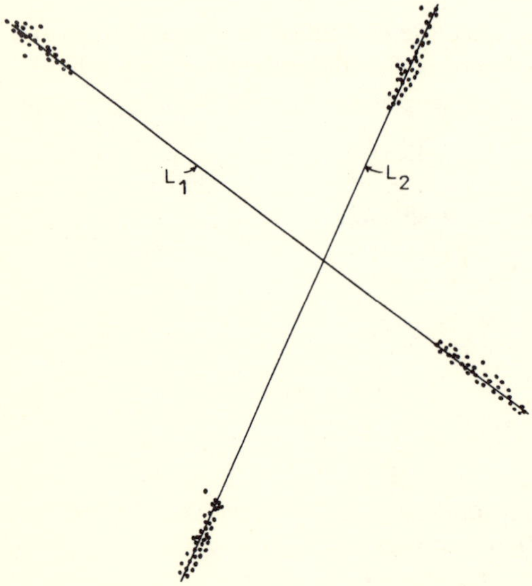

Figure 24.1. Linear varieties can be "too large." Two one-dimensional linear varieties, L_1 and L_2, fit $c = 4$ (co)linear clusters.

$\{\alpha_i\}$ be t numbers in (0, 1) with $\sum_{i=1}^{t} \alpha_i = 1$ (24.1b)

$\{\mathbf{V}_{r_i}\}$ be t sets of c linear varieties as in (D23.1), except here,

$$\mathbf{V}_{r_i} = (V_{r_{i1}}, \ldots, V_{r_{ic}}); \qquad \dim(V_{r_{ij}}) = r_i \;\forall j \text{ at each } i \tag{24.1c}$$

Define the functional which is a convex combination of the t r_i-dimensional $J_{V_{r_i}}$'s as

$$_{\{\alpha_i\}}J_{\{V_{r_i}\}m}(U, \mathbf{V}_{r_1}, \ldots, \mathbf{V}_{r_t}) = \sum_{i=1}^{t} \alpha_i J_{V_{r_i m}}(U, \mathbf{V}_{r_i}) \tag{24.2}$$

Our remarks above concerning this functional are substantiated in the following theorem

(T24.1) *Theorem 24.1.* Let $\hat{U} \in M_{fc}$ be fixed. For fixed $m > 1$ and fixed $\{r_i\}$, $\{\alpha_i\}$, and $\{\mathbf{V}_{r_i}\}$ as in (24.1), $(\mathbf{V}_{r_1}, \mathbf{V}_{r_2}, \ldots, \mathbf{V}_{r_t})$ is a local minimum of $_{\{\alpha_i\}}J_{\{V_{r_i}\}m}$ $(\hat{U}, \mathbf{V}_{r_1}, \ldots, \mathbf{V}_{r_t})$ if and only if $\mathbf{V}_{r_i}$ is determined, for *every* r_i, by (23.10a), (23.10b), and (23.10c) using only $\mathbf{V}_{r_q}$, where $r_q = \max_i\{r_i\}$.

Analysis. The key idea here is that each term of the functional at (24.2) utilizes r_i eigenvectors from the *same pool*, i.e., generated by the same S_{fjA} for cluster $\hat{u}_j \in \hat{U}$. Thus, adding on more "degrees of freedom" to the same prototypical varieties used for lower dimensions simply adds to the dimensions used previously.

Proof.

$$\min\left[\sum_{i=1}^{t} \alpha_i J_{V_{r_i m}}(\hat{U}, \mathbf{V}_{r_i})\right] \geq \sum_{i=1}^{t} \alpha_i\{\min[J_{\mathbf{V}_{r_i m}}(\hat{U}, \mathbf{V}_{r_i})]\}$$

$$\geq \left(\sum_{i=1}^{t} \alpha_i\right) J_{V_{r_q m}}(\hat{U}, \mathbf{V}_{r_q}) = J_{V_{r_q m}}(\hat{U}, \mathbf{V}_{r_q})$$

The reasoning behind the inequalities above is this: the "center of mass" of cluster j in variety $V_{r_{ij}}$ [a vector $\mathbf{v}_{r_{ij}}$ in place of $\mathbf{v}$ in (23.1)] lies in $V_{r_{qj}}$ for *every* r_i; moreover, choosing the r_q eigenvectors of S_{fjA} corresponding to its r_q largest eigenvalues *automatically* minimizes all of the lower-dimensional terms over choices of orthonormal spanning sets, because lower-dimensional FCV functionals simply use less eigenvectors from the same set. In other words, the varieties needed to minimize each of the $J_{V_{r_i m}}$'s are contained in the varieties needed to minimize $J_{V_{r_q m}}$, so to minimize (24.2) for fixed $\hat{U} \in M_{fc}$ is to minimize $J_{V_{r_q m}}$ alone. •

It remains to be seen how, for fixed $\{\hat{\mathbf{V}}_{r_i}\}$, one should choose $\hat{U} \in M_{fc}$ to minimize (24.2) over its first argument. There is a practical consideration, however, which dampens our enthusiasm for the general result, and it is this: convex combinations of FCV functionals which do not include J_{V0m} will not eliminate the "size" problem exemplified in Fig. 24.1. Accordingly, our attention is focused here on the case of immediate practical interest in connection with Fig. 24.2—a convex combination of J_{V0m} and J_{V1m}: let $\alpha \in [0, 1]$, and define

$$_{\alpha}J_{01m}(U, \mathbf{V}_0, \mathbf{V}_1) = (1-\alpha)J_{V0m}(U, \mathbf{V}_0) + \alpha J_{V1m}(U, \mathbf{V}_1) \tag{24.3}$$

where $\mathbf{V}_0 = (\mathbf{v}_1, \mathbf{v}_2, \ldots, \mathbf{v}_c) = \mathbf{v} \in \mathbb{R}^{cp}$ are c zero-dimensional prototypes (fuzzy c-means); and $\mathbf{V}_1 = (V_1, V_2, \ldots, V_c)$ are c one-dimensional varieties (fuzzy c-lines), say $V_i = L(\mathbf{v}_i; \mathbf{s}_i)$ with $\mathbf{v}_i$ the point through which line V_i passes in direction $\mathbf{s}_i$. Note that (T24.1) guarantees that the vectors $\{\mathbf{v}_i\}$ of $\mathbf{V}_0$ are the *same* vectors as the centers of mass $\{\mathbf{v}_i\}$ for the lines comprising $\mathbf{V}_1$. Heuristically, J_{01m} "penalizes" membership u_{ik} of $\mathbf{x}_k$ in cluster i if $\mathbf{x}_k$ is far from the center of mass $\mathbf{v}_i$, even though it may be quite close to the prototypical line V_i. Since it is no more trouble to find $\hat{U}$ for $_{\alpha}J_{rqm}$, where, say $0 \le r < q \le p-1$, consider the functional

$$\phi(U) = (1-\alpha)J_{Vrm}(U, \hat{\mathbf{V}}_r) + \alpha J_{Vqm}(U, \hat{\mathbf{V}}_q)$$

Assuming that $\sum_1^0 (\cdot) \doteq 0$, we have, since $\hat{V}_r \subset \hat{V}_q$:

$$\begin{aligned}
\phi(U) &= (1-\alpha) \sum_{i=1}^{c} \sum_{k=1}^{n} (u_{ik})^m \Big(\langle \mathbf{x}_k - \hat{\mathbf{v}}_i, \mathbf{x}_k - \hat{\mathbf{v}}_i \rangle_A - \sum_{j=1}^{r} \langle \mathbf{x}_k - \hat{\mathbf{v}}_i, \hat{\mathbf{s}}_{ij} \rangle_A^2 \Big) \\
&\quad + \alpha \sum_{i=1}^{c} \sum_{k=1}^{n} (u_{ik})^m \Big(\langle \mathbf{x}_k - \hat{\mathbf{v}}_i, \mathbf{x}_k - \hat{\mathbf{v}}_i \rangle_A - \sum_{j=1}^{q} \langle \mathbf{x}_k - \hat{\mathbf{v}}_i, \hat{\mathbf{s}}_{ij} \rangle_A^2 \Big) \\
&= \sum_{i=1}^{c} \sum_{k=1}^{n} (u_{ik})^m \Big(\langle \mathbf{x}_k - \hat{\mathbf{v}}_i, \mathbf{x}_k - \hat{\mathbf{v}}_i \rangle_A - \sum_{j=1}^{r} \langle \mathbf{x}_k - \hat{\mathbf{v}}_i, \hat{\mathbf{s}}_{ij} \rangle_A^2 \\
&\quad - \alpha \sum_{j=r+1}^{q} \langle \mathbf{x}_k - \hat{\mathbf{v}}_i, \hat{\mathbf{s}}_{ij} \rangle_A^2 \Big)
\end{aligned} \tag{24.4}$$

Setting

$$\hat{\gamma}_{ik} = \Big(\langle \mathbf{x}_k - \hat{\mathbf{v}}_i, \mathbf{x}_k - \hat{\mathbf{v}}_i \rangle_A - \sum_{j=1}^{r} \langle \mathbf{x}_k - \hat{\mathbf{v}}_i, \hat{\mathbf{s}}_{ij} \rangle_A^2 - \alpha \sum_{j=r+1}^{q} \langle \mathbf{x}_k - \hat{\mathbf{v}}_i, \hat{\mathbf{s}}_{ij} \rangle_A^2 \Big) \tag{24.5}$$

We get

$$\phi(U) = \sum_{i=1}^{c} \sum_{k=1}^{n} (u_{ik})^m \hat{\gamma}_{ik}$$

$\hat{\gamma}_{ik}$ is always ≥ 0 since it is a linear combination of distance from a point $\mathbf{x}_k$ to each of two varieties. With the result of (P23.1) applied to the numbers $\{\hat{\gamma}_{ik}\}$ instead of $\{D_{ik}\}$, we have proven the following theorem.

(T24.2) *Theorem 24.2.* If $0 \le r < q \le p-1$; $\hat{\mathbf{V}}_q$ is fixed in $(R^{cp})^{q+1}$; and $\hat{\gamma}_{ik} > 0 \ \forall i, k$, then $\hat{U} \in M_{fc}$ is a strict local minimum of ${}_\alpha J_{rqm}(U, \hat{\mathbf{V}}_p, \hat{\mathbf{V}}_q) = (1-\alpha)J_{V_{rm}}(U, \hat{\mathbf{V}}_r) + \alpha J_{V_{qm}}(U, \hat{\mathbf{V}}_q)$ if and only if for $m > 1$,

$$\hat{u}_{ik} = \left[\sum_{j=1}^{c}\left(\frac{\hat{\gamma}_{ik}}{\hat{\gamma}_{jk}}\right)\right]^{-1/(m-1)} \qquad \forall i, k \tag{24.6}$$

where $\hat{\gamma}_{ik}$ is calculated via (24.5) with parameters from $\hat{\mathbf{V}}_q$.

In practice, the use of convex combinations which do not include J_{V0m} will not eliminate the "size" problem exemplified in Fig. 24.1, hence we single out the special case of (24.4) and (24.5) for which $r = 0$, which yields

$$\hat{\gamma}_{ik} = \langle \mathbf{x}_k - \hat{\mathbf{v}}_i, \mathbf{x}_k - \hat{\mathbf{v}}_i \rangle_A - \alpha \sum_{j=1}^{q} \langle \mathbf{x}_k - \hat{\mathbf{v}}_i, \hat{\mathbf{s}}_{ij} \rangle_A^2 \tag{24.7}$$

A further simplification, setting $q = 1$, yields

$$\hat{\gamma}_{ik} = \langle \mathbf{x}_k - \hat{\mathbf{v}}_i, \mathbf{x}_k - \hat{\mathbf{v}}_i \rangle_A - \alpha \langle \mathbf{x}_k - \hat{\mathbf{v}}_i, \hat{\mathbf{s}}_i \rangle_A^2 \tag{24.8}$$

as the necessary "distances" for calculation of $\hat{U}$ via (24.6) for minimizing ${}_\alpha J_{01m}$ over M_{fc}. Theorems 24.1 and 24.2 provide necessary conditions for minimization via Picard iteration of any two-term convex combination ${}_\alpha J_{rqm}$ of fuzzy c-varieties functionals: in particular, this is the case for ${}_\alpha J_{01m}$, a convex combination of FCM and FCL. Before this algorithm is formalized, we consider the geometric property of clusters for which ${}_\alpha J_{01m}$ is a seemingly natural criterion.

A geometric interpretation of ${}_\alpha J_{01m}$ can be made by considering level surfaces of convex combinations. This functional uses a convex combination of the square of the distance from a data point to the "center of mass" and the square of the distance from the same point to a linear prototype. Let us examine level surfaces of such combinations. Specifically, let the point prototype be the origin in p-space, and the line prototype (through the origin of $\mathbb{R}^p$) be the axis specified by the vector

$$\mathbf{s}^T = (x_1, 0, \ldots, 0) \qquad \text{in } \mathbb{R}^p$$

Consider the set of points satisfying

$$(1-\alpha)\langle \mathbf{x}, \mathbf{x} \rangle_A + \alpha \langle \mathbf{y}, \mathbf{y} \rangle_A = k$$

where

$$\mathbf{x}^T = (x_1, \ldots, x_p)$$
$$\mathbf{y}^T = (0, x_2, \ldots, x_p)$$

For the case $A = I$ we have

$$(1-\alpha)(x_1^2 + x_2^2 + \cdots + x_p^2) + \alpha(x_2^2 + \cdots + x_p^2) = k$$

or

$$(1-\alpha)(x_1^2) + (x_2^2 + \cdots + x_p^2) = k$$

Hence level surfaces of ${}_\alpha J_{01m}$ are ellipsoids of a special kind, wherein a hypersphere has been stretched along one axis. In other words, ${}_\alpha J_{01m}$ measures the "hyperellipticity" of clusters; J_{V0m} assesses their "spherical part"; J_{V1m} their "linear part." As α changes from 1 to 0, J_{01m} has level surfaces in $\mathbb{R}^p$ which deform from linear to hyperspherical in shape, and are hyperellipsoidal for $0 < \alpha < 1$.

Since the *prototypes* generated by ${}_\alpha J_{01m}$ are neither lines nor hyperellipses for $0 < \alpha < 1$, these surfaces will be called *elliptotypes*; and the family of Picard algorithms using necessary conditions above for minimization of ${}_\alpha J_{01m}$ we call the *fuzzy c-elliptotypes* (FCE) family. Note that the vectors $\{\hat{\mathbf{s}}_{ij}\}$ derived from the fuzzy scatter matrices $\{\hat{S}_{fiA}\}$ associated with the prototypes of the lines portion of ${}_\alpha J_{01m}$ are *principal directions of the elliptotypes.* With this terminology established, we formalize the infinite family of fuzzy c-elliptotypes (FCE) algorithms based on (T24.1) and (T24.2) for minimization of ${}_\alpha J_{01m}$.

(A24.1) *Algorithm 24.1* [*Fuzzy c-Elliptotypes* (*FCE*), *Bezdek et al.*[17]]

(A24.1a) Fix c, $2 \leqslant c < n$; $r = 0$; $q = 1 < p$; $m, 1 < m < \infty$. Choose any positive-definite matrix A in V_{pp}. Choose lines weight α, $0 \leqslant \alpha \leqslant 1$. Initialize $U^{(0)} \in M_{fc}$. Then at step l, $l = 0, 1, 2, \ldots$:

(A24.1b) Calculate the c fuzzy cluster centers $\{\mathbf{v}_i^{(l)}\}$ using $U^{(l)}$ and (23.10a).

(A24.1c) Calculate the c generalized fuzzy scatter matrices $\{S_{fiA}^{(l)}\}$ using $U^{(l)}$, $\{\mathbf{v}_i^{(l)}\}$, A, and (23.10c).

(A24.1d) Extract from each $S_{fiA}^{(l)}$ its ($q = 1$) first principal unit eigenvector $\mathbf{y}_i^{(l)}$, and from it construct direction vector $\mathbf{s}_i^{(l)}$ using (23.10b).

(A24.1e) Update $U^{(l)}$ to $U^{(l+1)}$ using (24.6) and "distances" $\{\gamma_{ik}^{(l)}\}$ via (24.8).

(A24.1f) Compare $U^{(l+1)}$ to $U^{(l)}$ in a convenient matrix norm. If $\|U^{(l+1)} - U^{(l)}\| \leqslant \varepsilon_L$, stop. Otherwise, put $l = l + 1$ and return to (A24.1b).

As noted above, algorithms for minimization of ${}_\alpha J_{rqm}$ for any r, q such that $0 \leqslant r < q < p$ are well defined through formulas (24.5) and (24.6), in combination with equations (23.10). The only changes needed in (A24.1) for this more general case occur in (A24.1a), where one specifies r and q; in (A24.1d), where one extracts q principal eigenvectors from $S_{fiA}^{(l)}$; and in (A24.1e), where (24.5) must be used for the $\{\gamma_{ik}^{(l)}\}$ instead of (24.8). Remarks concerning the convergence properties of (A24.1) below obtain without change for this more general class of elliptotypes algorithms.

Convergence properties of (A24.1) are identical to those of (A23.1). Detailed proof of this fact is given in (17), where it is shown that *every* sequence generated by (A24.1) either converges to a strict local minimum of ${}_{\alpha}J_{rqm}$, or every convergent subsequence of such a sequence does so. This is true for every $0 \leqslant r < q \leqslant p-1$, not just $r = 0$, $q = 1$; and for every α, $0 \leqslant \alpha \leqslant 1$. If $\alpha = r = 0$, this reduces to (T12.5), convergence of FCM for J_m; if $\alpha = 1$ and $q \geqslant 1$, (T23.4) for convergence of FCV (A23.1) obtains.

(E24.1) *Example 24.1.* Figure 24.2a depicts a set of $n = 60$ data points $X \subset \mathbb{R}^2$. There are three visually apparent linear clusters: however, two of them lie along a common axis, so that X contains substructure which presumably causes the difficulty for J_{V1m} illustrated in Fig. 24.1. The purpose of this example is to test the supposition that a FCE algorithm using ${}_{\alpha}J_{01m}$ will rectify the dilemma caused by this type of structure.

First, observe that the problem alluded to really exists, by looking at Fig. 24.2b, which depicts a fuzzy 2-lines solution for X arrived at using (A23.1) with $r = 1$, $c = 2$, the Euclidean norm for J_{V1m}, $\varepsilon_L = 0.05$, and the max norm on $(U^{(l+1)} - U^{(l)})$ as a cutoff criterion. This corresponds to (A24.1) with $q = 1 = \alpha$ (the weight for J_{Vqm}). Several initializations for $U^{(0)}$ and exponents m resulted in the same prototypes $\{L(\mathbf{v}_i; \mathbf{s}_i)\}$. The plotted centers of mass $\{\mathbf{v}_i\}$ occupy exactly the positions theory predicts; and $L(\mathbf{v}_2, \mathbf{s}_2)$ passes through the collinear clusters as expected. These lines are in fact an excellent 2-lines fit to X, and constitute a nice 2-cluster solution with linear prototypes. [The truncation of $L(\mathbf{v}_i; \mathbf{s}_i)$ was done automatically, as in (E23.1).]

In view of the separation of the two linear segments along the line of negative slope, it is reasonable to hope that for $c = 3$, an algorithm will detect the visually appealing 3-cluster solution. Application of fuzzy c-lines algorithm (A23.1) with $r = 1$ or (A24.1) with $q = \alpha = 1$) using the protocols above with $c = 3$ resulted in the 3-lines solution shown in Fig. 24.3a, which represents the terminal prototypes for $l = 9$ iterations with $m = 2$.

The third center of mass $\mathbf{v}_3$ stations itself in the lower notch of the X, while $\mathbf{v}_2$ remains virtually unchanged, resulting in a third line which bears little relation to the apparent substructure in the data. The reason for this is simple: *all* of the points ranged along $L(\mathbf{v}_2; \mathbf{s}_2)$ through $\mathbf{v}_2$ are very close to it, and the functional J_{V1m} rewards them with very high membership in cluster u_2, even though there is a large separation between two subsets of them.

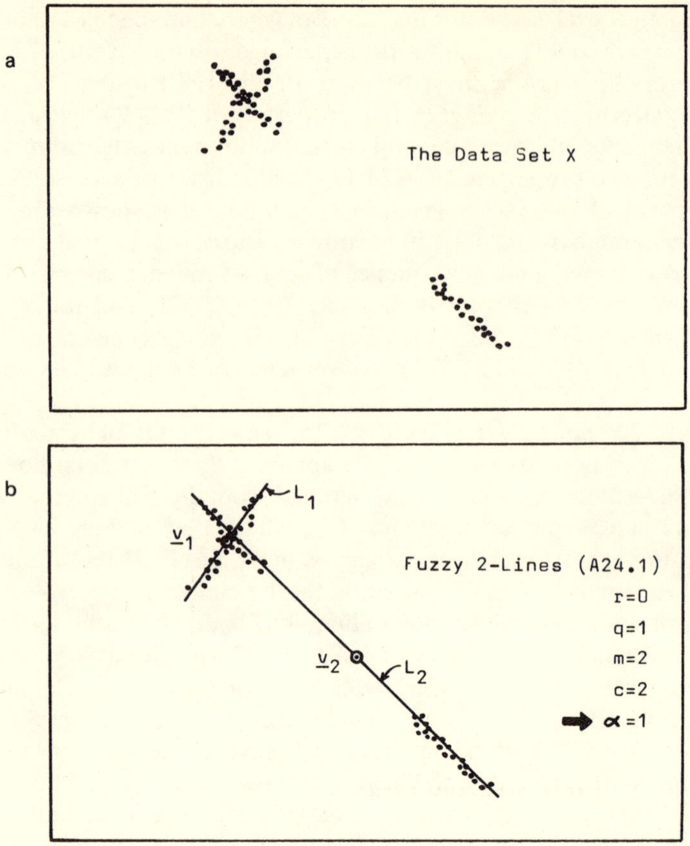

Figure 24.2. Collinear clusters: (E24.1).

Loosely speaking, c-lines functionals cannot "see" this separation because of the "size" problem of linear varieties alluded to above: the line $L(\mathbf{v}_2, \mathbf{s}_2)$ has infinite extent in $\mathbb{R}^2$ in direction $\mathbf{s}_2$, so any point close to this linear variety will readily attach itself to L_2, no matter how far it is from other points on the same line.

To remedy this defect, the term $(1 - \alpha)J_{V0m}$ in ${}_{\alpha}J_{01m}$ will presumably *penalize* points close to $L(\mathbf{v}_2; \mathbf{s}_2)$ but far from the point $\mathbf{v}_2$. In other words, fuzzy c-elliptotypes should mollify the propensity of J_{V1m} for lines, and that of J_{V0m} for points. The *prototypes*, of course, are neither lines nor points. Nevertheless, if α is kept close to unity, the principal directions of the generated elliptotypes seem to provide very reasonable estimates of the linear tendencies of the data; this seems to be corroborated by the 3-

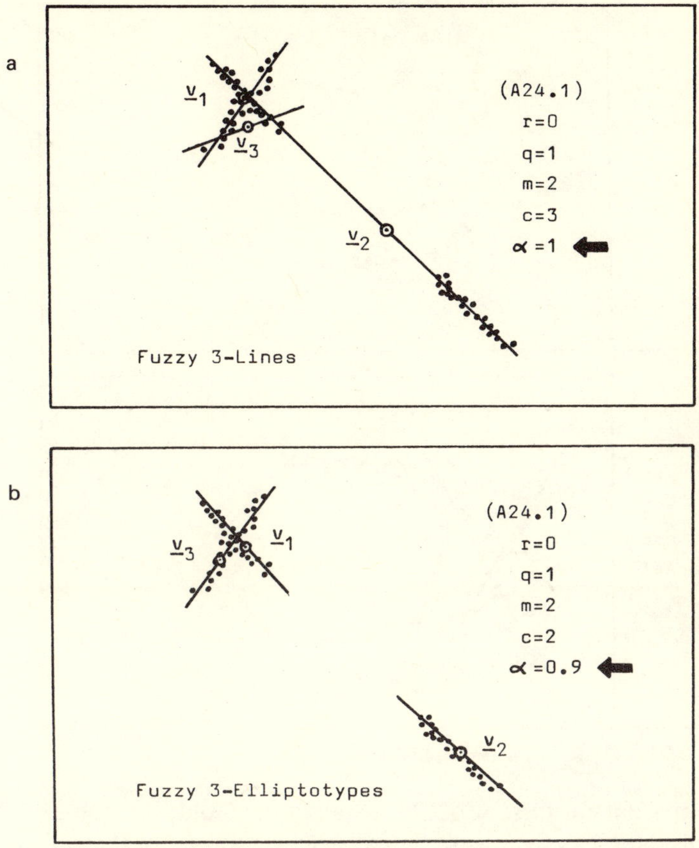

Figure 24.3. An elliptotypes solution for X: (E24.1).

elliptotypes solution displayed in Fig. 24.3b, which depicts the centers of mass and lines generated by the principal directions of elliptotypes obtained in 16 iterations (A24.1) using $c = 3$; $m = 2$; line weight $\alpha = 0.90$; point weight $(1 - \alpha) = 0.10$; $r = 0$; $q = 1$; and the remaining algorithmic parameters as specified above. Apparently the term $(1 - \alpha)J_{V0m}$ forces $\mathbf{v}_2$ away from the centralized position between the two collinear clusters it occupied in both previous instances: in turn, cluster center $\mathbf{v}_3$ becomes more effective as a locator for crossing line $L(\mathbf{v}_3; \mathbf{s}_3)$. Thus, FCE seems to behave as predicted, by mixing the advantages of c-means and c-lines via ${}_{\alpha}J_{01m}$. •

(E24.2) *Example 24.2* (*The Goldfish*). As a final example of fuzzy clustering in Chapter 5, consider the data set X of Fig. 24.4a. There are two

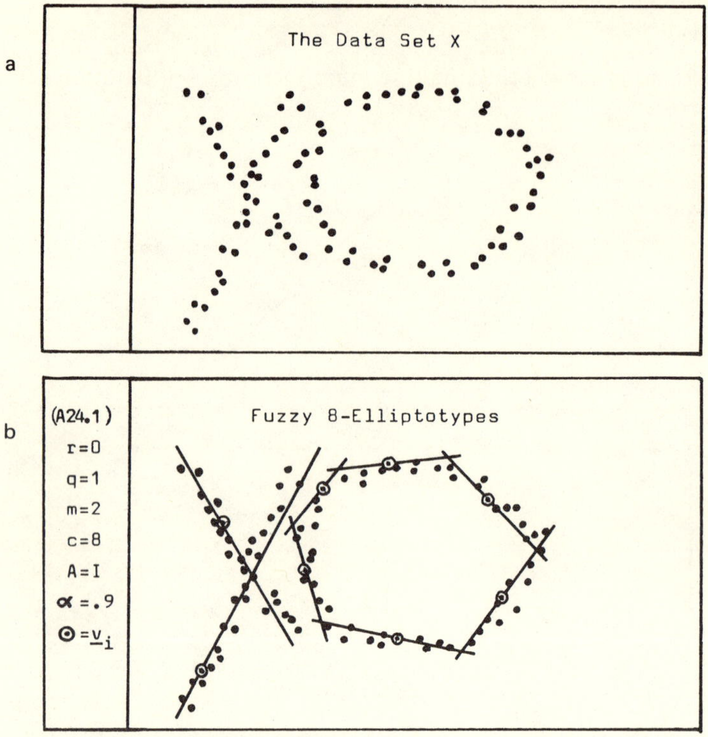

Figure 24.4. The goldfish: (E24.2).

visually apparent clusters: an "X" (the tail of the Goldfish); and an "O" (its body). This data set was constructed to embody several of the difficulties of earlier examples (cf. Fig. 8.1f): small cluster separation; linearity; curvilinearity; mixing; and noise. It is, although visually well structured, a truly difficult data set to recognize algorithmically. Both FCM ($\alpha = 0$) and FCL ($\alpha = 1$) produced expectedly poor results, independent of initialization, m, c, and the norm $\|\cdot\|_A$.

Shown in Fig. 24.4b are the means $\{\mathbf{v}_i\}$ and (clipped) principal directions $\{L(\mathbf{v}_i, \mathbf{s}_i)\}$ derived from (A24.1) in $l = 63$ iterations using ${}_\alpha J_{01m}$ with $\alpha = 0.90$; $c = 8$; $m = 2$; $\varepsilon_L = 0.05$; and the Euclidean norm. There is little doubt that the Goldfish has been successfully detected and characterized. •

Several final notes: Experimental evidence to date indicates that the FCV algorithms of S23 become decidedly sensitive to singularities ($D_{ik} \to 0$) for $r \geqslant 1$; further, the dimension of the fitting varieties seems to greatly increase the number of "poor" clustering solutions which are local minima of J_{Vrm}. These remarks seem equally applicable to algorithms based on ${}_{\{\alpha_i\}}J_{\{Vr_i\}m}$ whenever J_{V0m} is *not* one of the convex factors. However, adding a fuzzy c-means component $\alpha_i J_{V0m}$ to any of the higher-dimensional FCV functionals seems to restore order to this situation. In other words, adding even the smallest component ($\alpha_i \approx 0.01$ often suffices!) of J_{V0m} seems to force its good numerical characteristics onto ${}_{\{\alpha_i\}}J_{\{Vr_i\}m}$.

Remarks

The algorithms discussed in Chapter 5 are in their infancy. Many questions concerning their theory, usage, and interpretation remain open. Cluster validity is clearly one of the most important issues, because we have in each *section* of this chapter *infinite families* of fuzzy clustering algorithms! Concerning (A24.1), one wonders about generalizations of ${}_{\{\alpha_i\}}J_{\{Vr_i\}m}$ which might generate varieties of various dimensions simultaneously, enabling more reliable detection of mixed data structures. Another possible direction might consider local variations in shape via an expedient, such as Gustafson and Kessel's method, applied to the functionals J_{Vrm}. Finally, it is not hard to envision generalizations of J_{Vrm} which use prototypes other than linear varieties—e.g., hyperquadric surfaces. From a practical viewpoint, more experience with and applications of all of these algorithms is certainly desirable and seems inevitable. Another theoretical issue which deserves mention here concerns the possible link between and/or usage of the FCV algorithms for some traditional statistical problems and procedures; in particular, (A23.1) is clearly a weighted (orthogonal) linear regression technique. There appears to be a wealth of applications of (A23.1) directed towards problems of this kind which are currently studied by more conventional methods, such as analysis of variance, multiple discriminant analysis, and linear regression. Finally, we mention here the work of Roubens,[88] which could logically be included in this chapter because his algorithm is in some sense a generalization of (A11.1). However, the basic premise in Rouben's algorithm is that the data $\{\mathbf{x}_k\} = X$ are unknown; rather, the distances or dissimilarities $\{d_{ik}\}$ are the point of departure (i.e., the raw data). Thus, Rouben's algorithm is a modification of fuzzy c-means which accommodates a change in data structure as described in S7. Consequently, this algorithm is an objective function method for relation-theoretic data.

Exercise

H24.1. An *affine* combination of two vectors $\mathbf{x}$ and $\mathbf{y}$ is $\alpha\mathbf{x} + (1-\alpha)\mathbf{y}$, where $\alpha \in \mathbb{R}$. (The combination is *convex* when $\alpha \in [0, 1]$.) With this, one can envision equation (24.3) for *arbitrary* α as an affine combination of the fuzzy c-means and fuzzy c-lines functionals. Are conditions (23.10a), (23.10b), (23.10c), (24.6), and (24.8) necessary to minimize such a function for fixed α?

6

Selected Applications in Classifier Design

In this final chapter we consider several fuzzy algorithms that effect partitions of feature space $\mathbb{R}^p$, enabling classification of unlabeled (future) observations, based on the decision functions which characterize the classifier. S25 describes the general problem in terms of a canonical classifier, and briefly discusses Bayesian statistical decision theory. In S26 estimation of the parameters of a mixed multivariate normal distribution via statistical (maximum likelihood) and fuzzy (c-means) methods is illustrated. Both methods generate very similar estimates of the optimal Bayesian classifier. S27 considers the utilization of the prototypical means generated by (A11.1) for characterization of a (single) nearest prototype classifier, and compares its empirical performance to the well-known k-nearest-neighbor family of deterministic classifiers. In S28, an implicit classifier design based on Ruspini's algorithm is discussed and exemplified.

S25. An Overview of Classifier Design: Bayesian Classifiers

This chapter begins with a brief description of classifiers in general, and Bayesian classifiers in particular. More extensive discussions are available in a number of standard works, of which Duda and Hart[34] is one of the best. Let $f_j : \mathbb{R}^p \to \mathbb{R}$ be any scalar field, $1 \leq j \leq c$, and define the set

$$\mathrm{DR}_j = \{\mathbf{x} \in \mathbb{R}^p \mid f_j(\mathbf{x}) = \max_{1 \leq k \leq c} \{f_k(\mathbf{x})\}\} \tag{25.1}$$

In the context of classifier design, $\{f_j\}$ are called decision or discriminant functions; $\{\mathrm{DR}_j\}$ are called decision regions in $\mathbb{R}^p$; and the pair $(\{f_j\}, \{\mathrm{DR}_j\})$ is called a maximum classifier, because it defines the decision rule

$$\text{decide } \mathbf{x} \in \text{class } j \Leftrightarrow \mathbf{x} \in \mathrm{DR}_j \Leftrightarrow f_j(\mathbf{x}) = \bigvee_{k=1}^{c} f_k(\mathbf{x}) \tag{25.2}$$

The set in $\mathbb{R}^p$ upon which ties occur between f_i and f_j, $i \neq j$, is the level surface in $\mathbb{R}^p$ of $(f_i - f_j)$ or $(f_j - f_i)$ for the constant zero. For $i \neq j$ let

$$DS_{ij} = \{\mathbf{x} \in \mathbb{R}^p | f_i(\mathbf{x}) = f_j(\mathbf{x})\} \tag{25.3}$$

In all there are $c(c-1)/2$ distinct DS_{ij}'s and their union is $(\mathbb{R}^p - \bigcup_k DR_k)$. The DS_{ij}'s are called decision surfaces or decision boundaries for $\mathbb{R}^p$ determined by the f_j's because they divide $\mathbb{R}^p$ into the c disjoint decision regions at (25.1).

(D25.1) *Definition 25.1* (*Canonical Classifier*). Let $\{f_j : \mathbb{R}^p \to \mathbb{R} | 1 \leq j \leq c\}$ be any c scalar fields. If $\{DR_j\}$ are c decision regions defined as in (25.1), then $(\{f_j\}, \{DR_j\})$ is said to determine a maximum classifier via decision rule (25.2) for $\mathbb{R}^p$.

Since the functions $\{f_j\}$ determine the decision regions and boundaries, it is customary to regard the $\{f_j\}$ themselves as the classifier. Furthermore, the term "machine" is often used synonymously with classifiers, connoting the computational implementation of (25.2) for automatic classification of observations ($\mathbf{x}$'s in $\mathbb{R}^p$) as members of one of the *c populations* represented by the $\{DR_j\}$. A schematic diagram of the canonical classifier appears in Fig. 25.1. The entire procedure can be realized in software or hardware: see (103) for a nice discussion of the latter case.

The term "classifier design" means here the selection of the functions $\{f_j\}$. Classifiers are (themselves) classified by various methods: for our purposes, it suffices to distinguish them by *generic functional form* and by *method of estimation of the specific functional form*. Beginning with the generic description, suppose the $\{f_j\}$ to be linear maps, so that one can realize their action using the Euclidean inner product and c fixed p-vectors—say $\{\boldsymbol{\alpha}_j\}$—where $\boldsymbol{\alpha}_j^T = (\alpha_{j1}, \alpha_{j2}, \ldots, \alpha_{jp})$ is regarded as the matrix representation of the linear map f_j with respect to the standard bases of $\mathbb{R}^p$ and $\mathbb{R}^1$. In this instance, we have

$$f_j(\mathbf{x}) = \boldsymbol{\alpha}_j^T\mathbf{x} = \langle \boldsymbol{\alpha}_j, \mathbf{x} \rangle, \qquad 1 \leq j \leq c \tag{25.4}$$

When the $\{f_j\}$ have the form (25.4), the classifier is called a linear machine; the decision surfaces $\{DS_{ij}\}$ in this case are pieces of subspaces through the origin of $\mathbb{R}^p$ which have as their normals the difference vectors $(\boldsymbol{\alpha}_i - \boldsymbol{\alpha}_j)$; and design of a linear machine amounts to selecting the vectors $\{\boldsymbol{\alpha}_j\}$.

If a constant a_j is added to each f_j in (25.4), the resultant functions are affine mappings on $\mathbb{R}^p$, i.e.,

$$f_j(\mathbf{x}) = \boldsymbol{\alpha}_j^T\mathbf{x} + a_j = \langle \boldsymbol{\alpha}_j, \mathbf{x} \rangle + a_j, \qquad 1 \leq j \leq c \tag{25.5}$$

In this case, it is proper to call the $\{f_j\}$ an affine classifier, since the decision surfaces $\{DS_{ij}\}$ are pieces of affine subspaces (hyperplanes in $\mathbb{R}^p$) normal to

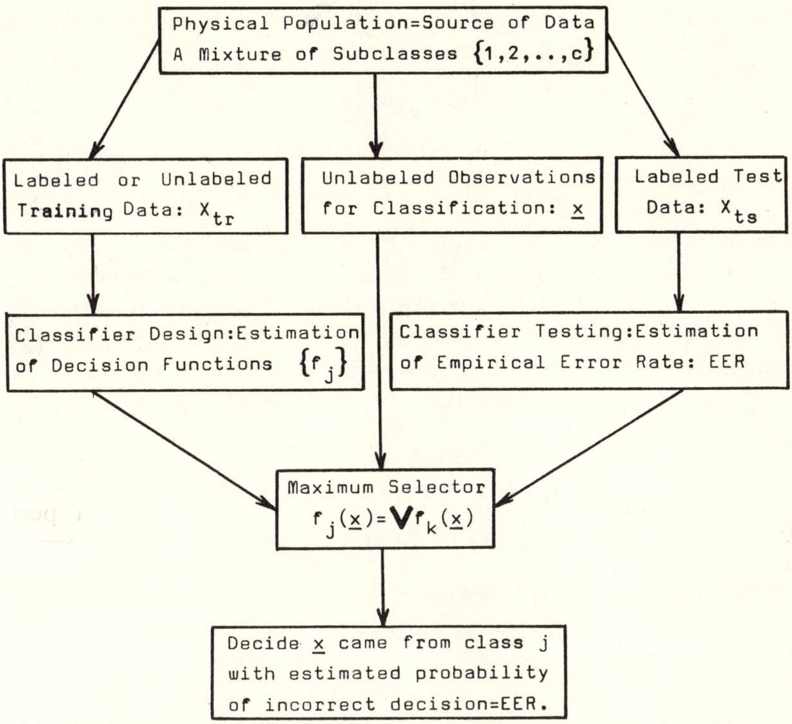

Figure 25.1. Design, test, and usage of a maximum classifier.

the vectors $(\boldsymbol{\alpha}_i - \boldsymbol{\alpha}_j)$ translated away from $\boldsymbol{\theta}$ unless $a_i = a_j$. However, it is customary to call a classifier designed with functions of type (25.5) a linear one. Any other form for the f_j's results in a nonlinear classifier, because the f_j's are nonlinear. For example, $f_j(\mathbf{x}) = \mathbf{x}^T A_j \, \mathbf{x} + \langle \boldsymbol{\alpha}_j, \mathbf{x} \rangle + a_j \forall j$ is said to define, with A_j a $p \times p$ matrix, a *quadratic* machine; the decision functions are quadratic; and the decision surfaces are hyperquadrics. Although a general classifier can be arbitrarily nonlinear, it is always possible to change the "nonlinear" partition of $\mathbb{R}^p$ induced by the $\{f_j\}$ into an affine one, by transforming the entire problem into $\mathbb{R}^{c+1}$ (cf. Tou and Gonzalez[103]), so (25.5) can be regarded as a canonical *functional form* for (D25.1). Since this device gains no advantage below, we shall not discuss it further.

(E25.1) *Example 25.1.* Consider the set of affine decision functions $f_j : \mathbb{R}^2 \to \mathbb{R}$ defined by

$$f_1(x, y) = 2x \qquad = \langle (x, y), (2, 0) \rangle + 0 \tag{25.6a}$$

$$f_2(x, y) = -3x - 2y = \langle (x, y), (-3, -2) \rangle + 0 \tag{25.6b}$$

$$f_3(x, y) = x - y + 1 = \langle (x, y), (1, -1) \rangle - 1 \tag{25.6c}$$

The $c(c-1)/2 = 3$ decision surfaces DS_{ij} defined by pairs (f_i, f_j) are the hyperplanes in $\mathbb{R}^2$ (straight lines) defined as

$$DS_{12} = \{(x, y) \in \mathbb{R}^2 | y = -(5/2)x\} \tag{25.7a}$$

$$DS_{13} = \{(x, y) \in \mathbb{R}^2 | y = -x + 1\} \tag{25.7b}$$

$$DS_{23} = \{(x, y) \in \mathbb{R}^2 | y = -4x - 1\} \tag{25.7c}$$

These decision boundaries divide $\mathbb{R}^2$ into three disjoint decision regions DR_1, DR_2, and DR_3 as shown in Fig. 25.2. The mutual intersection of all three DS_{ij}'s at $(-2/3, 5/3)$ is merely by chance: in general, decision regions need not be connected subsets of $\mathbb{R}^p$. Although one can perform classification here by geometric inspection, the f_j's in (25.6) determine its action algebraically. If, for example, observation $\mathbf{x}^* = (1, -1)$, then

$$f_1(\mathbf{x}^*) = 2(1) = 2$$

$$f_2(\mathbf{x}^*) = -3(1) - 2(-1) = -1$$

$$f_3(\mathbf{x}^*) = 1 - (-1) + 1 = 3$$

Consequently, $\mathbf{x}^* \in DR_3 \Rightarrow$ decide $\mathbf{x}^*$ is a class 3 object.

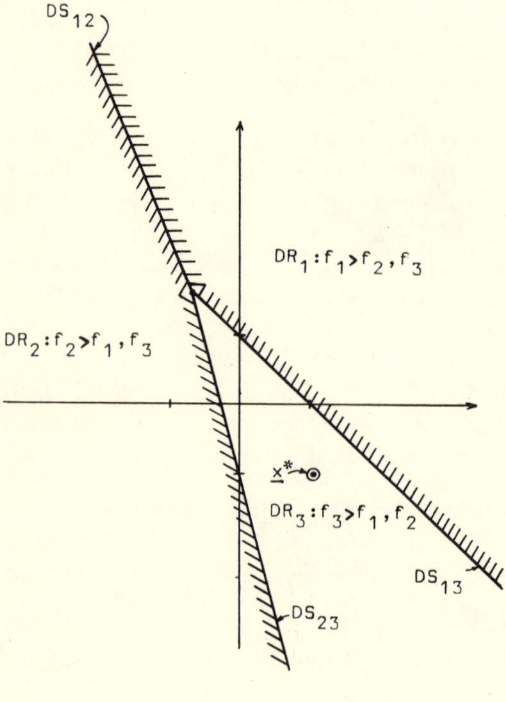

Figure 25.2. Decision surfaces and regions: (E25.1).

In general, the parameters needed to specify a particular functional form can be estimated by a variety of methods. In the present context data set $X \subset \mathbb{R}^p$ is often called a "training set," as it is used to train (design, estimate) the actual functions $\{f_i\}$ which delimit the classifier. X may be *labeled* data, in which case one knows a priori both c and the observationally correct hard $U \in M_c$ which specifies the class origin of each of the n data points: training a classifier with labeled data is often called "supervised learning." If X is not labeled, then c may or may not be known, and must be selected: classifier design using unlabeled data is referred to as "unsupervised learning." In this latter enterprise, a hard or fuzzy c-partition $U \in M_{fc}$ of X *may or may not* be required to train the classifier. In many methods, clustering algorithms generate necessary and/or useful information for classifier design: in this case, clustering is regarded as a "preprocessing" operation preparatory to estimation of the f_i's. In any event, classifiers are presumably designed to facilitate automatic recognition of unclassified data generated by a physical process which has c classes, so it is almost mandatory that one have *some* "correctly" labeled data—called a *test set*—with which to test and predict the level of classifier performance. These concepts are depicted schematically in Fig. 25.1.

There are essentially four broad categories of methods for classifier design: heuristic, deterministic, statistical, and fuzzy. For example, if $f_j(\mathbf{x}) = \langle \boldsymbol{\alpha}_j, \mathbf{x} \rangle$, one may define a linear machine simply by choosing c $\boldsymbol{\alpha}_j$'s according to any intuitively plausible heuristic. More objectively, the $\boldsymbol{\alpha}_j$'s might be estimated by a model of any of the afore-mentioned types. It thus makes perfect sense to describe one classifier as a statistical linear machine, another as a fuzzy quadratic machine, and so on. Regardless of the functional type or method of estimation, one needs a standard of performance against which a given classifier can be measured. One of the most popular *theoretical* criteria in this regard is the Bayesian classifier, a short description of which now follows.

To formulate the notion of "probability of misclassification," some distributional assumptions must be made. To begin, we suppose X to be one sample of n observations drawn from a mixture of c p-variate probability density functions. That is, the population from which X is drawn is assumed to consist of c subclasses, each having prior probability p_i, $0 < p_i < 1$, $\sum_{i=1}^{c} p_i = 1$; and each class i is distributed as a vector-valued random variable whose probability density (or mass) function we denote by $g(\mathbf{x}|i)$. This notation does not indicate "conditioning" in the usual sense; however, $g(\mathbf{x}|i)$ is often called a (class)-conditional density. The data set X is constructed as follows: on (independent) trial k, choose class i with probability p_i; and then sample $g(\mathbf{x}|i)$, yielding $\mathbf{x}_k \in X$. To simplify notation, $(\mathbf{x}_k \in i)$ shall mean *either* $\mathbf{x}_k$ was *drawn* from $g(\mathbf{x}|i)$; or we have *decided* that such was the case.

With these assumptions, X is a sample of n independent and identically distributed (i.i.d.) observations from the joint mixture density

$$F(\mathbf{x}) = \sum_{i=1}^{c} p_i\, g(\mathbf{x}|i) \tag{25.8}$$

Letting $h(i|\mathbf{x}) = \Pr(\mathbf{x} \in i|\mathbf{x})$ denote the posterior probability that $\mathbf{x} \in i$ given $\mathbf{x}$, we have by Bayes' rule

$$h(i|\mathbf{x}) = p_i g(\mathbf{x}|i)/F(\mathbf{x}) \tag{25.9}$$

It is customary in statistical decision theory to assign a loss, say l_{ij}, to the event "decide $\mathbf{x} \in j$ when in fact $\mathbf{x} \in i$." Then, the average conditional risk in deciding $\mathbf{x} \in j$ given $\mathbf{x}$ can be written

$$r(j|\mathbf{x}) = \sum_{i=1}^{c} l_{ij} h(i|\mathbf{x}), \qquad 1 \leq j \leq c \tag{25.10}$$

Suppose we put $f_{jb}(\mathbf{x}) = r(j|\mathbf{x})$. The resultant set of scalar fields $\{f_{jb}\}$ is called a *Bayesian classifier*, denoted hereafter as $\{f_{jb}\}$, because of the appearance of $\{h(i|\mathbf{x})\}$ in (25.10) via Bayes' rule. If $\{f_{jb}\}$ is used to define the statistical decision rule

$$\text{decide } \mathbf{x} \in i \Leftrightarrow f_{ib}(\mathbf{x}) = \min_j \{f_{jb}(\mathbf{x})\} \tag{25.11}$$

then (25.11) minimizes the expected loss (relative to $\{l_{ij}\}$) incurred when $\mathbf{x}$ is assigned to some class. If the loss function is $(l_{ij} = 1 - \delta_{ij})$, $1 \leq i, j \leq c$, then it is easy to check that

$$f_{jb}(\mathbf{x}) = \bigwedge_{j=1}^{c} f_{jb}(\mathbf{x}) \Leftrightarrow h(i|\mathbf{x}) = \bigvee_{j=1}^{c} h(j|\mathbf{x}) \tag{25.12}$$

In this case, minimizing the expected loss is equivalent to maximizing the posterior probability that $\mathbf{x} \in i$ given $\mathbf{x}$.

Comparing (25.12) with (25.1), we see that $\{h(j|\mathbf{x})\}$ is a maximum Bayesian classifier. More generally, since any monotone increasing function $\phi : \mathbb{R} \to \mathbb{R}$ does not alter the ordering of the numbers $\{h(j|\mathbf{x})\}$ for a given $\mathbf{x}$. $\{\phi(h(j|\mathbf{x}))\} = \{\phi(p_j g(\mathbf{x}|j))\} = \{f_{jb}\}$ is a maximum Bayesian classifier. Note that $F(\mathbf{x})$ is usually dropped, since it does not alter the decision implied by (25.12).

(E25.2) *Example 25.2.* Suppose F is a mixture of two univariate normal distributions $g(x|1) \sim n(1, 2)$; $g(x|2) \sim n(3, 1)$. If the prior probabilities of classes 1 and 2 are $p_1 = 0.4$, $p_2 = 0.6$, respectively, choosing

$$f_{1b}(x) = p_1 g(x|1),\ f_{2b}(x) = p_2 g(|2),$$

$$h(1|x) = h(2|x) \Leftrightarrow f_{1b}(x) = f_{2b}(x)$$

$$\Leftrightarrow (2^{-(1/2)})(0.4\, e^{-(x-1)^2/4}) = (0.6\, e^{-(x-3)^2/2})$$

$$\Leftrightarrow x = 1.68 \text{ or } 8.32.$$

Here then, $\mathbb{R}^p = \mathbb{R}$, and the decision surfaces are the $(p-1)$-dimensional (0-dimensional) hyperplanes (points) $DS_{1b} = 1.68$, $DS_{2b} = 8.32$. At these points, ties occur, and decisions are arbitrary. Otherwise, the Bayesian classifier has for its decision regions

$$DR_{1b} = (-\infty, 1.68) \cup (8.32, \infty) \quad \text{and} \quad DR_{2b} = (1.68, 8.32)$$

Given any x, deciding $x \in i \Leftrightarrow x \in DR_{ib} \Leftrightarrow h(i|x) > h(j|x) \Leftrightarrow f_{ib}(x) > f_{jb}(x)$, corresponds to the Bayes decision rule specified via (25.12). The situation is depicted geometrically in Fig. 25.3a. •

There are in (E25.2) two ways for the Bayesian classifier to err: decide $x \in 1$ when it is not, and vice versa. The probabilities of these two events are, respectively,

$$\Pr(\text{decide } x \in 1, x \in 2) = \int_{DR_{1b}} p_2 g(x|2)\, dx = A_1$$

and

$$\Pr(\text{decide } x \in 2, x \in 1) = \int_{DR_{2b}} p_1 g(x|1)\, dx = A_2$$

Each integral is the area under the tail of the smaller density; their sum is a measure of the overall probability of error (also called the error rate, or Bayes risk) in making decisions using $\{f_{jb}\}$. Figure 25.3b illustrates the components of this error rate. The integral of $p_2 g(x|2)$ from 8.32 to ∞ is essentially zero since the density at 8.32 is about 2×10^{-7}. For the classifier of (E25.2), the overall Bayes risk is 0.18, i.e., one expects, on the average, to misclassify about 18% of all observations drawn from $F(x)$ in (E25.2) when using $\{f_{jb}\}$.

Now let $\{f_1, f_2\}$ be *any other* classifier, defining regions $\{DR_i\} \neq \{DR_{ib}\}$. Under the assumption that $F(x)$ really governs observations being classified, it is reasonable to call

$$\int_{DR_1} p_2\, g(x|2)\, dx + \int_{DR_2} p_1\, g(x|1)\, dx$$

the error rate of $\{f_i\}$. Using this notion of error, one can visualize the effect of moving DR_{jb} to DR_j in Fig. 25.3b: the error can only go *up*! For example,

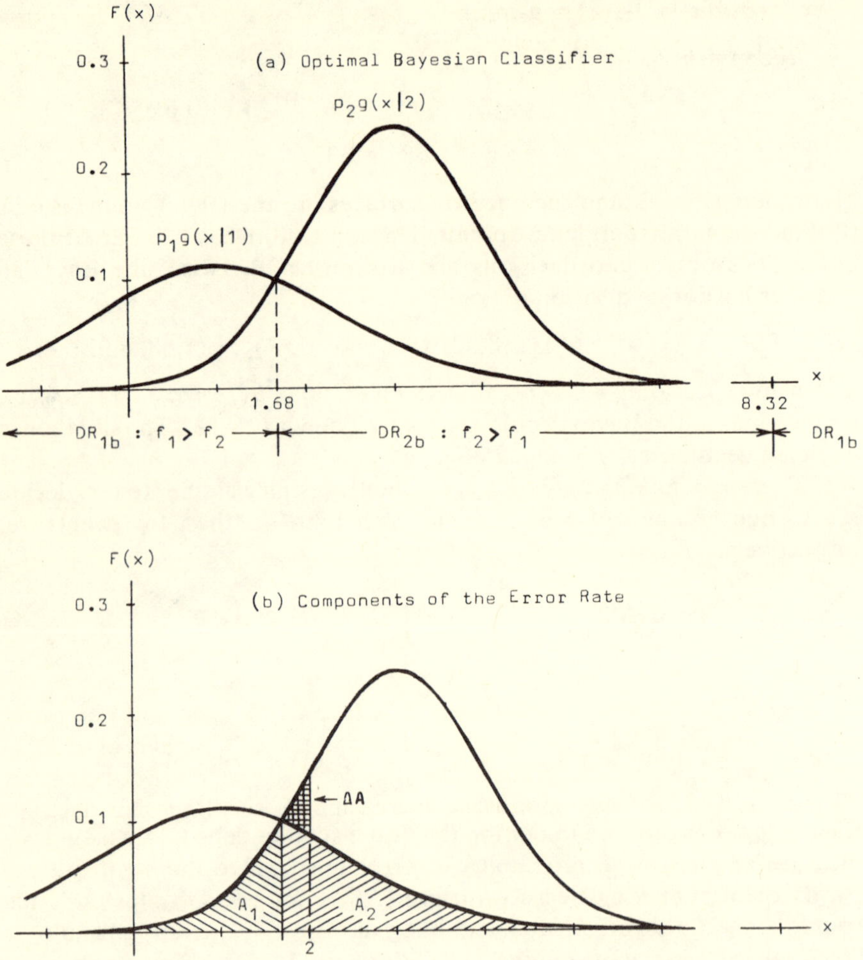

Figure 25.3. A mixture of two univariate normals: (E25.2).

moving $DS_{1b} = 1.68$ to $DS_1 = 2.00$ while keeping $DS_{2b} = DS_2 = 8.32$ results in the increase of area called ΔA in Fig. 25.3b. The area involved, using the values of (E25.2), is $\Delta A = 0.01$, so the error rate for $\{f_j\}$ is about 19%.

More generally, if we assume X to be drawn from $F(\mathbf{x})$ at (25.8), and if $\{f_j : \mathbb{R}^p \to \mathbb{R} | 1 \leq j \leq c\}$ is *any* maximum classifier defining c disjoint decision regions $\{DR_j\}$ for $\mathbb{R}^p$, then

$$\Pr(\mathbf{x} \in k, \mathbf{x} \in k) = \int\int_{DR_k}\cdots\int p_k\, g(\mathbf{x}|k)\, d\mathbf{x}; \qquad 1 \leq k \leq c \quad (25.13)$$

is the probability of deciding *correctly* that $\mathbf{x}$ came from class k when it did. Summing over k and subtracting from 1 yields the Neyman–Pearson error rate of the classifier in question:

(D25.2) *Definition 25.2* (*Neyman–Pearson Error Rate*). Let $F(\mathbf{x})$ be a mixture of c p-variate probability density functions, class j having prior probability $p_j \forall j$ as in (25.8). Let $\{f_j : \mathbb{R}^p \to \mathbb{R} | 1 \leq j \leq c\}$ be *any* maximum classifier defining decision regions $\{\mathrm{DR}_j\}$ as in (D25.1). The error rate or probability of misclassification using $\{f_j\}$ (or $\{\mathrm{DR}_j\}$) is

$$\mathrm{ER}(\{f_j\}) = 1 - \left(\sum_{k=1}^{c} \left[\int\int_{\mathrm{DR}_k} \cdots \int p_k g(\mathbf{x}|k)\, d\mathbf{x}\right]\right) \tag{25.14}$$

As written, (25.14) is applicable to continuous mixtures; for discrete mixtures, the integrals become sums as usual. In view of the discussion following (E25.2), the result below is not surprising.

(T25.1) *Theorem 25.1.* Let $\{f_{jb}\}$ be a Bayesian classifier; and let $\{f_j\}$ be any other classifier. Then

$$\mathrm{ER}(\{f_{jb}\}) \leq \mathrm{ER}(\{f_j\}) \tag{25.15}$$

Analysis and Proof. This theorem is true for a very simple reason: every Bayesian classifier uses *functions of the integrand functions* in (D25.2) to define the regions of integration $\{\mathrm{DR}_{jb}\}$. Since $\phi(p_j g(\mathbf{x}|j))$ defines the same DR_{jb} as $p_j g(\mathbf{x}|j)$ for any monotone increasing ϕ, the Bayesian decision regions maximize the integrands in (25.14). Thus, classifiers whose decision functions are not functions of $p_j g(x|j)$ define decision regions $\{\mathrm{DR}_j\}$ for (25.14) which reduce the probability of correct classification, or, equivalently, increase the theoretical error rate due to $\{f_j\}$. Thus $\{f_{jb}\}$ minimizes *both* the expected loss and probability of Neyman–Pearson error when $l_{ij} = 1 - \delta_{ij}$. •

(T25.1) is probably responsible for the oft-quoted phrase to the effect that "Bayesian classifiers are optimal." This is true provided that the assumptions underlying (25.15) are sustained. In practice, however, the theoretical substance of this assertion may lose its force. In the first place, $\mathrm{ER}(\{f_{jb}\})$ exists only in theory, since it cannot be calculated without knowing the $p_j g(\mathbf{x}|j)$'s, in which case it is known for certain that samples of the real process are actually distributed as mixture (25.8) and $\{f_{jb}\}$ would of course be used. Next, consider the number $\mathrm{ER}(\{f_j\})$ at (25.14): since the integrands are the (true but unknown) $p_j g(\mathbf{x}|j)$'s, $\mathrm{ER}(\{f_j\})$ still cannot be calculated, even over the decision regions $\{DR_j\}$ which approximate $\{DR_{jb}\}$. Thus, the first *estimate* for (25.14) that can be calculated—in principle—uses the approximations

$$\iint_{DR_k} \cdots \int p_k g(\mathbf{x}|k)\, d\mathbf{x} \approx \iint_{\widehat{DR}_{kb}} \cdots \int \hat{p}_k \hat{g}(\mathbf{x}|k)\, d\mathbf{x}$$

where $\{\hat{f}_{jb}\} = \{\phi(\hat{p}_j \hat{g}(\mathbf{x}|j))\}$ is an estimate of the true but unknown Bayesian classifier $\{f_{jb}\}$ produced with some method of training using data set X. Using these last integrals in (25.14) would yield what might be called an "empirical error rate" for $\{f_{jb}\}$—if it could be calculated; however, actually *performing* the integrations is usually impossible, so this concept seems somewhat vacuous. One might, of course, use numerical integration to approximate these estimates, but the results obtained may bear little connection to a useful estimate of the expected error rate.

Another point to be emphasized in connection with the optimality of $\{f_{jb}\}$ implied by (25.15) is that this inequality is *not* necessarily valid for $\{\hat{f}_{jb}\}$, i.e., *approximate* Bayesian classifiers do not, in theory, yield lower error rates than non-Bayesian ones. Finally, one should recall that theoretical optimality of $\{f_{jb}\}$ assumes X is really distributed as $F(\mathbf{x})$ in (25.8); failing this, (25.14) itself seems inappropriate.

The implication of all these remarks is this: Theorem (T25.1) provides some *psychological* justification for deciding to use X to design an approximate Bayesian classifier $\{\hat{f}_{jb}\}$, on the presumption that trying to estimate $\{f_{jb}\}$—the theoretically optimal design—may at least lead one to good approximations of the optimal decision regions $\{DR_{jb}\}$. There is, however, no theoretical reason which precludes the possibility (just to emphasize the point) of simply *guessing* regions $\{DR_j\}$ which are a better estimate of $\{DR_{jb}\}$ than are $\{\widehat{DR}_{jb}\}$! Further, since $\{DR_{jb}\}$ are unknown, there is no way to know *after the fact* which of $\{DR_j\}$ or $\{\widehat{DR}_{jb}\}$ is a better estimate of the assumed optimal design! Consequently, the most viable measure of classifier performance *in practice* seems to be an empirical error rate given by observations of its performance using trial data:

(D25.3) *Definition 25.3* (*Empirical Error Rate*). Let $\{f_j\}$ be any maximum classifier as in (D25.1). Let $X_{\text{test}} = X_{\text{ts}} \subset \mathbb{R}^p$ be a test set for $\{f_j\}$, i.e., a set of n_{ts} correctly labeled test vectors for the classifier under test. The empirical error rate of $\{f_j\}$ relative to X_{ts} is the number of observed errors in n_{ts} tries:

$$\text{EER}(\{f_j\};X_{\text{ts}}) = \frac{\text{No. of wrong decisions made by } \{f_j\}}{n_{\text{ts}} \text{ trials of } \{f_j\}} \tag{25.16}$$

Emphasis is placed on X_{ts} in (25.16) because different test sets of the same size may yield different EERs. A less obvious reason lies with the design of $\{f_j\}$: the relationship between X_{ts} and the *training data* X_{tr} with which $\{f_j\}$ is derived affects (25.16). For example, if one uses a labeled set $X = \{\mathbf{x}_1, \mathbf{x}_2, \ldots, \mathbf{x}_n\} = X_{tr} \subset \mathbb{R}^p$ to find the functions $\{f_i\}$, and then tests $\{f_i\}$ with the same data, i.e., uses the training data as the test set, it seems clear that (25.16) will produce an unduly optimisitic estimate of the expected error rate. The interaction between and selection of X_{tr} and X_{ts} for design and testing is an area of great current interest: readers anxious for more details should consult Toussaint,[106] who provides an excellent summary and comparison of several popular techniques. Among these, a method of estimating empirical error rates due to Toussaint himself[107] appears to be an excellent compromise between methods which have optimistic and pessimistic biases. Let $X = \{\mathbf{x}_1, \mathbf{x}_2, \ldots, \mathbf{x}_n\} \subset \mathbb{R}^p$ be a set of labeled data. Choose an integer z so that $n/z = n_t$ is an integer, and $z/n < 1/2$. Randomly partition X into n_t subsets, say $X_1, \ldots, X_{n_t}$, $|X_i| = z \forall i$. Set

$$X_{tr,i} = X_i, \qquad X_{ts,i} = X - X_i, \qquad 1 \leqslant i \leqslant n_t \tag{25.17}$$

On trial i, design classifier $\{f_{ji}\}$ using $X_{tr,i}$, and test it with $X_{ts,i}$. Calculate the average observed error rate for n_t rotations (i.e., n_t training sessions using a fixed scheme of classifier design):

$$\overline{\text{EER}}(\pi) = \frac{\sum_{i=1}^{n_t} \text{EER}(\{f_{ji}\}, X_{ts,i})}{n_t} \tag{25.18}$$

This number is conjectured to be a pessimistic estimate.[107] To compensate, Toussaint suggested taking a convex combination of it with an optimistic one, the R or resubstitution estimate, which uses all of X to design, and then all of X to test $\{f_j\}$—once. Letting $\overline{\text{EER}}(R)$ denote the R-estimate. Toussaint's method is to compute

$$\text{eer}(\alpha) = \alpha\overline{\text{EER}}(\pi) + (1 - \alpha)\overline{\text{EER}}(R) \tag{25.19}$$

where $\alpha \in [0, 1]$. In the absence of evidence to suggest otherwise, Toussaint recommends $\alpha = (1 - \alpha) = 0.50$, and conjectures that something like $n_t = 10$ rotations through X suffices to pin down eer(α). We return to (25.19)—which seems like a very reasonable compromise—in (S27).

Remarks

The discussion above is cursory at best: classifier design has become an enormous field in recent years, due mainly to high-speed, large data processing capabilities. Bayesian classifiers and estimates thereof are an increasingly popular type, due to the physical appeal of the mixture assumption embodied in (25.8). They are by no means, however, the only approach: Duda and Hart(34) and Tou and Gonzalez(103) exemplify many other strategies. Theoretical and empirical error rates, methods of training and testing, and error rate bounds are all topics of great current interest; as mentioned above, Toussaint(106) provides a particularly readable account of and bibliography for many papers concerning these topics. The remaining sections of Chapter 6 describe methods of constructing maximum classifiers $\{f_i\}$ using fuzzy algorithms. We mention that equation (25.14) produces an "empirical error rate" which is independent of test sets. This would follow by inserting the *estimated* $p_j g(\mathbf{x}|j)$ in each integral, and integrating over the decision region to be used; however, as mentioned above, the integration itself usually precludes using this as an empirical error rate.

Finally, note that procedures for estimating an approximate Bayes classifier $\{\hat{f}_{jb}\}$ can be heuristic, deterministic, statistical, or fuzzy. No philosophical dilemma arises, e.g., when using (A11.1)—FCM—to estimate (the parameters of) the functions $\{\phi(\hat{p}_j\hat{g}(\mathbf{x}|j))\}$; such a design would therefore be called a fuzzy approximate Bayesian classifier.

Exercises

H25.1. Let α, $\beta > 0$. Use the inequality $\alpha \wedge \beta \leqslant (\alpha\beta)^{1/2}$ to prove that the error rate of the optimal Bayesian classifier for any mixture of $g(\mathbf{x}|1)$ and $g(\mathbf{x}|2)$ satisfies

$$\mathrm{ER}(\{f_{jb}\}) \leqslant (p_1 p_2)^{1/2} k \leqslant k/2$$

where

$$k = \int\int_{\mathbb{R}^p}\cdots\int [g(\mathbf{x}|1)g(\mathbf{x}|2)]^{1/2}\,d\mathbf{x}$$

is the Bhattacharrya coefficient.

H25.2. $g(x|i) \sim n(\mu_i, \sigma^2)$ for $x \in \mathbb{R}$, $i = 1, 2$. If $p_1 = p_2 = 1/2$, show that with $K = (|\mu_2 - \mu_1|)/2\sigma$,

(i) $\mathrm{ER}(\{f_{jb}\}) = \int_K^\infty e^{-t^2/2}\,dt/(2\pi)^{1/2}$

(ii) $\lim[\mathrm{ER}(\{f_{jb}\})] = 0$ as $(|\mu_2 - \mu_1|)/2\sigma \to \infty$.

(iii) Both (i) and (ii) hold for $\mathbf{x} \in \mathbb{R}^p$ if $\mu_i = \mu_i$,

$$\Sigma = \sigma^2 I, \text{ and } K = \|\boldsymbol{\mu}_2 - \boldsymbol{\mu}_1\|/2\sigma \quad [\text{cf. } (26.1)]$$

H25.3. Let $f(\mathbf{x}) = \langle \boldsymbol{\alpha}, \mathbf{x} \rangle + a = (\boldsymbol{\alpha}^T \mathbf{x} + a)$ be an affine function, $\mathbf{x}, \boldsymbol{\alpha} \in \mathbb{R}^p$, $a \in \mathbb{R}$. Let H be the hyperplane $H = \{\mathbf{x} \in \mathbb{R}^p | f(\mathbf{x}) = 0\}$. Show that the orthogonal distance from $\boldsymbol{\theta}$ to H is $|a|/\|\boldsymbol{\alpha}\|$; more generally, show that the orthogonal distance from $\mathbf{x}$ to H is $|f(\mathbf{x})|/\|\boldsymbol{\alpha}\|$. What is the projection of $\mathbf{x}$ onto H? (Answer: $\mathbf{x} - [f(\mathbf{x})\boldsymbol{\alpha}/\|\boldsymbol{\alpha}\|^2)]$.}

H25.4.

$$g(x|1) = \begin{cases} 1/(b-a), & a \leq x \leq b \\ 0, & \text{otherwise} \end{cases}$$

$$g(x|2) = \begin{cases} 1/(d-c), & c \leq x \leq d \\ 0, & \text{otherwise} \end{cases}$$

Suppose $F(x) = p_1 g(x|1) + p_2 g(x|2)$ is a mixture of these two uniform densities. Find $\{f_{jb}\}$, ER($\{f_{jb}\}$) if

(i) $[a, b] \cap [c, d] = \varnothing$ [Answer: $b < c \Rightarrow x^* \in (b, c)$, ER $= 0$, $d < a \Rightarrow x^* \in (d, a)$, ER $= 0$.]

(ii) $[a, b] \subset [c, d]$

(iii) $[c, d] \subset [a, b]$

(iv) $[a, b] \cap [c, d] = [c, b]$ [Answer:
Let $p_1/(d-c) = k_1$ and
$p_2/(b-a) = k_2$.
Then
$k_1 = k_2 \Rightarrow x^* \in [c, b]$, ER $= k_1(b-c)$
$k_1 > k_2 \Rightarrow x^* = b$, ER $= k_2(b-c)$
$k_1 < k_2 \Rightarrow x^* = c$, ER $= k_1(b-c)$.]

H25.5. In Fig. 25.3, let $u_1(x) = p_1 g(x|1)$ and $u_2(x) = p_2 g(x|2)$. Then $u_1, u_2: \mathbb{R} \to [0, 1]$ are fuzzy sets in $\mathbb{R}$. Show that

$$\text{ER}(\{f_{jb}\}) = \int_{-\infty}^{\infty} (u_1 \wedge u_2)(x)\,dx$$

H25.6. Let $\{f_j\}$ be an affine classifier, $f_j(\mathbf{x}) = \langle \boldsymbol{\alpha}_j, \mathbf{x} \rangle + a_j$, $1 \leq j \leq c$; and let $\{\text{DR}_j\}$ be the decision regions in $\mathbb{R}^p$ determined by $\{f_j\}$. Prove that DR_j is a convex subset of $\mathbb{R}^p \forall j$.

H25.7. $X = X_1 \cup X_2$ is a 2-class labeled data set in $\mathbb{R}^p$. X is *linearly separable* if there is a hyperplane $H_{\boldsymbol{\alpha}} = \{\mathbf{x} | f(\mathbf{x}) = \langle \boldsymbol{\alpha}, \mathbf{x} \rangle + a = 0\}$ that "separates" X_1 and X_2, i.e., $f(\mathbf{x}) > 0 \forall \mathbf{x} \in X_1$, $f(\mathbf{x}) < 0 \forall \mathbf{x} \in X_2$. Given *any* $X = X_1 \cup X_2$, show that either X_1 and X_2 are linearly separable, or $\text{conv}(X_1) \cap \text{conv}(X_2) \neq \phi$.

H25.8. Let u be a fuzzy set, and H be the jth coordinate hyperplane in $\mathbb{R}^p$, $H_j = \{\mathbf{x} | x_j = 0\}$. The *projection* of u onto H_j is the fuzzy set u_{sj} in $\mathbb{R}^{p-1}$ whose membership function is[122]

$$u_{sj}(x_1, \ldots, x_{j-1}, x_{j+1}, \ldots, x_p) = \sup_{x_j} \{u(\mathbf{x})\}$$

(i) If u is a convex fuzzy set (cf. H6.14), show that u_{sj} is also.

(ii) If u, w are convex fuzzy sets and have equal projections on all p H_j's, prove that $u = w$.

H25.9. $H_{\boldsymbol{\alpha}}$ in H25.7 is called a *separating hyperplane*: if X_1 and X_2 are disjoint convex sets in $\mathbb{R}^p$, such an $H_{\boldsymbol{\alpha}}$ always exists. Let u and w be *bounded* fuzzy convex sets in $\mathbb{R}^p$ [u is bounded $\Leftrightarrow \{\mathbf{x}|u(\mathbf{x}) \geq \beta\}$ are bounded in $\mathbb{R}^p \forall \beta > 0$]. For a fixed hyperplane H_α in $\mathbb{R}^p$, let K_α be any number in [0, 1] such that $u(\mathbf{x}) \leq K_\alpha \forall \mathbf{x}$ on one side of H_α, and $w(\mathbf{x}) \leq K_\alpha$ for $\mathbf{x}$ on the other side of H_α. The *degree of separation* of u and w by H_α is $D_\alpha = 1 - \inf\{K_\alpha\}$; and the maximal degree of separation of u and w by a hyperplane is[122]

$$D = 1 - \inf_\alpha \{\inf\{K_\alpha\}\}$$

Let M_u, M_w, $M_{u \wedge w}$ be the maximal grades of membership of the bounded fuzzy sets u, w, and $u \wedge w$, respectively, [i.e., $M_u = \sup_\mathbf{x} \{u(x)\}$, etc.]. Prove that $D = 1 - M_{u \wedge w}$.

S26. A Heuristic Approach to Parametric Estimation: Bayesian Classifiers for Mixed Normal Distributions

This section considers the design of an approximate Bayesian classifier $\{\hat{f}_{jb}\}$ in the particular case of a mixture of c p-variate normal distributions. In this instance, the class conditional density $g(\mathbf{x}|j)$ in mixture (25.8) has the form

$$g(\mathbf{x}|j) = \frac{\exp\{-[(\mathbf{x} - \boldsymbol{\mu}_j)^T \Sigma_j^{-1} (\mathbf{x} - \boldsymbol{\mu}_j)/2]\}}{(2\pi)^{(p/2)}(\det \Sigma_j)^{(1/2)}} \tag{26.1}$$

where $\boldsymbol{\mu}_j$ is the (theoretically) expected mean vector [$\boldsymbol{\mu}_j = E(\mathbf{X}_j)$]; and Σ_j is the covariance matrix $\{\Sigma_j = E[\mathbf{X}_j - \boldsymbol{\mu}_j)(\mathbf{X}_j - \boldsymbol{\mu}_j)^T]\}$ of the normal random vector $\mathbf{X}_j$ whose density is (26.1). We assume Σ_j to be positive definite. If a sample X of n i.i.d. observations is drawn from the mixture $F(\mathbf{x}) = \Sigma_i p_i g(\mathbf{x}|i)$, and $(p_j, \boldsymbol{\mu}_j, \Sigma_j)$ is known for each j, it was argued in S25 that the theoretical Bayesian classifier $\{f_{jb}\} = \{\phi(h(j|\mathbf{x}))\}$ would minimize the expected loss and probability of error over all maximum classifiers $\{f_j\}$ of form (25.1).

If the parameters $(p_j, \boldsymbol{\mu}_j, \Sigma_j)$ of each class are *not* known, one may choose to estimate them using various methods of statistical or nonstatistical inference; this leads, in turn, to various estimates of $\{f_{jb}\}$—the optimal but unknown Bayesian classifier designed with $(p_j, \boldsymbol{\mu}_j, \Sigma_j)$. Our goal in this section is to compare maximum likelihood and fuzzy c-means estimates of the classifier $\{f_{jb}\}$. To formalize the parametric estimation problem, let $\boldsymbol{\theta}_j = (\boldsymbol{\mu}_j, \Sigma_j)$ denote the parameters of class j which characterize function $f_{jb}(\mathbf{x}) = \phi(p_j g(\mathbf{x}|j))$. Each $g(\mathbf{x}|j)$ is parametric in $(\boldsymbol{\mu}_j, \Sigma_j)$, which we denote herein by writing $g(\mathbf{x}|j; \boldsymbol{\theta}_j)$. Then f_{jb} is parametric in $\boldsymbol{\theta}_j$, and perhaps p_j as well. Letting $\boldsymbol{\omega}_j = (p_j, \boldsymbol{\theta}_j)$, the parametric form of the Bayesian decision functions is

$$f_{jb}(\mathbf{x}; \boldsymbol{\omega}_j) = \phi(p_j g(\mathbf{x}|j; \boldsymbol{\theta}_j)) \tag{26.2}$$

where ϕ is any monotone increasing real function. The form of mixture density $F(\mathbf{x})$ at (25.8) now becomes

$$F(\mathbf{x}; \boldsymbol{\omega}) = \sum_{i=1}^{c} p_i g(\mathbf{x}|i; \boldsymbol{\theta}_i) \tag{26.3}$$

where $\boldsymbol{\omega} = (\boldsymbol{\omega}_1, \boldsymbol{\omega}_2, \ldots, \boldsymbol{\omega}_c)$ is a parameter vector in parameter space Ω. Equation (26.3) actually represents the parametric mixture density for any c densities $\{g(\mathbf{x}|j; \boldsymbol{\theta}_j)\}$, not just the ones at (26.1). If $\mathbf{p} = (p_1, p_2, \ldots, p_c)$ is the vector of priors, $\boldsymbol{\mu} = (\boldsymbol{\mu}_1, \boldsymbol{\mu}_2, \ldots, \boldsymbol{\mu}_c)$ is the vector of means, and $\boldsymbol{\Sigma} = (\Sigma_1, \Sigma_2, \ldots, \Sigma_c)$ is the vector of $p \times p$ positive-definite covariance matrices, then $\boldsymbol{\omega} = (\mathbf{p}, \boldsymbol{\mu}, \boldsymbol{\Sigma}) \in \operatorname{conv}(B_c) \times \mathbb{R}^{cp} \times (\mathrm{PD})^c = \Omega$ is the parameter space for the densities at hand.

Let $X = \{\mathbf{x}_1, \mathbf{x}_2, \ldots, \mathbf{x}_n\} \subset \mathbb{R}^p$ be a sample of n observations drawn i.i.d. from (26.3), by selection on trial k of class j with probability p_j and then drawing $\mathbf{x}_k$ from $g(\mathbf{x}|j; \boldsymbol{\theta}_j)$. We want to use X to estimate $\boldsymbol{\omega}$. If X is labeled, the problem reduces to one of "supervised learning," wherein the n_j observations known to have come from class j are used (separately) to estimate $\boldsymbol{\omega}_j$, $1 \leq j \leq c$. In the more difficult situation (unsupervised learning), X is unlabeled, and all n observations in X are used to simultaneously estimate $\boldsymbol{\omega}$. In the first place, one must inquire whether X contains sufficient information about F to accomplish the task. This is the case whenever $F(\mathbf{x}; \boldsymbol{\omega})$ is *identifiable*:

(D26.1) *Definition 26.1* (*Identifiability*). The mixture $F(\mathbf{x}; \boldsymbol{\omega})$ at (26.3) is an identifiable mixture of the $g(\mathbf{x}|j; \boldsymbol{\theta}_j)$'s if for every $\boldsymbol{\omega}_1, \boldsymbol{\omega}_2 \in \Omega$ with $\boldsymbol{\omega}_1 \neq \boldsymbol{\omega}_2$, there is an $\mathbf{x}^* \in \mathbb{R}^p$ so that $F(\mathbf{x}^*; \boldsymbol{\omega}_1) \neq F(\mathbf{x}^*; \boldsymbol{\omega}_2)$.

The assumption of identifiability is made to ensure that $\boldsymbol{\omega}$ is at least *in principle* recoverable from a sample of F. Yakowitz and Spragins[121] demonstrated that identifiability is equivalent to linear independence of the component densities.

Assuming X to be an i.i.d. sample from (26.3), the parametric joint sample density is

$$H(X; \boldsymbol{\omega}) = \prod_{k=1}^{n} F(\mathbf{x}_k; \boldsymbol{\omega}) \tag{26.4}$$

From H we have, by regarding it as a function of $\boldsymbol{\omega}$, the likelihood function of X:

$$L(\boldsymbol{\omega}; X) = H(X; \boldsymbol{\omega}) = \prod_{k=1}^{n} F(\mathbf{x}_k; \boldsymbol{\omega}) \tag{26.5}$$

In the usual manner, we take natural logarithms of L, obtaining

$$\log L(\boldsymbol{\omega}; \mathbf{X}) = \sum_{k=1}^{n} \log \mathrm{F}(\mathbf{x}_k; \boldsymbol{\omega}) \tag{26.6}$$

The inferential principle of maximum likelihood asks for an $\hat{\boldsymbol{\omega}}$ which maximizes the probability of obtaining the sample X observed. Since maximizing L[or $\log(L)$] accomplishes this, we are led to the optimization problem

$$\underset{\boldsymbol{\omega} \in \Omega}{\text{maximize}} \{\log L(\boldsymbol{\omega}; X)\} \tag{26.7}$$

In the general case, existence, uniqueness, and solutions of (26.7) are not well known. Even for the case at hand, the covariance matrices $\{\Sigma_i\}$ need to be restricted before (26.7) necessarily has a finite solution.[34] However, classifiers designed with maximum likelihood estimates (MLE) obtained via (26.7) have enjoyed empirical success, so finite local maxima of $\log(L)$ seem to have practical utility. Assuming sufficient differentiability of the component densities in $F(\mathbf{x}; \boldsymbol{\omega})$, it is routine to check that $\hat{\boldsymbol{\omega}}$ may be a local maximum of $\log(L)$ only if, for $1 \leq i \leq c$,

$$\hat{p}_i = \frac{1}{n} \sum_{k=1}^{n} h(i|\mathbf{x}_k; \hat{\boldsymbol{\theta}}_i) \tag{26.8a}$$

$$\sum_{k=1}^{n} h(i|\mathbf{x}_k; \hat{\boldsymbol{\theta}}_i) \nabla_{\boldsymbol{\theta}_i}[\log (g(\mathbf{x}_k|i; \hat{\boldsymbol{\theta}}_i))] = 0 \tag{26.8b}$$

$$h(i|\mathbf{x}_k; \hat{\boldsymbol{\theta}}_i) = \hat{p}_i g(\mathbf{x}_k|i; \hat{\boldsymbol{\theta}}_i)/F(\mathbf{x}_k; \hat{\boldsymbol{\omega}}) \tag{26.8c}$$

In general, the only hope for solving (26.8) is by approximation via numerical means. Even for the case at hand, when the component densities are multinormals, system (26.8) is not amenable to closed form solutions. In 1969 Day[31] exhibited the explicit form of (MLE) for the parameters $(\hat{p}_i, \hat{\boldsymbol{\mu}}_i, \hat{\Sigma}_i) = \hat{\boldsymbol{\omega}}_i$ of (26.8)when $g(\mathbf{x}|i; \hat{\boldsymbol{\theta}}_i)$ has form (26.1). They are

$$\hat{p}_i = \frac{1}{n} \sum_{k=1}^{n} h(i|\mathbf{x}_k; \hat{\boldsymbol{\theta}}_i) \tag{26.9a}$$

$$\hat{\boldsymbol{\mu}}_i = \frac{\sum_{k=1}^{n} h(i|\mathbf{x}_k; \hat{\boldsymbol{\theta}}_i)\mathbf{x}_k}{\sum_{k=1}^{n} h(i|\mathbf{x}_k; \hat{\boldsymbol{\theta}}_i)} \tag{26.9b}$$

$$\hat{\Sigma}_i = \frac{\sum_{k=1}^{n} h(i|\mathbf{x}_k; \hat{\boldsymbol{\theta}}_i)(\mathbf{x}_k - \hat{\boldsymbol{\mu}}_i)(\mathbf{x}_k - \hat{\boldsymbol{\mu}}_i)^T}{\sum_{k=1}^{n} h(i|\mathbf{x}_k, \hat{\boldsymbol{\theta}}_i)} \tag{26.9c}$$

$$h(i|\mathbf{x}_k; \hat{\boldsymbol{\theta}}_i) = \frac{\hat{p}_i g(\mathbf{x}_k|i; \hat{\boldsymbol{\theta}}_i)}{F(\mathbf{x}_k; \hat{\boldsymbol{\omega}})} \tag{26.9d}$$

Using (26.1), the explicit form of Baye's rule (26.9d) can be written as

$$h(i|\mathbf{x}_k; \hat{\boldsymbol{\theta}}_i) = \left[\sum_{j=1}^{c} (\hat{w}_{ik}/\hat{w}_{jk})\right]^{-1} \qquad \forall i, k \tag{26.9d'}$$

where

$$\hat{w}_{ik} = (\hat{p}_i)^{-1}[\det(\hat{\Sigma}_i)]^{(1/2)} \exp[\tfrac{1}{2}(\mathbf{x}_k - \hat{\boldsymbol{\mu}}_i)^T \hat{\Sigma}_i^{-1} (\mathbf{x}_k - \hat{\boldsymbol{\mu}}_i)] \qquad \forall i, k$$

These four equations provide necessary conditions for local maxima of $\log(L)$, and are used to define a Picard iteration process for numerically approximating local solutions of (26.7). We call this the unlabeled maximum likelihood estimation (UMLE) algorithm for normal mixtures. For brevity, we let $H = [h_{ik}] = [h(i|\mathbf{x}_k; \boldsymbol{\theta}_i)]$; note that $H \in M_{fc}$, because the posterior probabilities $\{h_{ik}\}$ must sum to unity over i at each fixed k.

(A26.1) *Algorithm 26.1* (*UMLE, Duda and Hart*[34])

- (A26.1a) Given X, $|X| = n$, choose c, $2 \leq c < n$. Initialize $H^{(0)} \in M_{fc}$. Then at step l, $l = 0, 1, 2, \ldots$:
- (A26.1b) Calculate estimates of $\{p_i^{(l)}\}$, then $\{\boldsymbol{\mu}_i^{(l)}\}$, and then $\{\Sigma_i^{(l)}\}$ using, respectively, (26.9a), (26.9b), and (26.9c).
- (A26.1c) Update $H^{(l)}$ to $H^{(l+1)}$ with (26.9d) and $\{p_i^{(l)}, \boldsymbol{\mu}_i^{(l)}, \Sigma_i^{(l)}\}$
- (A26.1d) Compare $H^{(l)}$ to $H^{(l+1)}$ in a convenient matrix norm. If $\|H^{(l+1)} - H^{(l)}\| \leq \varepsilon_L$, stop. Otherwise, put $l = l + 1$ and return to (A26.1b).

Numerical convergence properties of the loop in (26.1) to local maxima of $\log(L)$ are not yet established [*stochastic* convergence of the MLE's in (26.9) to the theoretical parameters of $F(\mathbf{x})$ as $n \to \infty$ are well known]. The UMLE algorithm is superficially quite similar to FCM (A11.1); note, however, that step (26.9c) requires the construction of cn $p \times p$ matrices; and (26.9d) necessitates computation of determinants and inverses of the c $p \times p$ $\Sigma_i^{(l)}$'s at *each* step! Thus, (A26.1) is, for all but trivial c, p, and n, a computationally and economically unattractive method. Before illustrating its use in a numerical setting, we consider the situation when X is labeled. Towards this end, for fixed $\hat{\boldsymbol{\omega}} \in \Omega$, let us denote the posterior probabilities $\{h(i|\mathbf{x}_k; \hat{\boldsymbol{\theta}}_i)$ as $\hat{h}_{ik}\}$. Because they are probabilities, $0 \leq \hat{h}_{ik} \leq 1 \forall i, k$; and $\forall k, \sum_{i=1}^{c} \hat{h}_{ik} = 1$. Arraying $[\hat{h}_{ik}]$ as a $c \times n$ matrix $\hat{H}$, we note that H, the posterior matrix, and MLE $\hat{H}$ of it, lies in M_{fc}, fuzzy c-partition space for X! Suppose the data set X to be *labeled* by, say, $\hat{U} \in M_c$. Then $\hat{u}_{ik} = \hat{u}_i(\mathbf{x}_k) = 1$ when $\mathbf{x}_k \in i$; and is zero otherwise. On the other hand, $h_{ik} = \text{Prob}(\mathbf{x}_k \in i|\mathbf{x}_k \in i) = 1$ when $\mathbf{x}_k \in i$; 0 when $\mathbf{x}_k \notin i$. Thus, for *labeled* data,

$\hat{U} = \hat{H} = H$ is unique, and equations (26.9) yield the unique MLEs non-iteratively:

$$\hat{p}_i = \frac{\sum_{k=1}^{n} \hat{u}_{ik}}{n} = \frac{n_i}{n} \tag{26.10a}$$

$$\hat{\boldsymbol{\mu}}_i = \frac{\sum_{k=1}^{n} \hat{\mu}_{ik}\mathbf{x}_k}{\sum_{k=1}^{n} \hat{u}_{ik}} = \frac{\sum_{\mathbf{x}_k \in u_i}}{n_i} \tag{26.10b}$$

$$\hat{\Sigma}_i = \frac{\sum_{k=1}^{n} \hat{u}_{ik}(\mathbf{x}_k - \hat{\boldsymbol{\mu}}_i)(\mathbf{x}_k - \hat{\boldsymbol{\mu}}_i)^T}{\sum_{k=1}^{n} \hat{u}_{ik}} = C_i \tag{26.10c}$$

In other words, the unique MLE of $(\hat{p}_i, \hat{\boldsymbol{\mu}}_i, \hat{\Sigma}_i)$ are just the (sub-) sample proportion, mean, and covariance matrix, respectively, of the n_i observations drawn from class *i*. For the heuristic justification given below, it is crucial to observe that the estimates (26.10b) are also the unique least-squares estimates of the unknown means $\{\boldsymbol{\mu}_i\}$ in the present statistical context which minimize $J_W(\hat{U}, \mathbf{v})$ over $\mathbb{R}^{cp}$. For *labeled* data, then, problems (9.9)—minimize J_W—and (26.7)—maximize $\log(L)$—have, for mixtures of p-variate normals, the same unique solution. This fact, coupled with the obvious functional similarities between necessary conditions (11.3) and (26.9), suggests that solutions of (11.2) and (26.7) using *unlabeled* data may be close (i.e., $\hat{U} \approx \hat{H}; \hat{\mathbf{v}} \approx \hat{\boldsymbol{\mu}}$). Table 26.1 compares the necessary conditions. Fuzzy c-means algorithm (A11.1) does not generate estimates of the $\hat{p}_i$'s or $\hat{\Sigma}_i$'s; however, if $\hat{U}$ is taken as an estimate of $\hat{H}$, and $\hat{\mathbf{v}}_i$ of $\hat{\boldsymbol{\mu}}_i \forall i$, then obvious estimates of these parameters can be made by using $\hat{U}$ for $\hat{H}$, $\hat{\mathbf{v}}$ for $\hat{\boldsymbol{\mu}}$ in (26.9a) and (26.9c), respectively. That is, we *define* fuzzy estimates of $\hat{p}_i$, $\hat{\Sigma}_i$ using (26.9) as a guide:

$$\hat{p}_{fi} = \frac{1}{n} \sum_{k=1}^{n} \hat{u}_{ik} \tag{26.11a}$$

$$\hat{\Sigma}_{fi} = \frac{\sum_{k=1}^{n} (\hat{u}_{ik})(\mathbf{x}_k - \hat{\mathbf{v}}_i)(\mathbf{x}_k - \hat{\mathbf{v}}_i)^T}{\sum_{k=1}^{n} (\hat{u}_{ik})} \tag{26.11b}$$

Note especially that $\hat{\Sigma}_{fi}$ in (26.11b) is *not* the fuzzy covariance matrix $\hat{C}_{fi}$

Table 26.1. Necessary Conditions for ML and FCM Parametric Estimates

Parameter	MLE to maximize $\log(L)$ over Ω	FCM to minimize J_m over $M_{fc} \times \mathbb{R}^{cp}$
Prior prob. $\hat{p}_i$	$\left(\frac{1}{n}\sum_{k=1}^{n}\hat{h}_{ik}\right)$	—
Class-i mean $\hat{\boldsymbol{\mu}}_i$	$\dfrac{\sum_{k=1}^{n}\hat{h}_{ik}\mathbf{x}_k}{\sum_{k=1}^{n}\hat{h}_{ik}}$	$\dfrac{\sum_{k=1}^{n}(\hat{u}_{ik})^m\mathbf{x}_k}{\sum_{k=1}^{n}(\hat{u}_{ik})^m}$
Class-i covariance matrix $\hat{\Sigma}_i$	$\dfrac{\sum_{k=1}^{n}\hat{h}_{ik}(\mathbf{x}_k-\hat{\boldsymbol{\mu}}_i)(\mathbf{x}_k-\hat{\boldsymbol{\mu}}_i)^T}{\sum_{k=1}^{n}\hat{h}_{ik}}$	—
Posterior prob. $\hat{h}_{ik}$	$\left(\sum_{j=1}^{c}\frac{\hat{w}_{ik}}{\hat{w}_{jk}}\right)^{-1}$	$\left(\sum_{j=1}^{c}\left(\frac{\hat{d}_{ik}}{\hat{d}_{jk}}\right)^{2/(m-1)}\right)^{-1}$

defined at (22.4), because the $\hat{u}_{ik}$'s in (26.11) are not raised to the power m. For $\hat{U} \in M_c$ *hard*, (26.11b) reduces to $\hat{C}_i$, as does $\hat{C}_{fi}$, i.e., $\hat{\Sigma}_{fi} = \hat{C}_{fi} = \hat{C}_i \doteq \hat{\Sigma}_i$ for *labeled* data. Otherwise, $\hat{\Sigma}_{fi}$ and $\hat{C}_{fi}$ are distinct, each arising as a legitimate generalization of covariance from different contexts. In fact, letting $\hat{C}_{fi}(m)$ be the matrix arising from (22.4), one has infinitely many different estimates for Σ_i for a fixed $\hat{U} \in M_{fc}$, in addition to $\hat{\Sigma}_{fi}$ from (26.11b). Furthermore, for fixed X, different clustering algorithms, e.g., (A11.1), (A22.1), and (A23.1), will produce different $\hat{U}$'s at fixed m, which further complicates resolution of which estimate of Σ_i is "best." This dilemma is a familiar one in fuzzy models, for it often happens that one can propose many generalizations of a fixed conventional concept, all of which extend the concept in the mathematical sense. We circumvent this with a teleological heuristic: choose a generalization which provides useful results for the application at hand. This should not discourage the search for a theoretically distinguished "best" extension; it simply reinforces our belief that different models of the same process may yield comparable and useful results.

(E26.1) *Example 26.1.*[18] This example is a condensation of an example presented in (18) which compared four methods of estimation of $\hat{\boldsymbol{\omega}}_i = (\hat{p}_i, \hat{\boldsymbol{\mu}}_i, \hat{\Sigma}_i)$, using 12 samples of size $n = 50$ drawn from a mixture of two bivariate normals. The four methods are: labeled MLE (LMLE), wherein $\hat{\boldsymbol{\omega}}_i$ is calculated noniteratively using labeled subsamples and equations (26.10); UMLE algorithm (A26.1); FCM (A11.1); and HCM (A9.2). The same 12 samples were used for all four algorithms; for the last three, the samples were processed as if unlabeled. The termination criterion ε_L was 0.001 for all algorithms using the max norm on H or U. The Euclidean norm was used for (A9.2) and (A11.1); $c = 2$ was fixed; $m = 2$ was fixed for FCM; and the same initialization $H^{(0)} = U^{(0)}$ was used for the three iterative procedures.

The 12 samples processed were drawn from the mixture

$$F(\mathbf{x}) = 0.4n(\boldsymbol{\mu}_1, \Sigma) + 0.6n(\boldsymbol{\mu}_2, \Sigma) \tag{26.12}$$

where $\boldsymbol{\mu}_1 = (-2, -2)$; $\boldsymbol{\mu}_2 = (0, 0)$; and $g(\mathbf{x}|1)$, $g(\mathbf{x}|2)$ shared the common covariance matrix

$$\Sigma = \begin{pmatrix} 3.062 & 1.531 \\ 1.531 & 3.062 \end{pmatrix} \tag{26.13}$$

The assumption $\Sigma_1 = \Sigma_2 = \Sigma$ simplifies somewhat the necessary conditions at (26.9); the results given below did not account for this modification. The geometric structure of a typical sample would be quite like that depicted in Fig (8.2d), where $\boldsymbol{\mu}_i$ roughly corresponds to the center of cluster i (hence also the name "touching Gaussian clusters"). Table 26.2 lists the *mean* (over 12 samples) estimates of each of the ten parameters involved in the (hypothetical) parametric mixture (26.1).

There are various ways to compare these estimates, several of which are discussed in (18). Loosely speaking, LMLE does the best job, which is hardly suprising, since labeled data provide "more information" than unlabeled data. Hard c-means (A9.2) does quite poorly compared to the intermediate values supplied by UMLE and FCM. Concentrating now on the latter two, the priors $\{\hat{p}_i\}$ are somewhat better (0.02) with maximum likelihood; the

Table 26.2. Mean Estimates of $(p_i, \boldsymbol{\mu}_i, \Sigma_i)$: (E26.1)

		Method of estimation			
Parameter estimated	Actual	LMLE (26.10)	UMLE (A26.1)	FCM (A11.1)	HCM (A9.2)
p_1	0.40	0.40	0.45	0.47	0.46
p_2	0.60	0.60	0.55	0.53	0.54
μ_{11}	−2.00	−1.90	−1.92	−1.94	−2.23
μ_{12}	−2.00	−1.94	−2.07	−1.99	−2.42
μ_{21}	0.00	−0.11	0.32	0.19	0.44
μ_{22}	0.00	−0.15	0.32	0.18	0.47
Σ_{11}	3.06	3.02	2.82	2.68	1.85
Σ_{12}	1.53	1.61	1.55	1.30	0.37
Σ_{21}	1.53	1.61	1.55	1.30	0.37
Σ_{22}	3.06	2.80	2.69	2.58	1.44

four means $\{\hat{\mu}_{ij}\}$ are estimated much more accurately by fuzzy c-means; and the covariance matrix $[\hat{\Sigma}_{ij}]$ supplied by UMLE is slightly more accurate than that provided by FCM. The effect of varying m in (A11.1) was not studied in (18). •

Aside from the obvious qualitative similarities between ML and FCM estimates for unlabeled data, several important points should be made. First: if the asymptotic statistical properties of ML estimates are persuasive enough to justify solutions of system (26.9), fuzzy c-means ostensibly produces very accurate starting guesses [take $U^{(\text{final})}$ from (A11.1) as $H^{(0)}$] for algorithm (A26.1) at a significant fraction of the computational cost involved in UMLE. Secondly: as an initialization generator, fuzzy c-means is probably much more efficient than hard c-means, which has been used for this purpose in previous investigations.[34] And finally: if the objective in producing the estimates in, say, Table 26.2, is to use them to design approximate Bayesian classifiers, there is *no* theoretical basis for assuming that one classifier will be, a priori, superior (in the sense of empirical error rates) to another! Example (E26.2) illustrates this.

(E26.2) *Example 26.2.* The decision surface DS_b which characterizes the optimal Bayesian classifier $\{f_{jb}\}$ for any mixture of two multivariate normals having a common covariance matrix Σ can be found by requiring $p_i n(\boldsymbol{\mu}_1, \Sigma) = p_2 n(\boldsymbol{\mu}_2, \Sigma)$:

$$p_1 n(\boldsymbol{\mu}_1, \Sigma) = p_2 n(\boldsymbol{\mu}_2, \Sigma)$$

$$\Leftrightarrow \ln p_1 - \tfrac{1}{2}(\|\mathbf{x} - \boldsymbol{\mu}_1\|^2_{\Sigma^{-1}}) = \ln p_2 - \tfrac{1}{2}(\|\mathbf{x} = \boldsymbol{\mu}_2\|^2_{\Sigma^{-1}})$$

$$\Leftrightarrow \langle \mathbf{x}, \boldsymbol{\mu}_2 - \boldsymbol{\mu}_1 \rangle_{\Sigma^{-1}} = \ln(p_1/p_2) + \tfrac{1}{2}(\|\boldsymbol{\mu}_2\|^2_{\Sigma^{-1}} - \|\boldsymbol{\mu}_1\|^2_{\Sigma^{-1}}) \qquad (26.14)$$

Evidently DS_b is in this case a hyperplane orthogonol to $(\boldsymbol{\mu}_2 - \boldsymbol{\mu}_1)$ (in the sense of $\langle \cdot, \cdot \rangle_{\Sigma^{-1}}$); and $\{f_{jb}(\mathbf{x}) = p_j n(\boldsymbol{\mu}_j, \Sigma)\}$ is an affine classifier, because the quadratic terms $\|\mathbf{x}\|^2_{\Sigma^{-1}}$ cancel out. Applying (26.14) to the theoretical parameters of (E26.1) yields the theoretically optimal decision hyperplane (a line in $\mathbb{R}^2$ for $p = 2$) DS_b. And each algorithm in (E26.1) provides via (26.14) an approximate Bayesian decision surface, say $(\text{DS}_b)_{\text{LMLE}}$; $(\widehat{\text{DS}}_b)_{\text{UMLE}}$; and $(\widehat{\text{DS}}_b)_{\text{FCM}}$ [$(\widehat{\text{DS}}_b)_{\text{HCM}}$ is clearly inferior to the other three estimates so it is not exhibited here]. Figure 26.1 illustrates these four decisions surfaces. The Cartesian equations of these four decision surfaces are

$$(\text{DS}_b) \sim y = -1.00x - 2.95 \qquad (26.15\text{a})$$

$$(\widehat{\text{DS}}_b)_{\text{LMLE}} \sim y = -1.00x - 3.51 \qquad (26.15\text{b})$$

$$(\widehat{\text{DS}}_b)_{\text{UMLE}} \sim y = -0.73x - 1.78 \qquad (26.15\text{c})$$

$$(\widehat{\text{DS}}_b)_{\text{FCM}} \sim y = -0.87x - 1.89 \qquad (26.15\text{d})$$

Note that all four surfaces are displaced away from mean $\boldsymbol{\mu}_2$, because prior probability $p_2 > p_1$ renders decision region DR_2 slightly larger than DR_1. These equations allow one to form error rate equation (25.14) for the true and three estimated Bayesian classifiers defined by the hyperplane (26.15). For example, using $\boldsymbol{\mu}_1 = (-2, -2)$; $\boldsymbol{\mu}_2 = (0, 0)$; $\Sigma_{11} = \Sigma_{22} = 3.06$; $\Sigma_{12} = \Sigma_{21} = 1.53$; and (26.15a), equation (25.14) yields

$$\mathrm{ER}\{f_{jb}\} = (1 - I_1 - I_2) \tag{26.16a}$$

where $\mathbf{x} = (x, y)$, and

$$I_1 = 0.036 \int_{-\infty}^{\infty} dx \int_{-\infty}^{-x-2.95} \exp\left[-\tfrac{1}{2}(\mathbf{x} - \boldsymbol{\mu}_2)^T \Sigma^{-1} (\mathbf{x} - \boldsymbol{\mu}_2)\right] dy \tag{26.16b}$$

$$I_2 = 0.024 \int_{-\infty}^{\infty} dx \int_{-x-2.95}^{\infty} \exp\left[-\tfrac{1}{2}(\mathbf{x} - \boldsymbol{\mu}_1)^T \Sigma^{-1} (\mathbf{x} - \boldsymbol{\mu}_1)\right] dy \tag{26.16c}$$

This number, when calculated, is the lower bound in (25.15), i.e., the theoretically minimal error rate one can expect when (26.15a) is used to classify samples drawn from (26.12). Displacement of the boundary $\widehat{\mathrm{DS}}_b$ away from DS_b as shown in Fig. 26.1 increases the error rate of the three approximate Bayesian classifiers defined by (26.15b), (26.15c), and

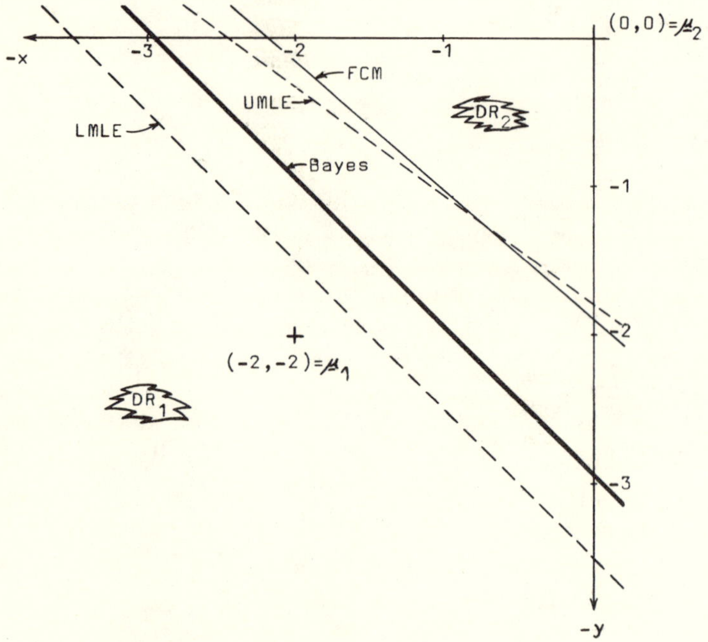

Figure 26.1. Bayesian decision surfaces: (E26.2).

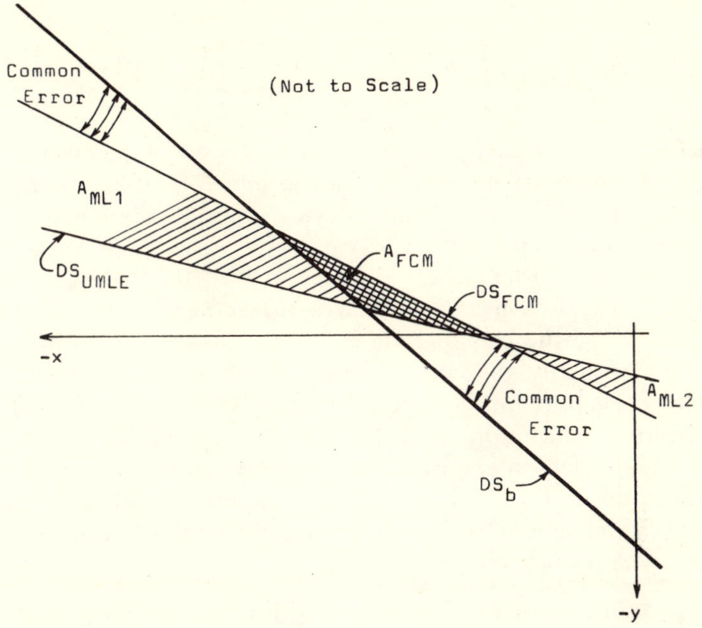

Figure 26.2. Error areas of approximate Bayesian machines: (E26.2).

(26.15d), by virtue of increasing the volume under the tail of the smaller density in one or both decision regions $\{\widehat{\text{DR}}_{jb}\}$. Since the true $p_j g(\mathbf{x}|j)$'s are known in this example, ER $\{\hat{f}_{jb}\}$ for each classifier is given by (26.16a) using the integrals at (26.16b) and (26.16c) but with limits of integration determined by the appropriate hyperplane in (26.15). Without calculating these numbers, it is quite probable that the LMLE classifier specified by (26.15b) enjoys the next best error rate, since its slope (– 1.00) is the same as DS_b's. The other two classifiers have decision boundaries that intersect DS_b, so the support of the integrals in (26.16b) and (26.16c) is "much larger" than for $(\widehat{\text{DS}}_b)_{\text{LMLE}}$. Figure 26.2 illustrates the regions of integration yielding error rates for the UMLE and FCM Bayesian classifiers. Since the integrand functions are fixed here, an incremental error analysis is possible. Figure 26.2 shows that the triangle bounded by vertices at (– 8.15, 5.20), (– 4.33, 1.38), and (– 0.79, – 1.21) is a region (A_{FCM}) of error for FCM-$\{\widehat{\text{DR}}_{jb}\}$ but not UMLE-$\{\widehat{\text{DR}}_{jb}\}$. Conversely, the two semiinfinite strips (labeled A_{ML1}, A_{ML2}) correspond to regions where the fuzzy Bayesian classifier is correct while the maximum likelihood Bayesian classifier will err. Let

$$I(A_{\text{FCM}}) = \iint_{A_{\text{FCM}}} p_1 g(\mathbf{x}|1)\, dx\, dy$$

and

$$I(A_{\mathrm{ML}}) = \iint_{A_{\mathrm{ML1}}} p_2 g(\mathbf{x}|2)\, dx\, dy + \iint_{A_{\mathrm{ML2}}} p_1 g(\mathbf{x}|1)\, dx\, dy$$

then $\mathrm{ER}(\{\hat{f}_{jb}\}_{\mathrm{FCM}}) < \mathrm{ER}(\{\hat{f}_{jb}\}_{\mathrm{UMLE}}) \Leftrightarrow I(A_{\mathrm{FCM}}) > I(A_{\mathrm{ML}})$. Without performing the integrations, one cannot order the three approximating classifiers by their Neyman–Pearson error rates. Readers with a flair for multiple integrals might enjoy verifying or disproving the conjecture that these rates are ordered as $\mathrm{ER(Bayes)} < \mathrm{ER(LMLE)} \leq \mathrm{ER(FCM)} \leq \mathrm{ER(UMLE)}$. The ordering actually makes little difference, for the geometric division of $\mathbb{R}^2$ by the classifiers in Fig. 26.1 shows that all of them will perform with roughly the same accuracy. The *order of magnitude* of these numbers is perhaps of more interest: are these 1%, 10% or 40% classifiers? The difficulty of calculating (25.14) makes this a hard question to answer without falling back on empirical error rates as exemplified in S27 below. The implication of Fig. 26.1 is that fuzzy parametric estimation can be as reliable as the maximum likelihood method, if not more so, and at a fraction of the computing costs required by (A26.1). •

Remarks

Unsupervised learning has become an important area for many applications: Cooper and Cooper[28] provide an excellent survey of many recent methods; Wolfe's paper[117] is exceedingly clear concerning mixtures of multinormals and multinomials. Examples 26.1 and 26.2 are discussed in greater detail in (18), where it was suggested that $\|\hat{H} - \hat{U}\|$ and $\|\hat{\boldsymbol{\mu}}_i - \hat{\mathbf{v}}_i\| \to 0$ as $n \to \infty$, that is, that FCM may converge (stochastically) to the same asymptotic (ML) estimates as (A26.1) under the assumptions of this section; this conjecture has not been investigated. Another question of interest concerns the differences between $\{C_{fi}(m)\}$ and Σ_{fi}; further research concerning the statistical properties of $J_m(U, \mathbf{v}, \mathbf{A})$ should provide new insights concerning the appropriateness of a specific form of "fuzzy covariance" in particular instances. A recent paper using fuzzy correlation defined via Sugeno's fuzzy integral[101] by Seif and Martin[97] describes an algorithm for unsupervised learning when the data are linguistic rather than numerical variables. This paper includes an example of prosthetic control (an artificial hand) using the fuzzy classifier so designed.

Exercises

H26.1. Suppose X to be a binary random variable distributed as $F(x;\mathbf{w}) = (w_1)^x(1-w_1)^{(1-x)}/2 + (w_2)^x(1-w_2)^{1-x}/2$, where x can be zero or one. Prove that F is not identifiable.

H26.2. Let $F(x;\mathbf{w}) = p_1 n(w_1, 1) + p_2 n(w_2, 1)$ be a mixture of two univariate normal densities. For what prior probabilities $\{p_1, p_2\}$ is F unidentifiable? (Answer: $p_1 = p_2 = 0.5$.)

H26.3. Suppose $F(\mathbf{x};\mathbf{w}) = \sum_{i=1}^{c} p_i g(\mathbf{x}|i;\mathbf{w}_i)$, where each g is a p-variate Bernoulli distribution, $g(\mathbf{x}|i,\mathbf{w}_i) = \prod_{j=1}^{p} w_{ij}^{x_j}(1-w_{ij})^{(1-x_j)}$, $x_j = 0$ or $1 \forall j$, $w_{ij} = \Pr(x_j = 1$ on trial $i) \forall i, j$. Find $\hat{\mathbf{w}}_i$, the maximum likelihood estimate for $\mathbf{w}_i$, using n draws from F. [Answer: $\hat{\mathbf{w}}_i = \sum_{k=1}^{n} h(i|\mathbf{x}_k;\hat{\mathbf{w}}_i)\mathbf{x}_k / \sum_{k=1}^{n} h(i|\mathbf{x}_k;\hat{\mathbf{w}}_i)$.]

H26.4. Let $F(x;w) = p_1 n(0, 1) + (1-p_1)n(0, \frac{1}{2})$. Find the maximum likelihood estimate of p_1 using one draw x_1 from F; (a) if $x_1^2 < \log(2)$; (b) if $x_1^2 > \log(2)$. [Answer: (a) $p_1 = 0$.]

H26.5. $A \in V_{pp}$, $\mathbf{x} \in \mathbb{R}^p$, $F(\mathbf{x}) = \mathbf{x}^T A \mathbf{x}$. Find

- (i) $\nabla_{\mathbf{x}} F$ for any $A \in V_{pp}$ [Answer: $(A + A^T)\mathbf{x}$]
- (ii) $\nabla_{\mathbf{x}} F$ for symmetric A (Answer: $2A\mathbf{x}$)
- (iii) $\nabla_{\mathbf{x}} F$ for A the identity matrix (Answer: $2\mathbf{x}$)

H26.6. $A \in V_{pp}$ invertible, $\mathbf{x}, \mathbf{x}_k \in \mathbb{R}^p$, $F(\mathbf{x}) = \sum_{k=1}^{n} (\mathbf{x}-\mathbf{x}_k)^T A^{-1}(\mathbf{x}-\mathbf{x}_k)$. Find the unique vector $\mathbf{x} \in \mathbb{R}^p$ that minimizes F over $\mathbb{R}^p$. (Answer: $\mathbf{x} = \sum_{k=1}^{n} \mathbf{x}_k/n$.)

H26.7. Use the results of H22.1 to show that $\sum_{k=1}^{n} (\mathbf{x}_k - \boldsymbol{\mu})(\mathbf{x}_k - \boldsymbol{\mu})^T/n = \hat{\Sigma}$ is the unique maximum likelihood estimate of the covariance matrix of $g(\mathbf{x};\mathbf{w}) = n(\boldsymbol{\mu}, \Sigma)$ when $\boldsymbol{\mu}$ is known, using n samples drawn from g.

H26.8. Let $F(\mathbf{x}) = \sum_{j=1}^{c} p_j n(\boldsymbol{\mu}_j, \Sigma_j)$ be a mixture of c p-variate normal densities, and let $\{f_{jb}\}$ denote the optimal Bayesian classifier relative to F determined with (25.12). If all the parameters of F are known, show that the decision function $f_{jb}(\mathbf{x})$ can be written as

- (i) $\langle \boldsymbol{\alpha}_j, \mathbf{x}\rangle + a_j$ when $\Sigma_j = \sigma^2 I \forall j$.
  [Answer: $\boldsymbol{\alpha}_j = \boldsymbol{\mu}_j/\sigma^2$, $a_j = \log(p_j) - (\boldsymbol{\mu}_j^T\boldsymbol{\mu}_j/2\sigma^2)$.]
- (ii) $\langle \boldsymbol{\beta}_j, \mathbf{x}\rangle + b_j$ when $\Sigma_j = \Sigma \forall j$.
  [Answer: $\boldsymbol{\beta}_j = \Sigma^{-1}\boldsymbol{\mu}_j$; $b_j = \log(p_j) - (\boldsymbol{\mu}_j^T\Sigma^{-1}\boldsymbol{\mu}_j/2)$.]
- (iii) $\mathbf{x}^T\Gamma_j\mathbf{x} + \langle \boldsymbol{\gamma}_j, \mathbf{x}\rangle + c_j$ when Σ_j = arbitrary. [Answer: $\Gamma_j = -\Sigma_j^{-1}/2$, $\boldsymbol{\gamma}_j = \Sigma_j^{-1}\boldsymbol{\mu}_j$, $c_j = \log(p_j) - \frac{1}{2}(\log\det\Sigma_j) - (\boldsymbol{\mu}_j^T\Sigma_j^{-1}\boldsymbol{\mu}_j/2)$.]

H26.9. $X = \{4, 1, 2, 1, 3, 5, 2, 2\} \cup \{6, 6, 8, 6, 7, 5, 9, 9\} \subset \mathbb{R}$ is a labeled sample of size 16 from the mixture $F(x;\mathbf{w}) = p_1 n(\mu_1, 1) + p_2 n(\mu_2, 1)$. Find the maximum likelihood estimates of $\mathbf{w}_1$ and $\mathbf{w}_2$ based on X. Find the estimated Bayesian decision surface for partitioning $\mathbb{R}$ based on these estimates. (Answer: $p_1 = p_2 = 0.5$; $\hat{\mu}_1 = 2.5$; $\hat{\mu}_2 = 7.0$; $\widehat{\mathrm{DS}}_b = 4.75$.)

H26.10. $X = \{(4, 1), (2, 1), (3, 5), (2, 2)\} \cup \{(6, 6), (8, 6), (7, 5), (9, 9)\} \subset \mathbb{R}^2$ is a labeled sample of size 8 from the mixture $F(\mathbf{x};\mathbf{w}) = p_1 n(\boldsymbol{\mu}_1, \Sigma_1) + p_2 n(\boldsymbol{\mu}_2, \Sigma_2)$. Find the maximum likelihood estimates $(\hat{p}_1, \hat{\boldsymbol{\mu}}_1, \hat{\Sigma}_1), (\hat{p}_2, \hat{\boldsymbol{\mu}}_2, \hat{\Sigma}_2)$ based on X, and sketch the decision surface $\widehat{\mathrm{DS}}_b$ in $\mathbb{R}^2$ which these parameters define.

S27. Nearest Prototype Classifiers: Diagnosis of Stomach Disease

In this section we discuss two popular non-Bayesian designs: nearest-prototype (1-NP) and nearest-neighbor (1-NN) classifiers. These machines are formulated via distances in data space between observations to be classified and paradigms of the c classes so represented. In the 1-NP case, the paradigms are c prototypical vectors, say $\{\boldsymbol{\alpha}_1, \boldsymbol{\alpha}_2, \ldots, \boldsymbol{\alpha}_c\} \subset \mathbb{R}^p$; in the 1-NN case, the paradigms are n *labeled* observations of the process, say $X = X_1 \cup X_2 \cup \cdots \cup X_c$; $|X_i| = n_i \forall i$; $\sum_i n_i = n = |X|$; $X_i \cap X_j = \emptyset \forall i \neq j$, and $\mathbf{x}_k \in i \Leftrightarrow \mathbf{x}_k \in X_i$. Considering first the 1-NP classifier, let d be a metric for $\mathbb{R}^p$. The decision to classify $\mathbf{x} \in j$ in case $\mathbf{x}$ is metrically closest to prototype $\boldsymbol{\alpha}_j$ defines the nearest prototype decision rule as follows:

$$\text{decide } \mathbf{x} \in j \Leftrightarrow d(\mathbf{x}, \boldsymbol{\alpha}_j) = \min_{1 \leq k \leq c} \{d(\mathbf{x}, \boldsymbol{\alpha}_k)\} \tag{27.1}$$

If d is an inner-product-induced norm metric on $\mathbb{R}^p$, squaring distances in (27.1) and expanding in the inner product leads to the equivalent decision rule

$$\text{decide } \mathbf{x} \in j \Leftrightarrow f_j(\mathbf{x}) = \max_{1 \leq k \leq c} \{f_k(\mathbf{x})\} \tag{27.2a}$$

where

$$f_j(\mathbf{x}) = \langle \mathbf{x}, \boldsymbol{\alpha}_j \rangle - (\|\boldsymbol{\alpha}_j\|^2/2), \qquad 1 \leq j \leq c \tag{27.2b}$$

Functional form (27.2b) is identical to (25.5) with $(-\|\boldsymbol{\alpha}_j\|^2/2) = a_j$: consequently, the 1-NP classifier is an affine machine; and the minimum distance (to NP $\boldsymbol{\alpha}_j$) strategy at (27.1) defines through the f_j's in (27.2b) a maximum classifier as in (D25.1).

(D27.1) *Definition 27.1* [*Nearest Prototype (1-NP) Classifier*]. Let $f_j(\mathbf{x}) = \langle \mathbf{x}, \boldsymbol{\alpha}_j \rangle - (\|\boldsymbol{\alpha}_j\|^2/2)$, $1 \leq j \leq c$, where $\{\boldsymbol{\alpha}_j\}$ are c prototypical vectors in $\mathbb{R}^p$. Then $\{f_j\}_{\text{NP}}$ with decision rule (25.1)—or equivalently, (27.1)—is said to determine a nearest-prototype (affine) maximum classifier via (D25.1) on $\mathbb{R}^p$. Ties are resolved arbitrarily.

One can, of course, use (27.1) with any measure of distance for $\mathbb{R}^p$, not just metrics derived from inner products. Decision rule (27.1) still defines a 1-NP machine, but it is not necessarily an affine maximum classifier as in (D27.1): and in either case, 1-NP classifiers are not "Bayesian," since the f_j's are not functions of either side of (25.9), except in very special circumstances.[34]

The nearest-neighbor classifier operates on the same principle as the 1-NP machine, by classifying each observation according to a minimizing distance; the difference is that distances are measured to labeled observations, not labeled prototypes. Given a labeled set X as above, with, say, $U \in M_c$ the hard labeling of X, the nearest-neighbor decision rule is to

$$\text{decide } \mathbf{x} \in j \Leftrightarrow d(\mathbf{x}, \mathbf{x}_i) = \min_{1 \leq k \leq n} \{d(\mathbf{x}, \mathbf{x}_k)\} \quad \text{and} \quad u_{ji} = 1 \tag{27.3}$$

where, as usual, $u_{ji} = u_j(\mathbf{x}_i)$ = membership of $\mathbf{x}_i$ in cluster j.

(D27.2) *Definition 27.2* [*Nearest Neighbor (1-NN) Classifier*]. Let

$$X = \bigcup_{i=1}^{c} X_i, \qquad |X| = n, \qquad |X_i| = n_i, \qquad \sum_{i=1}^{c} n_i = n,$$
$$X_i \cap X_j = \emptyset, \qquad i \neq j$$

be a labeled data set, with $U \in M_c$ its labeling (hard c-partitioning). The decision rule (27.3) is said to define a nearest-neighbor (1-NN) classifier for $\mathbb{R}^p$. Ties are resolved arbitrarily.

Because n f_j's are defined by (27.3), the 1-NN classifier, say $\{f_{j,n}\}_{\text{NN}}$, does not have the canonical form (25.2) unless $c = n$. In this case, the labeled observations comprising X can be thought of as c prototypes, $\{\mathbf{x}_j \in X_j\} = \{\boldsymbol{\alpha}_j\}$, and the 1-NN classifier is a special 1-NP classifier. Ordinarily, however, n is much larger than c. In fact, the efficiency of the 1-NN classifier is asymptotically less than twice the theoretically optimal Bayes risk: Cover and Hart showed in (29) that

$$\text{ER}(\{f_{jb}\}) \leq \lim_{n \to \infty} \{\text{ER}(\{f_{j,n}\}_{1\text{-NN}})\} \leq 2\text{ER}(\{f_{jb}\}) \tag{27.4}$$

Although (27.4) is vacuous in the sense that $\text{ER}(\{f_{jb}\})$ is always unknown, this result affords some psychological justification for using the 1-NN classifier.

More generally, it is easy to extend definitions 27.1 and 27.2 to k-NP and k-NN classifiers, where $k > 1$. In the first case, one uses *multiple* prototypes of each class, finds the k nearest prototypes to $\mathbf{x}$ in the metric d, and assigns $\mathbf{x}$ to the class which is closest (in prototypes) to $\mathbf{x}$ most often in k tries. The k-NN rule works exactly the same way, by assigning $\mathbf{x}$ to the class which has the most among its k nearest labeled neighbors. It seems intuitively clear that as *both* k and $n \to \infty$, the k-NN error rate should approach $\text{ER}(\{f_{jb}\})$. This is precisely the case, as was demonstrated by Fukunaga and Hostetler;[43] however, one should interpret this cautiously, because asymptotic results have little connection to real applications involving finite data sets! Nonetheless, the fact that k-NN classifiers are asymptotically optimal in the sense of Neyman–Pearson error rate lends them a certain credibility which often leads to their use.

Although nearest-prototype and nearest-neighbor classifiers are conceptually quite similar, their designs call for rather different strategies. Considering first the NN design, one needs the labeling partition $U \in M_c$ of X, and a distance measure for $\mathbb{R}^p$. If U is obtained during collection of the sample X, so that the labels are known to be correct on *physical* grounds, it is hard to fault the logic of using a k-NN classifier, as long as the process itself also recommends a "natural" measure of distance with which to implement the design. In some sense, this brings all of the available "information" in X to bear on each decision rendered, and thus constitutes an efficient use of the data. One could, of course, use (X, U) to estimate $\{\hat{f}_{jb}\}$ as in S26, but there is no *a priori* reason to expect $\{\hat{f}_{jb}\}$ to perform more accurately than $\{f_{j,n}\}_{k\text{-NN}}$. The k-NN classifier, given U, is designed simply by choosing k and d, whereas numerical estimation of the parameters of $F(\mathbf{x}; \boldsymbol{\omega})$ is time-consuming, costly, and (possibly) intractable; these are far better reasons than asymptotic optimality for using k-NN designs!

If X is unlabeled, it can be used for k-NN classification after being itself partitioned. The labeling of X can be estimated by *any* of the hard or fuzzy clustering algorithms discussed here or elsewhere: since (D27.2) is made with $U \in M_c$, fuzzy c-partitions of X will have to be converted to hard ones in order to implement the k-NN rule as we have defined it. Two fairly obvious generalizations of (D27.2) which have yet to be investigated are these. First: (27.3) could be implemented with fuzzy c-partitions of X simply by altering it to assign $\mathbf{x}$ to the class j in which $\mathbf{x}_i$ has maximum membership; and for $k > 1$, the majority class could be defined as the one maximizing the sum of the k nearest-neighbor memberships. Such a classifier renders hard decisions based on fuzzy labels. An even more general situation would allocate fuzzy class memberships to $\mathbf{x}$ based on a variety of proximity notions, thus affording fuzzy decisions based on fuzzy labels. Both of these generalizations should yield interesting and potentially useful algorithms; since the example below involves (hard) 1-NP and k-NN designs, further explorations in these directions are left to future researchers.

Nearest prototype classifiers obviously require the prototypes $\{\boldsymbol{\alpha}_j\}$, and d, a measure of distance. The choice for d is similar in every respect to that above; error rates of $\{f_j\}_{\text{NP}}$ and $\{f_{j,n}\}_{\text{NN}}$ clearly depend on the notion of distance used. The choice of the $\boldsymbol{\alpha}_j$'s may or may not depend on X being labeled: for example, given $U \in M_c$ labeling X, one may (but need not!) take for $\boldsymbol{\alpha}_j$ the centroid of u_j. Conversely, if X is not labeled, it is not necessary to label it in order to find prototypes, although some algorithms, e.g., (A9.2), produce both. If it seems desirable, NN and NP designs may be categorized according to the method of estimating the labels U(NN), or prototypes $\{\boldsymbol{\alpha}_j\}$(NP) as heuristic, deterministic, statistical, or fuzzy. For example, using the $\mathbf{v}_i$'s from (E26.1) as the $\boldsymbol{\alpha}_i$'s yields with (D27.1) a fuzzy 1-NP machine; using the $\boldsymbol{\mu}_i$'s from (E26.1) as the $\boldsymbol{\alpha}_i$'s yields a statistical maximum likelihood 1-NP machine; and so on. The following example compares the empirical error rate of a fuzzy 1-NP design to the rate observed for k-NN classifiers designed with labeled data.

(E27.1) *Example 27.1.* The data for this example are a set of $n = 300$ feature vectors $\mathbf{x}_k \in \mathbb{R}^{11}$, representing 300 patients with one of $c = 6$ abdominal disorders: hiatal hernia, duodenal ulcer, gastric ulcer, cancer, gallstones, and functional disease. The data are binary, each feature representing the presence or absence of a measured symptom (the example of S13 used the subset of these data consisting of all hernia and gallstone cases). These data were collected by Rinaldo, Scheinok, and Rupe[83] in 1963; each patient was labeled unequivocally using conventional medical diagnosis, so the data affords a natural basis for k-NN designs. A subsequent study of classifier design using these data was done by Scheinok and Rinaldo,[94] who gave empirical error rates based on various subsets of selected features. Later, Toussaint and Sharpe[107] studied different methods of estimation of empirical error rates using k-NN designs as the classifier basis. Using the Euclidean norm for d in (D27.2) and the medically correct labels for each patient, it was their conclusion that the most credible estimate for the empirical error rate of *all* k-NN machines based on this fixed data set was eer(0.5) $\approx 48\%$, where eer (0.5) is given by formula (25.19) with $\alpha = 0.5$. In other words, Toussaint and Sharpe suggest that the best one can hope for, on the average, using Euclidean k-NN designs based on the 300 labels supplied, is about 52 correct diagnoses per 100 tries.

Bezdek and Castelaz[15] contrasted this to the empirical error rate predicted by (25.19) using the same data and methods for calculating eer(α), but using a fuzzy 1-NP classifier designed as follows: for each trial i, $1 \leq i \leq 20$, a training set of 230 vectors drawn randomly from X was used with (A11.1) to find six prototypes $\{\boldsymbol{\alpha}_j\}_i = \{\mathbf{v}_j\}_i$, the fuzzy cluster centers at (11.3b). This method of rotation through X is a slight modification of Toussaint's π method above; it presumably secures a somewhat more stabilized average for $\overline{\text{EER}}(\pi)$. For each of these runs, (A11.1) used $m = 1.4$, $\|\cdot\| =$ Euclidean, $c = 6$, $\varepsilon_L = 0.01$, and the max norm on iterates $U^{(l)}$. The cluster centers $\{\mathbf{v}_j\}_i$ were then substituted for $\{\boldsymbol{\alpha}_j\}$ in (27.2b), resulting in a 1-NP classifier based on fuzzy prototypes. The inner product and norm at (27.2a) were also Euclidean. The resubstitution estimate $\overline{\text{EER}}(R)$ for eer(α) was made as described in (S25). The overall result was this: based on formula (25.19) with $\alpha = 0.5$, the expected empirical error rate of the fuzzy 1-NP classifier was approximately 38%. That is, using the empirical method (25.19) for error rates defined by Toussaint, 1-NP classifiers designed with (27.2) using cluster centers from (A11.1) for the prototypes seems to improve the probability of correct automatic diagnosis of the six abdominal diseases represented by X by approximately 10% when compared to *all* k-NN designs based on this 300-point data set.

Several additional observations should be made. First, the labels for k-NN designs are unequivocal, so testing always yields an "absolute" number of mistakes. At the end of a training session with (A11.1), prototypes $\{\mathbf{v}_j\}_i$ have numerical—but not *physical*—labels. That is, the computer does not know whether $(\mathbf{v}_3)_i$ represents, e.g., gallstones or hernia. The numerator of (25.16)—the number of wrong decisions in n_{ts} trials—was calculated by assigning to each $(\mathbf{v}_j)_i$ that *physical* label which resulted in the minimum number of testing errors on trial i. At the end of *testing*, then, the fuzzy 1-NP design has physical labels which minimize EER($\{f_j\}$, $X_{ts,i}$), and which render it operable (i.e., ready to classify *physically* incoming observations). If the classes are not well represented in X, it is entirely conceivable that the random sample $X_{tr,i}$ is insufficient to characterize all c physical classes by mathematical prototypes. Thus, it may happen that several distinct $(\mathbf{v}_j)_i$'s are assigned the same physical label, and some physical classes end up without prototypical representatives! This enigmatic possibility still optimizes classifier performance under the circumstances that cause it: judicious sampling methods should eliminate this occurrence in practice.

Secondly, one must always guard against generalizing the percentages above with undue haste. The error rates above are for one sample of size 300: a different sample of the same size and process might reverse the above figures!

Finally, the asymptotic properties of k-NN designs may tempt one to assert that k-NN machines will, for "large enough" samples, yield better EERs than any other classifier. One should remember, however, that "large enough" is not accepted as the upper limit of a "Do Loop"! •

Remarks

S26 and S27 show that neither estimated Bayesian nor asymptotically Bayesian designs *necessarily* yield better classifiers *in practice* than other decision-theoretic strategies. The basic reason for this is simple: design and testing of all classifiers is ultimately limited by the finiteness of the sample used. Accordingly, classifier performance is ultimately linked to the usual vagaries of sampling. For this reason, it seems best to approach classifier design with a view towards finding a machine whose observed performance is good on samples of the process being studied, rather than choosing the design a priori, on the basis of some attractive theoretical property (unless, of course, the theoretical property is explicitly linked to the physical process). This should not, however, dull one's appetite for theoretical reassurance. In particular, no analysis such as Cover and Hart's[29] has been formulated for fuzzy classifier designs; whether fuzzy classifiers such as $\{\hat{f}_{jb}\}_{\text{FCM}}$ in (S26) and $\{f_j\}_{1\text{-NP}}$ in (S27) have nice asymptotic relations to $\{f_{jb}\}$ or others remains to be discovered. That they can perform at least as well as more conventional designs in some circumstances seems empirically substantiated by the examples of (S26) and (S27). Reference (19) contains an example similar to (E27.1) using hypertension data which further corroborates this supposition.

Exercises

H27.1. Suppose X is a labeled sample from a mixture of c known p-variate normal densities. Under what circumstances is the optimal Bayesian classifier a 1-NP classifier? What are the prototypes? [Answer: cases (i), (ii) of H26.9 when $p_j = (1/c)\forall j$.]

H27.2. Let $\{\boldsymbol{\alpha}_k | 1 \leq k \leq c\} \subset \mathbb{R}^p$, be c fixed prototypes, and suppose $\mathbf{x} \in \mathbb{R}^p$ is fixed. Let d be (any fixed) metric on $\mathbb{R}^p$, and suppose $d(\mathbf{x}, \boldsymbol{\alpha}_j)$ to be the minimum among the distances $\{d(\mathbf{x}, \boldsymbol{\alpha}_k) | 1 \leq k \leq c\}$. Is 1-NP classification of $\mathbf{x} \in j$ via (27.1) invariant if d is changed to

(i) an arbitrary metric for $\mathbb{R}^p$?
(ii) an arbitrary norm-induced metric for $\mathbb{R}^p$?
(iii) an arbitrary inner-product-induced metric for $\mathbb{R}^p$?
(iv) an arbitrary matrix-induced inner-product-induced metric for $\mathbb{R}^p$?

H27.3. $X = \{(0, 0), (2, 0), (2, 2), (0, 2)\} \cup \{(1, 0), (0, 1), (-1, 0), (0, -1)\} \subset \mathbb{R}^2$ is a 2-class labeled sample. Determine and sketch the decision boundaries of the

(i) 1-NP classifier designed with subsample centroids as prototypes (Answer: $y = 1 - x$);
(ii) 1-NN classifier based on the given labels, assuming for d the Euclidean distance in both cases.

H27.4. Repeat H27.3, using the norm N_2 in H22.2 to define d.

H27.5. Repeat H27.3, using the norm N_3 in H22.2 to define d.

H27.6. The upper bound in (27.4) can be improved: the tighter upper bound derived in (29) is

$$\text{ER}(\{f_{jb}\})[2 - (c\text{ER}(\{f_{jb}\})/(c-1)].$$

Under what circumstances is the 1-NN classifier *optimal* independent of the number of labeled neighbors used? See (34) for an example wherein the 1-NN and Bayes risks are equal and not extremes. [Answer: ER = 0; 1 − (1/c).]

H27.7. The fact that $\{f_{i,n}\}_{k\text{-NN}}$ is asymptotically optimal suggests that its error rate decreases monotonically as k increases. Argue against this by considering a $c = 2$ class mixture, $p_1 = p_2 = 0.5$, $g(\mathbf{x}|j)$ uniform, k odd. Show that for n labeled observations drawn from this mixture

(i) $\text{ER}(\{f_{i,n}\}_{1\text{-NN}}) = 2^{-n}$

(ii) $\text{ER}(\{f_{i,n}\}_{k\text{-NN}}) = 2^{-n}\left[\sum_{i=0}^{(k-1)/2} \binom{n}{i}\right] > 2^{-n}$ if $k > 1$

(iii) Generalize this result: under what circumstances is 1-NN theoretically superior to k-NN for $k > 1$, $c > 2$?

H27.8. Let $X = \{\mathbf{x}_1, \ldots, \mathbf{x}_n\}$, $Y = \{\mathbf{y}_1, \ldots, \mathbf{y}_m\}$ be labeled data sets drawn from classes 1 and 2 in $\mathbb{R}^p$. Let

$$H_{ij} = \{\mathbf{z} \in \mathbb{R}^p | f(\mathbf{z}) = \langle [\mathbf{z} - (\mathbf{x}_i + \mathbf{y}_j)/2], (\mathbf{x}_i - \mathbf{y}_j) \rangle = 0\}.$$

(i) Show that H_{ij} is a separating hyperplane for $\mathbf{x}_i$ and $\mathbf{y}_j$ unless $\mathbf{x}_i = \mathbf{y}_j$ (cf. H25.7); sketch one such H_{ij} schematically.

(ii) Let

$$H_{ij}^+ = \{\mathbf{z}|f(\mathbf{z}) > 0\rangle, \qquad H_{ij}^- = \{\mathbf{z}|f(\mathbf{z}) < 0\}$$

$$H^+ = \bigcap_{j=1}^{m} \bigcup_{i=1}^{n} H_{ij}^+, \qquad H^- = \bigcap_{i=1}^{n} \bigcup_{j=1}^{m} H_{ij}^-$$

Consider the decision rule: decide $\mathbf{z} \in 1 \Leftrightarrow \mathbf{z} \in H^+$; decide $\mathbf{z} \in 2 \Leftrightarrow \mathbf{z} \in H^-$; ties are arbitrary. Show that this is equivalent to the 1-NN rule (27.3).

H27.9. Sketch the data set of H9.9. Let X be the first eight vectors and Y be the last 12 vectors of this set, labels as in H27.8.

(i) Are X and Y linearly separable? (Answer: Yes.)

(ii) Sketch $H_{1,12}$; $H_{1,12}^+$; $H_{1,12}^-$. How many other H_{ij}'s are there parallel to $H_{1,12}$? (Answer: 11.)

(iii) Sketch H_{81}; H_{81}^+; H_{81}^-. This hyperplane separates $\mathbf{x}_8$ and $\mathbf{y}_1$ as well as X and Y. Does *every* H_{ij} determined as in H27.8 separate all of X and Y? (Answer: Yes.)

(iv) Which H_{ij} separates X and Y most optimistically for class-X objects (i.e., is closest to Y)? (Answer: $H_{8,11}$.)

(v) Which H_{ij} maximizes the distances from it to both X and Y?

(vi) Try to sketch H^+ and H^-, retaining your sense of humor in the process.

H27.10. Consider the metric derived from N_1 in H22.2:

$$d_1(\mathbf{x}, \boldsymbol{\alpha}) = \sum_{i=1}^{p} |x_i - \alpha_i| = \|\mathbf{x} - \boldsymbol{\alpha}\|_1$$

where $\mathbf{x}, \boldsymbol{\alpha} \in \mathbb{R}^p$. N_1 is not inner-product induced, so the 1-NP rule of (27.1) is, in general, not equivalent to the form displayed in (27.2). If, however, the prototypes $\{\boldsymbol{\alpha}_j\}$ and $\mathbf{x}$ are always *binary* valued, decision rule (27.1) can be expressed as an affine classifier. For this special case, prove that

$d_1(\mathbf{x}, \boldsymbol{\alpha}_j) = \bigwedge_{k=1}^{c} d_1(\mathbf{x}, \boldsymbol{\alpha}_k) \Leftrightarrow f_j(\mathbf{x}) = \bigvee_{k=1}^{c} f_k(\mathbf{x})$, where $f_j(\mathbf{x}) = \langle \mathbf{x}, \boldsymbol{\alpha}_j \rangle - (\|\boldsymbol{\alpha}_j\|_1/2)$, the dot product being Euclidean. This is called a 1-NP Minkowski metric classifier.

S28. Fuzzy Classification with Nearest Memberships

This section describes an interesting combination of several ideas which is due to Ruspini.[92] The classifier involved is not of the type described in previous sections, because it does not partition feature space $\mathbb{R}^p$ into decision regions as in (D25.1). Instead, it yields, upon application, simultaneous *fuzzy* classifications of a finite number of unlabeled points, based on the notion of membership proximity between the desired labels and a set of given labels. To describe this method, suppose that $(X, U) \in \mathbb{R}^p \times M_{fc}$ is given, where $|X| = n$. $X = \{\mathbf{x}_1, \mathbf{x}_2, \ldots, \mathbf{x}_n\}$ is regarded here as a set of n training vectors in $\mathbb{R}^p$. If X is labeled, $U \in M_c$ is given, and the input to Ruspini's classifier coincides with that required for the k-NN classifier in S27. Otherwise, a clustering algorithm (hard or fuzzy) is applied to X, resulting in the labels $U \in M_{fc}$ for X.

Next, let $\hat{X} = \{\hat{\mathbf{x}}_1, \hat{\mathbf{x}}_2, \ldots, \hat{\mathbf{x}}_q\} \subset \mathbb{R}^p$ be a set of q *unlabeled* vectors in $\mathbb{R}^p$. Imagine $\hat{X}$ to be an incoming, unlabeled sample generated by the same process that produced X; we seek a $(c \times q)$ matrix $\hat{U} \in M_{fc}$ which labels the vectors in $\hat{X}$ with hard or fuzzy memberships in each of the c classes represented by memberships in the "training partition" U. Ruspini's idea was to incorporate as unknowns the desired memberships $\{\hat{u}_{ij} = \hat{u}_i(\hat{\mathbf{x}}_j)\} \forall \hat{\mathbf{x}}_j \in \hat{X}$ in the functional J_R at (9.2), while leaving the given memberships $\{u_{ik} = u_i(\mathbf{x}_k)\} \forall \mathbf{x}_k \in X$ fixed, in such a way that subsequent minimization of this augmented functional, say $\hat{J}_R$, would assign memberships $\hat{u}_{ij}$ to $\hat{\mathbf{x}}_j$ that are similar to u_{ik} for $\mathbf{x}_k$ when the dissimilarity between $\hat{\mathbf{x}}_j$ and $\mathbf{x}_k$, say $d(\hat{\mathbf{x}}_j, \mathbf{x}_k)$, is small (ordinarily d is a metric for $\mathbb{R}^p$). To formalize this procedure, let us denote the kth column of U as $\mathbf{U}^{(k)} \in \mathbb{R}^c$, and the jth column of $\hat{U}$ as $\hat{\mathbf{U}}^{(j)} \in \mathbb{R}^c$, $1 \leq k \leq n$; $1 \leq j \leq q$. The inner sum in J_R, equation (9.2), can be rewritten in this notation as

$$\sum_{i=1}^{c} \sigma(u_{ij} - u_{ik})^2 = \sigma\|\mathbf{U}^{(j)} - \mathbf{U}^{(k)}\|_{\mathbb{R}^c}^2 \tag{28.1}$$

where $\mathbf{U}^{(j)}$ appears instead of $\hat{\mathbf{U}}^{(j)}$ because in S9 the $c \times n$ matrix $U \in M_{fc}$ was the variable. The norm in (28.1) is just the Euclidean norm on $\mathbb{R}^c$. Letting $d_{jk}^2 = d^2(\mathbf{x}_j, \mathbf{x}_k)$ be the dissimilarity between $\mathbf{x}_j$ and $\mathbf{x}_k$ in X, Ruspini's original functional minimized a sum of squares of form

$$(\sigma\|\mathbf{U}^{(j)} - \mathbf{U}^{(k)}\|_{\mathbb{R}^c}^2 - d_{jk}^2)^2 \tag{28.2}$$

in the unknown columns of U as the criterion for finding U. Equation (28.2) can, when d is a norm metric for $\mathbb{R}^p$, be written much more suggestively as

$$(\sigma\|\mathbf{U}^{(j)} - \mathbf{U}^{(k)}\|^2_{\mathbb{R}^c} - \|\mathbf{x}_j - \mathbf{x}_k\|^2_{\mathbb{R}^p})^2 \tag{28.3}$$

A simple alteration which makes it possible to search for classifications of $\hat{X}$ by $\hat{U}$ given (X, U), is to replace $\mathbf{U}^{(j)}$ by $\hat{\mathbf{U}}^{(j)}$ and $\mathbf{x}_j$ by $\hat{\mathbf{x}}_j$ in (28.3), sum over j and k, and attempt to minimize the resultant sum of squares. Although Ruspini discussed the more general form (28.2) in principle, his examples used for d the Euclidean norm on $\mathbb{R}^p$. Accordingly, let $\hat{J}_R: M_{fco} \to \mathbb{R}^+$ be defined as

$$\hat{J}_R(\hat{U}) = \sum_{j=1}^{q} \sum_{k=1}^{n} (\sigma\|\hat{\mathbf{U}}^{(J)} - \mathbf{U}^{(k)}\|^2_{\mathbb{R}^c} - \|\hat{\mathbf{x}}_j - \mathbf{x}_k\|^2_{\mathbb{R}^p})^2 \tag{28.4}$$

where X, $\hat{X}$, and U are given as described above, and the $(c \times q)$ matrix $\hat{U}$ provides hard or fuzzy memberships for the vectors in $\hat{X}$ to be classified. Evidently, a U^* in M_{fco} that solves

$$\underset{\hat{U} \in M_{fco}}{\text{minimize}} \{\hat{J}_R(\hat{U})\} \tag{28.5}$$

will provide labels for the $\hat{\mathbf{x}}_j$'s that are based on a combination of nearest-neighbor and nearest-membership goals: (28.4) asks for a U^* which makes $(\mathbf{U}^*)^{(j)}$ close to $\mathbf{U}^{(k)}$ when $\hat{\mathbf{x}}_j$ is close to $\mathbf{x}_k$. This is an eminently plausible criterion, which seems to have the selection of distance measures for $\mathbb{R}^c$ and $\mathbb{R}^p$ as its major algorithmic parameters. Ruspini correctly points out that the efficacy of using (28.4) depends entirely on having "correct" labels $U \in M_{fc}$ for X: in particular, labeled data, as used, e.g., by the k-NN classifier above, should provide much more reliable classifications for X than an "artificial" labeling of X would. Further, it seems entirely reasonable that Ruspini is also correct in assuming that if X is labeled by an algorithm, $\hat{J}_R$ will be more compatible with substructures identified by J_R via (A9.1) than otherwise.

Ruspini describes a procedure for iterative optimization of $\hat{J}_R$ which is nothing more than the obvious modification of (A9.1) to accommodate $\hat{J}_R$ instead of J_R. Loosely speaking, minimization of $\hat{J}_R$ over M_{fco} is reduced to sequential minimization for each $\hat{\mathbf{x}}_j$ in $\hat{X}$, holding fixed all parameters except $\hat{\mathbf{U}}^{(j)}$, with (A9.1) providing the necessary steps for each fixed column of $\hat{U}$. The output (U^*) of this procedure is actually a fuzzy c-partition of the vectors in $\hat{X}$; the guidance of (A9.1) towards U^* by a *labeled* training partition U of X as incorporated by $\hat{J}_R$ renders this a legitimate "classifier" in the context of Chapter 6, rather than a clustering algorithm in the sense of Chapters 3 or 5. If, however, X is initially unlabeled, this procedure breaks down into first clustering X with J_R and then clustering $\hat{X}$ with $\hat{J}_R$. In this latter instance, one wonders whether the partitioned matrix $(U \mid U^*)$ arrived at via the "composition" $\hat{J}_R \circ J_R$ would be substantially different from the $[c \times (n+q)]$ matrix, say $\hat{U}^* \in M_{fc}$, derived by simply clustering $X \cup \hat{X}$ with (A9.1) and J_R. In any case if one has the information (X, U) needed to implement (D27.2) for a k-NN *hard* classifier, this same information can be used in equation (28.4) with (A9.1) to provide a *fuzzy* classifier for incoming data sets $\hat{X} \subset \mathbb{R}^p$. As an example of this method, we give the following example.

(E28.1) *Example 28.1* (*Ruspini*[92]). Figure 28.1 depicts $n = 95$ training vectors $X \subset \mathbb{R}^2$, together with the hard $c = 4$ partition $U \in M_4$ of X which renders the pair (X, U) a labeled data set. Taking these as fixed, Ruspini then generated the $q = 11$ unlabeled vectors $\{\hat{\mathbf{x}}_j\} = \hat{X} \subset \mathbb{R}^2$ shown in Fig. 28.1, and applied (A9.1) in its modified form to $\hat{J}_R(\hat{U})$, using the Euclidean norm on $\mathbb{R}^2$ as the dissimilarity measure between pairs of points $(\hat{\mathbf{x}}_j, \mathbf{x}_k) \in \hat{X} \times X$. Other computing protocols were similar to those used in (E9.1). The fuzzy 4-partition U^* of X at which (A9.1) terminated is listed as Table (28.1), columns 2 through 5.

Since the 11 points in X were not assigned "correct" labels independently of their algorithmic labels in Table 28.1, it is not possible to calculate here an empirical error rate [Equation (25.16), of course, is not even applicable to fuzzy labels]. We can, however, make a comparison of sorts by assuming that each column of U^* exhibits the *values* of a set of four scalar

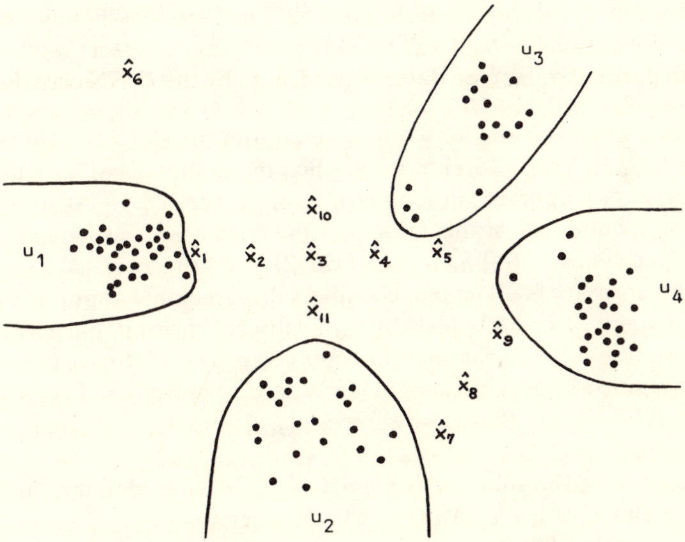

Figure 28.1. Labeled (●) and unlabeled ($\hat{x}$) data: (E28.1).

Table 28.1. Fuzzy Classifications Using $\hat{J}_R$: (E28.1)

Data point	Memberships via (A9.1) ~ $\tilde{J}_R$ u_1^*	u_2^*	u_3^*	u_4^*	$\{f_{jR}\}$ class	$\{f_{j,nNN}\}$ class
$\hat{\mathbf{x}}_1$	1.00	—	—	—	1	1
$\hat{\mathbf{x}}_2$	0.89	0.11	—	—	1	1
$\hat{\mathbf{x}}_3$	0.28	0.72	—	—	2	2
$\hat{\mathbf{x}}_4$	—	0.11	0.89	—	3	3
$\hat{\mathbf{x}}_5$	—	—	0.77	0.23	3	3
$\hat{\mathbf{x}}_6$	1.00	—	—	—	1	1
$\hat{\mathbf{x}}_7$	—	0.22	—	0.78	4 ←	→ 2
$\hat{\mathbf{x}}_8$	—	0.15	—	0.85	4 ←	→ 2
$\hat{\mathbf{x}}_9$	—	0.05	0.01	0.94	4	4
$\hat{\mathbf{x}}_{10}$	0.19	—	0.81	—	3	3
$\hat{\mathbf{x}}_{11}$	0.13	0.87	—	—	2	2

fields—let us call them $\{f_{jR}\}$—which are defined implicitly through their values on the vectors in $\hat{X}$. Thus, we imagine that

$$f_{jR}(\hat{\mathbf{x}}_k) = \hat{u}_{jk}^*, \qquad 1 \leq j \leq 4, \qquad 1 \leq k \leq 11 \tag{28.6}$$

In the general context of fuzzy c-partitions, f_{jR} is just the membership function or fuzzy set $u_j^*: \hat{X} \to [0, 1]$, which is always defined implicitly by its values on $\hat{X}$—i.e., the jth *row* of U^*. In other words, Ruspini's algorithm can be interpreted as defining an IMPLICIT classifier, $\{f_{jR}\} = \{u_j^*\}$. The functions themselves are unknown, but their values are specified, so that hard decisions via the maximum decision rule (25.2) can in fact be rendered exactly as in previous sections. Interpreting the rows of U^* (which are the columns $\{u_j^*\}$ of Table 28.1) as an implicit maximum classifier leads to the hard decisions reported in column 6 of Table 28.1. As a basis for comparison, column 7 of this table lists the hard decisions emanating from the 1-NN classifier $\{f_{j,n}\}_{NN}$ using (D27.2) with the pair (X, U) and the Euclidean norm for $\mathbb{R}^2$. The two classifiers disagree only at points $\hat{\mathbf{x}}_7$ and $\hat{\mathbf{x}}_8$. The assignment of $\hat{\mathbf{x}}_7$ to cluster 4 by $\hat{J}_R$ is rather surprising, since it is visually quite close to cluster 2; the membership vector $(\mathbf{U}^*)^{(8)}$ for $\hat{\mathbf{x}}_8$ is even more surprising, since $\hat{\mathbf{x}}_7$ appears most clearly to be "more of" a class-2 object than $\hat{\mathbf{x}}_8$. Aside from these differences, $\{f_{jR}\}$ and $\{f_{j,n}\}_{NN}$ yield identical results. •

Interpreting the numerical results of (E28.1) as defining an implicit maximum classifier leads naturally to a comparison of $\{f_{jR}\}$ with $\{f_{jb}(\hat{\mathbf{x}}_k) = h(\hat{\mathbf{x}}_k | j: \hat{\boldsymbol{\omega}}_j)\}$ as in S26. The c numbers $\{f_{jb}(\hat{\mathbf{x}}_k) | 1 \leq j \leq c\}$ have exactly the same properties as $\{f_{jR}(\hat{\mathbf{x}}_k) = \hat{u}_{jk}^* | 1 \leq j \leq c\}$, namely, they lie in [0, 1] and sum to unity. There are two ways to look at this observation. On the one hand, the posterior probabilities $\{f_{jb}(\hat{\mathbf{x}}_k)\}$ may be viewed as a "probabilistic classification" of $\hat{\mathbf{x}}_k$, just as the $\{f_{jR}(\hat{\mathbf{x}}_k)\}$ are regarded as a "fuzzy classification" of $\hat{\mathbf{x}}_k$. From this point of view, the two c-vectors provide exactly the same *operational* options for rendering decisions about $\mathbf{x}_k$; but their "information" content has a different philosophical connotation (cf. sections 3 and 16). On the other hand, the *use* of the $f_{jb}(\hat{\mathbf{x}}_k)$'s in decision rule (25.2) is rooted through the notions of expected loss and error rate to a sound theoretical basis, whereas the implicit nature of $f_{jR} = \hat{u}_j^*$ renders the criterion being optimized when using (25.2) with the $\hat{u}_{jk}^*$'s obscure at best. From this latter viewpoint, it is hard to accord $\{f_{jR}\}$ the same analytical status as $\{f_{jb}\}$; nonetheless, there is still no a priori reason to suspect therefore that this scheme will generate less reliable classifications than $\{f_{jb}\}$ in practice!

Remarks

Ruspini's classifier entails a number of innovations that deserve attention. First, the idea of *fuzzy* classifications: rather than decide $\hat{\mathbf{x}}_j \in i$, this algorithm provides a membership vector for $\hat{\mathbf{x}}_j$. One may, of course, use $(U^*)^{(j)}$ to render a hard decision as in (E28.1) if this is desirable; on the other hand, memberships of $\hat{\mathbf{x}}_j$ in all c classes may provide more useful information about it than an unequivocal classification. Other classifiers almost certainly can be modified to provide fuzzy decisions. Secondly, a relationship between (28.4) and k-NN machines seems implicit, but has yet to be investigated. Further, a study of the theoretical relationship between $\{f_{jR}\}$ and $\{f_{jb}\}$, the optimal Bayes' design, would be most welcome. This will be difficult (perhaps impossible) because the functions $\{f_{jR}\}$ are defined implicitly through the rows of U^*. If these could be found, it would relieve an obvious computational disadvantage of Ruspini's algorithm, viz, processing iteratively all of $\hat{X}$ to find U^*, the calculations involving (U, X) and an initialization $U^{(0)}$ for $\hat{J}_R$ which might lead to possibly different U^*'s for the same $\hat{X}$, as compared to simply evaluating c scalar fields on each $\hat{\mathbf{x}} \in \hat{X}$.

Exercise

H28.1. Let X be a set of n *unlabeled* observations from the mixture described in H27.7. Suppose X to then be labeled by any hard clustering algorithm, and let s be the number of incorrectly labeled points in n tries. Derive generalizations of H27.7(i) and H27.7(ii) for this case.

References

1. Anderberg, M. R., *Cluster Analysis for Applications*, Academic Press, New York (1973).
2. Anderson, E., The Irises of the Gaspe Peninsula, *Bull. Amer. Iris Soc.*, Vol. 59 (1935), pp. 2–5.
3. Arbib, M. A., and Manes, E. G., A Category-Theoretic Approach to System in a Fuzzy World, *Synthese*, Vol. 30 (1975), pp. 381–406.
4. Backer, E., *Cluster Analysis by Optimal Decomposition of Induced Fuzzy Sets*, Delft Univ. Press, Delft, The Netherlands (1978).
5. Backer, E., Cluster Analysis Formalized as a Process of Fuzzy Identification based on Fuzzy Relations, Report No. IT-78-15, Delft Univ. of Technology (1978).
6. Backer, E., and Jain, A., A Clustering Performance Measure based on Fuzzy Set Decomposition, *IEEE Trans. PAMI*, Vol. PAMI-3(1) (1981), pp. 66–74.
7. Balas, E., and Padberg, M. W., Set Partitioning: A Survey, *SIAM Rev.*, Vol. 18-4 (1976), pp. 710–760.
8. Ball, G. H., and Hall, D. J., A Clustering Technique for Summarizing Multivariate Data, *Behav. Sci.*, Vol. 12 (1967), pp. 153–155.
9. Bellman, R., Kalaba, R., and Zadeh, L., Abstraction and Pattern Classification, *J. Math. Anal. Appl.*, Vol. 13 (1966), pp. 1–7.
10. Bezdek, J. C., *Fuzzy Mathematics in Pattern Classification*, Ph.D. Thesis, Applied Math. Center, Cornell University, Ithaca (1973).
11. Bezdek, J. C., Numerical Taxonomy with Fuzzy Sets, *J. Math. Biol.*, Vol. 1-1 (1974), pp. 57–71.
12. Bezdek, J., Mathematical Models for Systematics and Taxonomy, in: *Proc. Eighth Int. Conf. on Numerical Taxonomy* (G. Estabrook, ed.), pp. 143–164, Freeman, San Francisco (1975).
13. Bezdek, J. C., A Physical Interpretation of Fuzzy ISODATA, *IEEE Trans. Syst. Man Cybern.*, Vol. SMC-6 (1976), pp. 387–390.
14. Bezdek, J. C., Feature Selection for Binary Data: Medical Diagnosis with Fuzzy Sets, in: *Proc. 25th National Comp. Conf.* (S. Winkler, ed.), pp. 1057–1068, AFIPS Press, Montvale, New Jersey (1976).

15. Bezdek, J. C., and Castelaz, P. F., Prototype Classification and Feature Selection with Fuzzy Sets, *IEEE Trans. Syst. Man Cybern.*, Vol. SMC-7 (1977), pp. 87–92.
16. Bezdek, J. C., Coray, C., Gunderson, R., and Watson, J., Detection and Characterization of Cluster Substructure. I. Linear Structure: Fuzzy c-Lines, *SIAM J. Appl. Math.*, Vol. 40 (1981), in press.
17. Bezdek, J. C., Coray, C., Gunderson, R., and Watson, J., Detection and Characterization of Cluster Substructure. II. Fuzzy c-Varieties and Convex Combinations Thereof, *SIAM J. Appl. Math.*, Vol. 40 (1981), in press.
18. Bezdek, J. C., and Dunn, J. C., Optimal Fuzzy Partitions: A Heuristic for Estimating the Parameters in a Mixture of Normal Distributions, *IEEE Trans. Comp.*, Vol. C-24 (1975), pp. 835–838.
19. Bezdek, J. C., and Fordon, W. A., Analysis of Hypertensive Patients by the Use of the Fuzzy ISODATA Algorithm, in: *Proc. 1978 JACC, Vol. 3*, (H. Perlis, Chm.), pp. 349–356, ISA Press, Philadelphia (1978).
20. Bezdek, J. C., and Harris, J., Convex Decompositions of Fuzzy Partitions, *J. Math. Anal. Appl.*, Vol. 67-2 (1979), pp. 490–512.
21. Bezdek, J. C., and Harris, J., Fuzzy Relations and Partitions: An Axiomatic Basis for Clustering, *J. Fuzzy Sets Systems*, Vol. 1 (1978), pp. 111–127.
22. Bezdek, J. C., Trivedi, M., Ehrlich, R., and Full, W., Fuzzy Clustering: A New Approach for Geostatistical Analysis, *J. Math. Geo.*, (1980), in review.
23. Bezdek, J., Windham, M., and Ehrlich, R., Statistical Parameters of Cluster Validity Functionals, *Int. J. Comp. Inf. Science*, Vol. 9-4, (1980), pp. 323–336.
24. Bremmerman, H. J., What Mathematics Can and Cannot Do for Pattern Recognition, in: *Zeichenerkennung durch Biologische und Technische Systeme* (O. J. Grusser and R. Klinke, eds.), pp. 27–39, Springer, Berlin (1971).
25. Bongard, N., *Pattern Recognition*, Spartan, New York (1970).
26. Chang, C. L., Some Properties of Fuzzy Sets in E^n, Report, Div. of Comp. Res. and Tech., NIH, Bethesda (1967).
27. Chen, Z., and Fu, K. S., On the Connectivity of Clusters, *Inf. Sci.*, Vol. 8 (1975), pp. 283–299.
28. Cooper, D. B., and Cooper, P. W., Nonsupervised Adaptive Signal Detection and Pattern Recognition, *Inf. Control*, Vol. 7 (1964), pp. 416–444.
29. Cover, T. M., and Hart, P. E., Nearest Neighbor Pattern Classification, *IEEE Trans. Inf. Theory*, Vol. IT-13 (1967), pp. 21–27.

30. Croft, D., Mathematical Models in Medical Diagnosis, *Ann. Biom. Eng.*, Vol. 2 (1974), pp. 68–69.
31. Day, N. E., Estimating the Components of a Mixture of Normal Distributions, *Biometrika*, Vol. 56 (1969), pp. 463–474.
32. DeLuca, A., and Termini, S., A Definition of a Nonprobabilistic Entropy in the Setting of Fuzzy Sets Theory, *Inf. Control*, Vol. 20 (1972), pp. 301–312.
33. Diday, E., and Simon, J. C., Clustering Analysis, in: *Communication and Cybernetics 10*: *Digital Pattern Recognition* (K. S. Fu, ed.), pp. 47–94, Springer-Verlag, Heidelberg (1976).
34. Duda, R., and Hart, P., *Pattern Classification and Scene Analysis*, Wiley, New York (1973).
35. Dunn, J. C., A Fuzzy Relative of the ISODATA Process and its Use in Detecting Compact, Well Separated Clusters, *J. Cybern.*, Vol. 3 (1974), pp. 32–57.
36. Dunn, J. C., Well Separated Clusters and Optimal Fuzzy Partitions, *J. Cybern.*, Vol. 4-1 (1974), pp. 95–104.
37. Dunn, J. C., Indices of Partition Fuzziness and the Detection of Clusters in Large Data Sets, in: *Fuzzy Automata and Decision Processes* (M. Gupta and G. Saridis, eds.), Elsevier, New York (1977).
38. Fisher, R. A., The Use of Multiple Measurements in Taxonomic Problems, *Ann. Eugenics*, Vol. 7 (1936), pp. 179–188.
39. Flake, R. H., and Turner, B. L., Numerical Classification for Taxonomic Problems, *J. Theoret. Biol.*, Vol. 20 (1968), pp. 260–270.
40. Friedman, H., and Rubin, J., On Some Invariant Criteria for Grouping Data, *J. Am. Stat. Assoc.* Vol. 62 (1967), pp. 1159–1178.
41. Fu, K. S., *Sequential Methods in Pattern Recognition and Machine Learning*, Academic Press, New York (1968).
42. Fukunaga, K., *Introduction to Statistical Pattern Recognition*, Academic Press, New York (1972).
43. Fukunaga, K., and Hostetler, L. D., *k*-Nearest Neighbor Bayes Risk Estimation, *IEEE Trans. Inf. Theory*, Vol. IT-21-3, (1975), pp. 285–293.
44. Gaines, B. R., Foundations of Fuzzy Reasoning, *Int. J. Man–Mach. Stud.*, Vol. 8 (1967), pp. 623–668.
45. Gaines, B. R., and Kohout, L. J., The Fuzzy Decade: A Bibliography of Fuzzy Systems and Closely Related Topics, *Int. J. Man–Mach. Stud.*, Vol. 9 (1977), pp. 1–68.
46. Gitman, I., and Levine, M., An Algorithm for Detecting Unimodal Fuzzy Sets and its Application as a Clustering Technique, *IEEE Trans. Comp.*, Vol. C-19 (1970), pp. 583–593.

47. Goguen, J. A., *L*-Fuzzy Sets, *J. Math. Anal. Appl.*, Vol. 18 (1967), pp. 145–174.
48. Gose, E. A., Introduction to Biological and Mechanical Pattern Recognition, in: *Methodologies of Pattern Recognition* (S. Watanabe, ed.), pp. 203–252, Academic Press, New York (1969).
49. Gower, J., and Ross, G., Minimum Spanning Trees and Single Linkage Cluster Analysis, *Appl. Stat.*, Vol. 18 (1969), pp. 54–64.
50. Grenander, U., Foundations of Pattern Analysis, *Q. Appl. Math.*, Vol. 27 (1969), pp. 2–55.
51. Grossman, S. I., and Turner, J. E., *Mathematics for the Biological Sciences*, MacMillan, New York (1974).
52. Gunderson, R., Application of Fuzzy ISODATA Algorithms to Star Tracker Pointing Systems, in: *Proc. 7th Triennial World IFAC Congress*, Helsinki (1978), pp. 1319–1323.
53. Gunderson, R., and Watson, J., Sampling and Interpretation of Atmospheric Science Experimental Data, in: *Fuzzy Sets: Theory and Applications to Policy Analysis and Information Systems* (S. Chang and P. Wang, eds.), Plenum, New York (1980).
54. Gustafson, D. E., and Kessel, W., Fuzzy Clustering with a Fuzzy Covariance Matrix, in *Proc. IEEE-CDC*, Vol. 2 (K. S. Fu, ed.), pp. 761–766, IEEE Press, Piscataway, New Jersey (1979).
55. Hadley, G., *Nonlinear and Dynamic Programming*, Addison-Wesley, Reading, Massachusetts (1964).
56. Halpern, J., Set Adjacency Measures in Fuzzy Graphs, *J. Cybern.*, Vol. 5-4 (1975), pp. 77–87.
57. Hartigan, J., *Clustering Algorithms*, Wiley, New York (1975).
58. Ho, Y. C., and Agrawala, A. K., On Pattern Classification Algorithms; Introduction and Survey, *Proc. IEEE*, Vol. 56-12 (1968), pp. 2101–2114.
59. Jardine, N., and Sibson, R., *Mathematical Taxonomy*, Wiley and Sons, New York (1971).
60. Johnson, S. C., Hierarchical Clustering Schemes, *Psychometrika*, Vol. 32, pp. 241–254.
61. Kanal, L. N., Interactive Pattern Analysis and Classification Systems: A Survey, *Proc. IEEE*, Vol. 60-12 (1972).
62. Kaufmann, A., *Introduction to the Theory of Fuzzy Subsets*, Vol. 1, Academic Press, New York (1975).
63. Kemeny, J. G., and Snell, J. L., *Mathematical Models in the Social Sciences*, Ginn-Blaisdell, Waltham, Massachusetts (1962).
64. Kendall, M. G., Discrimination and Classification, in: *Multivariate Analysis* (P. Krishnaiah, ed.), pp. 165–185, Academic Press, New York (1966).

65. Khinchin, A. I., *Mathematical Foundations of Information Theory* (R. Silverman and M. Friedman, trans.), Dover, New York (1957).
66. Klaua, D., Zum Kardinalzahlbegriff in der Mehrwertigen Mengenlehre, in: *Theory of Sets and Topology*, pp. 313–325, Deutscher Verlag der Wissenschaften, Berlin (1972).
67. Knopfmacher, J., On Measures of Fuzziness, *J. Math. Anal. Appl.*, Vol. 49 (1975), pp. 529–534.
68. Kokawa, M., Nakamura, K., and Oda, M., Experimental Approach to Fuzzy Simulation of Memorizing, Forgetting and Inference Process, in: *Fuzzy Sets and Their Applications to Cognitive and Decision Processes* (L. Zadeh, K. S. Fu, K. Tanaka, and M. Shimura, eds.), pp. 409–428, Academic Press, New York (1975).
69. Lance, G., and Williams, W., A General Theory of Classificatory Sorting Strategies I, *Comput. J.*, Vol. 9 (1966), pp. 337–380.
70. Lance, G., and Williams, W., A General Theory of Classificatory Sorting Strategies II, *Comput. J.*, Vol. 10 (1967), pp. 271–277.
71. Larsen, L. E., Ruspini, E. H., McNew, J. J., Walter, D. O., and Adey, W. R., A Test of Sleep Staging Systems in the Unrestrained Chimpanzee, *Brain Res.*, Vol. 40 (1972), pp. 319–343.
72. Lee, R. C. T., Application of Information Theory to Select Relevant Variables, *Math. Biosciences*, Vol. 11 (1971), pp. 153–161.
73. Levine, M. D., Feature Extraction: A Survey, *Proc. IEEE*, Vol. 57 (1969), pp. 1391–1407.
74. Ling, R. F., *Cluster Analysis*, Ph.D. Thesis, Yale Univ., New Haven, Connecticut (1971).
75. Luenberger, D. L., *An Introduction to Linear and Non-Linear Programming*, Addison-Wesley, Reading, Massachusetts (1973).
76. MacQueen, J., Some Methods for Classification and Analysis of Multivariate Observations, in: *Proc. 5th Berkeley Symp. on Math. Stat. and Prob.* (L. M. LeCam and J. Neyman, eds.), pp. 281–297, Univ. of California Press, Berkeley and Los Angeles (1967).
77. Maki, D. P., and Thompson, M., *Mathematical Models and Applications*, Prentice-Hall, Englewood Cliffs, New Jersey (1973).
78. Matula, D. W., Graph Theoretic Techniques for Cluster Analysis Algorithms, in: *Classification and Clustering* (J. Van Ryzin, ed.), pp. 95–129, Academic Press, New York (1977).
79. Meisel, W. S., *Computer-Oriented Approaches to Pattern Recognition*, Academic Press, New York (1972).
80. Nagy, G., State of the Art in Pattern Recognition, *Proc. IEEE*, Vol. 56-5 (1968), pp. 836–862.
81. Patrick, E. A., *Fundamentals of Pattern Recognition*, Prentice-Hall, Englewood Cliffs, New Jersey (1972).

82. Ragade, R., Fuzzy Sets in Communications Systems and in Consensus Formation Systems, *J. Cybern.*, Vol. 6 (1976), pp. 1–13.
83. Rinaldo, J., Scheinok, P., and Rupe, C., Symptom Diagnosis: A Mathematical Analysis of Epigastric Pain, *Ann. Int. Med.*, Vol. 59-2 (1963), pp. 145–154.
84. Roberts, A. W., and Varberg, D. E., *Convex Functions*, Academic Press, New York (1973).
85. Roberts, F. S., *Discrete Mathematical Models*, Prentice-Hall, Englewood Cliffs, New Jersey (1976).
86. Ross, S., *A First Course in Probability*, MacMillan, New York (1976).
87. Rossini, L., Martorana, F., and Periti, P., Clustering Cholinergic Receptors by Muscarine and Muscarone Analogs, in: *Proc. 2nd Int. Meeting of Medical Advisors in the Pharmaceutical Industry* (1975), Firenze, Italy.
88. Roubens, M., Pattern Classification Problems and Fuzzy Sets, *J. Fuzzy Sets Syst.*, Vol. 1 (1978), pp. 239–253.
89. Ruspini, E., Numerical Methods for Fuzzy Clustering, *Inf. Sci.*, Vol. 2 (1970), pp. 319–350.
90. Ruspini, E., New Experimental Results in Fuzzy Clustering, *Inf. Sci.*, Vol. 6 (1972), pp. 273–284.
91. Ruspini, E., Optimization in Sample Descriptions: Data Reductions and Pattern Recognition using Fuzzy Clustering, *IEEE Trans. SMC*, Vol. SMC-2 (1972), p. 541.
92. Ruspini, E., A New Approach to Clustering, *Inf. Control*, Vol. 15 (1969), pp. 22–32.
93. Ruspini, E., Applications of Fuzzy Clustering to Pattern Recognition, Internal Report, Brain Research Institute, UCLA, Los Angeles (1972).
94. Scheinok, P., and Rinaldo, J., Symptom Diagnosis: Optimal Subsets for Upper Abdominal Pain, *Comp. Bio. Res.*, Vol. 1 (1967), pp. 221–236.
95. Scott, A., and Symons, M., Clustering Methods Based on Likelihood Ratio Criteria, *Biometrics*, Vol. 27 (1971), pp. 387–397.
96. Sebestyen, A. S., *Decision Making Processes in Pattern Recognition*, MacMillan, New York (1962).
97. Seif, A., and Martin, J., Multigroup Classification using Fuzzy Correlation, Fuzzy Sets and Systems, *J. Fuzzy Sets Syst.*, Vol. 3 pp. 109–122 (1980).
98. Shannon, C. E., A Mathematical Theory of Communication, *Bell Syst. Tech. J.*, Vol. XXVII-3 (1948), pp. 379–423.
99. Sneath, P. H. A., and Sokal, R., *Numerical Taxonomy*, Freeman, San Francisco (1973).

100. Sokal, R., and Michener, C. D., A Statistical Method for Evaluating Systematic Relationships, *Univ. Kansas Sci. Bull.*, Vol. 38 (1958), pp. 1409–1438.
101. Sugeno, M., *Theory of Fuzzy Integrals and Its Applications*, Ph.D. thesis, Tokyo Institute of Technology, Tokyo (1974).
102. Thorndike, R. L., Who Belongs in the Family?, *Psychometrika*, Vol. 18 (1953), pp. 267–276.
103. Tou, J. T., and Gonzalez, R., *Pattern Recognition Principles*, Addison-Wesley, Reading, Massachusetts (1974).
104. Tou, J. T., and Wilcox, R. H., *Computer and Information Sciences*, Spartan, Washington (1964).
105. Toussaint, G. T., Subjective Clustering and Bibliography of Books on Pattern Recognition, *Inf. Sci.*, Vol. 8 (1975), pp. 251–257.
106. Toussaint, G. T., Bibliography on Estimation of Misclassification, *IEEE Trans. Inf. Theory*, Vol. IT-20-4 (1974), pp. 472–479.
107. Toussaint, G. T., and Sharpe, P., An Efficient Method for Estimating the Probability of Misclassification Applied to a Problem in Medical Diagnosis, *Comp. Biol. Med.*, Vol. 4 (1975), pp. 269–278.
108. Trillas, E., and Riera, T., Entropies in Finite Fuzzy Sets, Internal Report, Dept. of Mathematics, Universitat Politecnica de Barcelona (1978).
109. Tryon, R. C., and Bailey, D. E., *Cluster Analysis*, McGraw-Hill, New York (1970).
110. Uhr, L., *Pattern Recognition*, Wiley, New York (1966).
111. Verhagen, C. J. D. M., Some General Remarks about Pattern Recognition; Its Definition; Its Relation with other Disciplines; A Literature Survey, *J. Patt. Recog.*, Vol. 8-3 (1975), pp. 109–116.
112. Watanabe, S., *Methodologies of Pattern Recognition*, Academic Press, New York (1969).
113. Watanabe, S., *Frontiers of Pattern Recognition*, Academic Press, New York (1972).
114. Wee, W. G., On Generalizations of Adaptive Algorithms and Application of the Fuzzy Sets Concept to Pattern Classification, Ph.D. thesis, Purdue Univ., Lafayette, Indiana (1967).
115. Windham, M. P., Cluster Validity for Fuzzy Clustering Algorithms, *J. Fuzzy Sets Syst.*, Vol. 3 (1980), pp. 1–9.
116. Wishart, D., Mode Analysis: A Generalization of Nearest Neighbor which reduces Chaining Effects, in: *Numerical Taxonomy* (A. J. Cole, ed.), pp. 282–308, Academic Press, New York (1969).
117. Wolfe, J., Pattern Clustering by Multivariate Mixture Analysis, *Multivar. Behav. Res.*, Vol. 5 (1970), pp. 329–350.

118. Woodbury, M. A., and Clive, J. A., Clinical Pure Types as a Fuzzy Partition, *J. Cybern.*, Vol. 4-3 (1974), pp. 111–121.
119. Woodbury, M. A., Clive, J. A., and Garson, A., Mathematical Typology: A Grade of Membership Technique for Obtaining Disease Definition, *Comp. Bio. Res.*, Vol. 11 (1978), pp. 277–298.
120. Woodbury, M. A., Clive, J. A., and Garson, A., A Generalized Ditto Algorithm for Initial Fuzzy Clusters, presented to 7th Annual Meeting of the Classification Society, Rochester (1976).
121. Yakowitz, S. J., and Spragins, J. D., On the Identifiability of Finite Mixtures, *Ann. Math. Stat.*, Vol. 29 (1968), pp. 209–214.
122. Zadeh, L. A., Fuzzy Sets, *Inf. Control.*, Vol. 8 (1965), pp. 338–353.
123. Zadeh, L. A., Probability Measures of Fuzzy Events, *J. Math. Anal. Appl.*, Vol. 23-2 (1968), pp. 421–427.
124. Zahn, C. T., Graph-Theoretical Methods for Detecting and Describing Gestalt Clusters, *IEEE Trans. Comp.*, Vol. C-20 (1971), pp. 68–86.
125. Zangwill, W., *Non-Linear Programming: A Unified Approach*, Prentice-Hall, Englewood Cliffs, New Jersey (1969).

Algorithm Index

The main algorithms described and/or discussed in this volume are collected here by function, type or name, and page of first appearance.

Author Index

Subject Index